FM 3-22.90

Mortars

★

Headquarters
Department of the Army

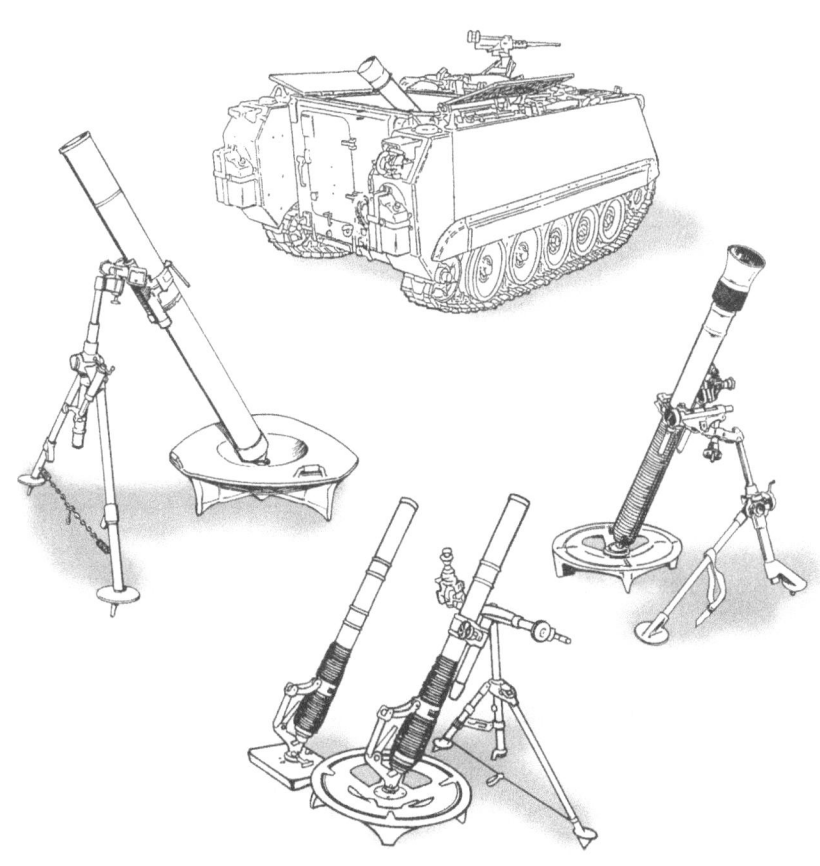

Published by Books Express Publishing
Copyright © Books Express, 2010
ISBN 978-1-907521-31-7
To purchase copies at discounted prices please contact
info@books-express.com

**Field Manual
No. 3-22.90**

*FM 3-22.90

Headquarters
Department of the Army
Washington, DC, 7 December 2007

Mortars

Contents

Page

	PREFACE xiii	
Chapter 1	INTRODUCTION 1-1	
	Section I. General Doctrine 1-1	
	Effective Mortar Fire 1-1	
	Mortar Positions 1-1	
	Section II. Indirect Fire Team 1-2	
	Applications 1-2	
	Team Mission 1-2	
	United States Mortars 1-3	
	Section III. Safety Procedures 1-5	
	Duties Of The Safety Officer and Supervisory Personnel 1-5	
	Safety Diagram and Safety "T" 1-8	
	Surface Danger Zones 1-12	
	Section IV. Ammunition 1-13	
	Ammunition Care and Handling 1-13	
	Ammunition Color Codes 1-15	
	Field Storage of Ammunition 1-15	
Chapter 2	SIGHTING AND FIRE CONTROL EQUIPMENT 2-1	
	Section I. Compass, M2 2-1	
	Characteristics 2-1	
	Description 2-2	
	Use 2-2	
	Section II. Aiming Circles, M2 and M2A2 2-5	
	Characteristics 2-5	
	Description 2-5	
	Use 2-6	
	Accessory Equipment 2-9	

DISTRIBUTION RESTRICTION: Approved for public release; distribution is unlimited.

*This publication supersedes FM 3-22.90, 31 December 2004.

Table of Contents

    Setup and Leveling of the Aiming Circle 2-10
    Declination Constant 2-11
    Orienting of the Instrument on Grid North to Measure Grid Azimuth to Objects 2-14
    Measuring of the Horizontal Angle Between Two Points 2-14
    Orienting of the 0-3200 Line on a Given Grid Azimuth 2-15
    Orienting of the 0-3200 Line on a Given Magnetic Azimuth 2-15
    Verification of the Lay of the Platoon 2-16
    Orienting by Orienting Angle 2-17
    Disassembly of the Aiming Circle 2-17
    Care and Maintenance 2-18
    Section III. Sightunits **2-18**
    Sightunit, M67 2-18
    Sightunit, M64-Series 2-19
    Operation of Sightunits 2-21
    Care and Maintenance of Sightunits 2-23
    Radioactive Tritium Gas 2-23
    Section IV. Boresights **2-24**
    Boresight, M45-Series 2-24
    Boresight, M115 2-25
    Sight Calibration 2-26
    Boresight Method Of Calibration 2-26
    Calibration for Deflection Using the M2 Aiming Circle 2-28
    Section V. Other Equipment **2-30**
    Aiming Posts, M14 and M1A2 2-30
    Aiming Post Lights, M58 and M59 2-31
    Section VI. Laying of the Section **2-32**
    Reciprocal Laying 2-33
    Placing Out Aiming Posts 2-39
    Section VII. Loading and Firing **2-42**
    Firing of the Ground-Mounted Mortar 2-42
    Target Engagement 2-42
    Execution of Fire Commands 2-43
    Arm-and-Hand Signals 2-45
    Subsequent Fire Commands 2-46
    Repeating and Correcting of Fire Commands 2-46
    Reporting of Errors in Firing 2-47
    Night Firing 2-47

Chapter 3   **60-mm MORTAR, M224** **3-1**
    Section I. Squad and Section Organization and Duties **3-1**
    Organization 3-1
    Duties 3-1

Table of Contents

	Section II. Components	3-2
	Tabulated Data	3-2
	Cannon Assembly, M225	3-3
	Bipod Assembly, M170	3-3
	Baseplate, M7	3-5
	Baseplate, M8	3-5
	Section III. Operation	3-6
	Premount Checks	3-6
	Mounting of the Mortar	3-6
	Safety Checks Before Firing	3-7
	Small Deflection and Elevation Changes	3-8
	Large Deflection and Elevation Changes	3-8
	Referring of the Sight and Realignment of Aiming Posts	3-9
	Malfunctions	3-10
	Removal of a Misfire	3-11
	Dismounting and Carrying of the Mortar	3-15
	Section IV. Ammunition	3-16
	Classification and Types of Ammunition	3-16
	Fuzes	3-19
	Cartridge Preparation	3-21
	Care And Handling	3-22
Chapter 4	81-mm MORTAR, M252	4-1
	Section I. Squad and Section Organization and Duties	4-1
	Organization	4-1
	Duties	4-1
	Section II. Components	4-3
	Tabulated Data	4-4
	Cannon Assembly, M253	4-5
	Mount, M177	4-5
	Baseplate, M3A1	4-6
	Section III. Operation	4-7
	Premount Checks	4-7
	Mounting of the Mortar	4-8
	Safety Checks Before Firing	4-10
	Small Deflection and Elevation Changes	4-10
	Large Deflection and Elevation Changes	4-11
	Referring of the Sight and Realignment of Aiming Posts Using the M64 Sight	4-12
	Malfunctions	4-12
	Removal of a Misfire	4-12
	Dismounting of the Mortar	4-16

Table of Contents

	Section IV. Ammunition	**4-16**
	Classification and Types of Ammunition	4-16
	Fuzes	4-21
	Cartridge Preparation	4-24
	Care and Handling	4-25
Chapter 5	**120-mm MORTARS, M120 AND M121**	**5-1**
	Section I. Squad Organization and Duties	**5-1**
	Organization	5-1
	Duties	5-1
	Section II. Components	**5-3**
	Tabulated Data For the 120-mm Mortar	5-4
	Cannon Assembly, M298	5-4
	Bipod Assembly, M191 (Carrier-/Ground-Mounted)	5-6
	Bipod Assembly, M190 (Ground-Mounted)	5-7
	Baseplate, M9	5-8
	Section III. Operations	**5-8**
	Premount Checks	5-8
	Placing a Ground-Mounted 120-mm Mortar Into Action	5-10
	Performing Safety Checks on a Ground-Mounted 120-mm Mortar	5-11
	Performing Small Deflection and Elevation Changes on a Ground-Mounted 120-mm Mortar	5-12
	Performing Large Deflection and Elevation Changes on a Ground-Mounted 120-mm Mortar	5-13
	Malfunctions on a Ground-Mounted 120-mm Mortar During Peacetime	5-14
	Referring of the Sight and Realignment of Aiming Posts During Peacetime	5-14
	Removal of a Misfire on a Ground-Mounted 120-mm Mortar	5-14
	Loading and Firing of the Ground-Mounted 120-mm Mortar	5-21
	Taking the 120-mm Mortar Out of Action	5-22
	Section IV. Mortar Carrier, M1064A3	**5-23**
	Description	5-23
	Tabulated Data for the M1064A3 Carrier	5-25
	Mortar and Vehicular Mount	5-25
	Maintenance	5-26
	Section V. Operation of a Carrier-Mounted 120-mm Mortar	**5-26**
	Premount Checks	5-26
	Placing a Carrier-Mounted 120-mm Mortar Into Action	5-26
	Mounting of the Mortar From a Carrier- to a Ground-Mounted Position	5-27
	Performing Safety Checks on a Carrier-Mounted 120-mm Mortar	5-28
	Performing Small Deflection and Elevation Changes on a Carrier-Mounted 120-mm Mortar	5-29
	Performing Large Deflection and Elevation Changes on a Carrier-Mounted 120-mm Mortar	5-29

Table of Contents

	Removal of a Misfire on a Carrier-Mounted 120-mm Mortar	5-30
	Taking the Mortar Out of Action (Ground-Mounted to M1064A3 Carrier-Mounted)	5-37
	Reciprocally Laying the Mortar Carrier Section	5-38
	Section VI. Ammunition	**5-38**
	Classification and Types of Ammunition	5-38
	Fuzes	5-40
	Cartridge Preparation	5-43
	Care And Handling of Cartridges	5-44
Chapter 6	**MORTAR FIRE CONTROL SYSTEM**	**6-1**
	Description	6-1
	Capabilities	6-8
	Soldier Graphic User Interface	6-9
	Startup	6-12
	Log-In Procedures	6-12
	Data Initialization and System Configuration	6-12
	Additional Functions	6-21
Chapter 7	**CONDUCT FIRE MISSIONS USING THE MORTAR FIRE CONTROL SYSTEM**	**7-1**
	Pointing Device	7-1
	Navigation and Emplacement	7-5
	Fire Commands	7-10
	Final Protective Fires	7-14
Chapter 8	**FIRE WITHOUT A FIRE DIRECTION CENTER**	**8-1**
	Section I. Fire Procedures	**8-1**
	Advantages and Disadvantages	8-1
	Firing Data	8-1
	Observer Corrections	8-2
	Initial Fire Commands	8-3
	Fire Commands	8-3
	Fire Control	8-5
	Movement to Alternate and Supplementary Positions	8-5
	Squad Conduct of Fire	8-5
	Reference Line	8-6
	Squad Use of Smoke and Illumination	8-6
	Attack of Wide Targets	8-6
	Attack of Deep Targets	8-8
	Section II. Direct-Lay Method	**8-9**
	Step 1: Initial Firing Data	8-9
	Step 2: Referring the Sight	8-9

Table of Contents

	Step 3: Bracketing the Target	8-10
	Step 4: Fire For Effect	8-10
	Section III. Direct-Alignment Method	**8-10**
	Mortar Dismounted	8-10
	Mortar Mounted	8-10
	Natural Object Method	8-11
	Section IV. Adjustment of Range	**8-11**
	Spottings	8-11
	Bracketing Method	8-12
	Creeping Method of Adjustment	8-13
	Ladder Method of Adjustment	8-13
	Establishment of a Reference Line and Shifting From That Line	8-15
Chapter 9	**GUNNER'S EXAMINATION**	**9-1**
	Section I. Preparation	**9-1**
	Methods of Instruction	9-1
	Prior Training	9-1
	Preparatory Exercises	9-1
	Examining Board	9-1
	Location and Date	9-2
	Eligible Personnel	9-2
	Qualification Scores	9-2
	General Rules	9-3
	Section II. Gunner's Examination with the Ground-Mounted Mortar	**9-4**
	Subjects and Credits	9-4
	Equipment	9-4
	Organization	9-4
	Procedure	9-4
	Mounting of the Mortar	9-4
	Small Deflection Change	9-9
	Referring of the Sight and Realignment of Aiming Posts	9-10
	Large Deflection and Elevation Changes	9-11
	Reciprocal Laying	9-13
	Section III. Gunner's Examination with the Track-Mounted Mortar, M121	**9-14**
	Subjects and Credits	9-14
	Equipment	9-14
	Organization	9-15
	Procedure	9-15
	Placement of Mortar into a Firing Position from Traveling Position, 120-mm Mortar	9-15
	Small Deflection Change	9-16
	Referring of the Sight and Realignment of Aiming Posts	9-17
	Large Deflection and Elevation Changes	9-19

Table of Contents

	Reciprocal Laying	9-20
	Support Squad	9-21
Appendix A	**MORTAR TRAINING STRATEGY**	**A-1**
	Training Philosophy	A-1
	Unit Mortar Training	A-1
	Mortar Training at Training Base	A-1
	Training in Units	A-3
	Training Evaluation	A-9
Appendix B	**TRAINING DEVICES**	**B-1**
	Section I. Full-Range Training Cartridge, M931	**B-1**
	Description	B-1
	Procedures	B-1
	Section II. Short-Range Training Round, M880	**B-2**
	Training with the Short-Range Training Round, M880	B-2
	Components	B-3
	Training Considerations	B-9
	Construction of a Scaled Map	B-9
	Safety	B-13
	Malfunctions and Removal of a Misfire	B-13
	Section III. Subcaliber Insert, M303	**B-14**
	Characteristics	B-14
	Maintenance	B-15
	Misfire Procedures	B-16
	Section IV. Subcaliber Trainer, M313	**B-16**
	Characteristics	B-16
	Maintenance	B-17
	Misfire Procedures	B-19
	Glossary	**Glossary-1**
	References	**References-1**
	Index	**Index-1**

Figures

Figure 1-1. Indirect fire team.	1-2
Figure 1-2. Example completed safety record or card.	1-9
Figure 1-3. Safety diagram for M821 HE and M853A1 ILLUM.	1-11
Figure 1-4. Safety "T" for M821 HE.	1-11
Figure 1-5. Safety "T" for M853A1 ILLUM.	1-11
Figure 1-6. Stacked ammunition.	1-16
Figure 2-1. Compass, M2 (top view).	2-1

Table of Contents

Figure 2-2. Compass, M2 (side view). 2-3
Figure 2-3. Compass, M2 (user's view). 2-4
Figure 2-4. Aiming circles, M2 and M2A2, and accessory equipment. 2-6
Figure 2-5. Aiming circle, M2. 2-7
Figure 2-6. Leveling screws. 2-10
Figure 2-7. Marginal data from a map. 2-13
Figure 2-8. Aiming circle oriented in desired direction of fire. 2-15
Figure 2-9. Method used to orient an aiming circle, M2. 2-16
Figure 2-10. Orienting by orienting angle. 2-17
Figure 2-11. Sightunit, M67. 2-18
Figure 2-12. Sightunit, M64-series. 2-20
Figure 2-13. Warning label for tritium gas (H^3). 2-23
Figure 2-14. Boresight, M45. 2-25
Figure 2-15. Boresight, M115. 2-26
Figure 2-16. Verifying proper alignment of the boresight device. 2-28
Figure 2-17. Calibration for deflection using the angle method. 2-29
Figure 2-18. Calibration for deflection using the distant aiming point method. 2-30
Figure 2-19. Aiming posts, M14 and M1A2. 2-30
Figure 2-20. Aiming post lights, M58 and M59. 2-31
Figure 2-21. Parallel sheaf. 2-32
Figure 2-22. Principle of reciprocal laying. 2-33
Figure 2-23. Mortar laid parallel with the aiming circle. 2-35
Figure 2-24. Mortars laid parallel in the desired azimuth. 2-36
Figure 2-25. Mortar laid parallel with sights. 2-37
Figure 2-26. Sighting on the mortar sight. 2-38
Figure 2-27. Arm-and-hand signals used in placing out aiming posts. 2-40
Figure 2-27. Arm-and-hand signals used in placing out aiming posts (continued). 2-41
Figure 2-28. Arm-and-hand-signals for ready, fire, and cease firing. 2-46
Figure 3-1. 60-mm mortar, M224, handheld and conventional mode. 3-2
Figure 3-2. Cannon assembly, M225. 3-3
Figure 3-3. Bipod assembly, M170. 3-4
Figure 3-4. Baseplate, M7. 3-5
Figure 3-5. Baseplate, M8. 3-5
Figure 3-6. Large deflection and elevation changes. 3-9
Figure 3-7. Compensated sight picture. 3-10
Figure 3-8. Kicking the mortar to clear a misfire. 3-12
Figure 3-9. Multioption fuze, M734. 3-19
Figure 4-1. Position of squad members. 4-2
Figure 4-2. 81-mm mortar, M252. 4-3
Figure 4-3. Cannon assembly, M253. 4-5
Figure 4-4. Mount, M177, in folded position. 4-5
Figure 4-5. Baseplate, M3A1. 4-6

Table of Contents

Figure 4-6. Layout of equipment. 4-8
Figure 4-7. Baseplate placed against baseplate stake. 4-9
Figure 4-8. Kicking the mortar to dislodge the round. 4-13
Figure 4-9. Removing the firing pin. 4-14
Figure 4-10. Raising the cannon to a horizontal position. 4-15
Figure 4-11. Removing the round from the cannon. 4-15
Figure 4-12. Correct way to open an ammunition box. 4-26
Figure 4-13. Floating firing pin. 4-26
Figure 5-1. Position of squad members. 5-2
Figure 5-2. 120-mm mortar. 5-3
Figure 5-3. Cannon, M298, with old and new styles of breech cap. 5-5
Figure 5-4. Bipod assembly, M191 (carrier-/ground-mounted). 5-6
Figure 5-5. Bipod assembly, M190 (ground-mounted). 5-7
Figure 5-6. Baseplate, M9. 5-8
Figure 5-7. Rotating the artillery cleaning staff. 5-16
Figure 5-8. Holes in the cartridge body. 5-17
Figure 5-9. Withdrawing the cartridge from the barrel, M120. 5-17
Figure 5-10. Pressing the extractor catches. 5-18
Figure 5-11. Removing the breech cap assembly from the barrel. 5-20
Figure 5-12. Mortar carrier, M1064A3, front and side view. 5-23
Figure 5-13. Mortar carrier, M1064A3, rear view. 5-24
Figure 5-14. Withdrawing the cartridge from the barrel, M121. 5-33
Figure 5-15. Mechanical time superquick fuze, M776. 5-41
Figure 5-16. Point-detonating fuze, M935. 5-41
Figure 5-17. Multioption fuze, M734. 5-42
Figure 5-18. Point-detonating fuze, M745. 5-43
Figure 6-1. Mortar Fire Control System. 6-2
Figure 6-2. Commander's interface. 6-3
Figure 6-3. Power distribution assembly. 6-5
Figure 6-4. Pointing device. 6-6
Figure 6-5. Gunner's display. 6-7
Figure 6-6. Driver's display. 6-7
Figure 6-7. Vehicle motion sensor. 6-8
Figure 6-8. Graphic user interface. 6-11
Figure 6-9. Log-in screen. 6-12
Figure 6-10. "Unit List" screen. 6-14
Figure 6-11. "Configuration" screen. 6-15
Figure 6-12. "Data" screen. 6-16
Figure 6-13. "Geographic Reference" screen. 6-17
Figure 6-14. "Position" screen. 6-18
Figure 6-15. "Mounting Azimuth and Reference" screen. 6-19
Figure 6-16. "Channel A" screen. 6-20

Table of Contents

Figure 6-17 "Ammo Fire Unit" screen. 6-22
Figure 6-18. "Ammo Roll-up" screen. 6-23
Figure 6-19. "Status Fire Unit" screen. 6-24
Figure 6-20. "Check Fire" screen. 6-25
Figure 6-21. "Read" screen. 6-26
Figure 6-22. "Send" screen. 6-27
Figure 6-23. "Alerts" screen. 6-28
Figure 6-24. "Plot" screen. 6-29
Figure 6-25. "Plot" screen legend. 6-30
Figure 7-1. Pointing device "Status" screen. 7-2
Figure 7-2. "Boresight" screen. 7-4
Figure 7-3. "Navigation/Emplacement" screen. 7-5
Figure 7-4. "Navigation/Emplacement" screen: Send status. 7-6
Figure 7-5. Driver's display showing steering directions, distance, and position. .. 7-6
Figure 7-6. "Navigation/Emplacement" screen: destination azimuth, destination range, and heading. 7-7
Figure 7-7. Driver's display showing arrival at the waypoint. 7-7
Figure 7-8. "Navigation/Emplacement" screen: message transmitted and received by the FDC. 7-8
Figure 7-9. "Navigation/Emplacement" screen: fire area. 7-9
Figure 7-10. Driver's display activated. 7-10
Figure 7-11. "Fire Command" screen. 7-11
Figure 7-12. Subsequent "Fire Command" screen. 7-12
Figure 7-13. "End of Mission" screen. 7-13
Figure 7-14. "Not in Mission" screen. 7-13
Figure 7-15. Fire command for an assigned FPF. 7-14
Figure 7-16. End of mission. 7-15
Figure 7-17. Fire the stored FPF. 7-16
Figure 8-1. Observer more than 100 meters from mortar but within 100 meters of GT line. 8-3
Figure 8-2. Traversing fire. 8-7
Figure 8-3. Searching fire. 8-9
Figure 8-4. Bracketing method. 8-13
Figure 8-5. Ladder method of fire adjustment. 8-14
Figure 8-6. Adjusting fire onto a new target with the observer within 100 meters of the GT line. 8-16
Figure 9-1. Example of completed DA Form 5964-R. 9-2
Figure 9-2. Diagram of equipment layout and position of personnel for the gunner's examination (60-mm mortar). 9-5
Figure 9-3. Diagram of equipment layout and position of personnel for the gunner's examination (81-mm mortar, M252). 9-6
Figure 9-4. Diagram of equipment layout and position of personnel for the gunner's examination (120-mm mortar). 9-7

Figure A-1. Integrated training strategy. A-4
Figure A-2. Example training program for IBCT battalion. A-7
Figure A-3. Example training program for HBCT battalion. A-8
Figure B-1. Scaled range for short-range training round, M880. .... B-3
Figure B-2. Short-range training round, M880—practice round. B-4
Figure B-3. Converting 1:50,000 grid to 1:5,000 grid. B-10
Figure B-4. Plotting targets on the 1:5,000-scale map. B-11
Figure B-5. Determining direction from mortar position to the registration point. B-12
Figure B-6. Example completed DA Form 2188-R. B-12
Figure B-7. Example completed DA Form 2399 showing SHELL AND FUZE entries in the FDC ORDER and INITIAL FIRE COMMAND columns. B-13
Figure B-8. Subcaliber insert, M303. .... B-14
Figure B-9. Subcaliber trainer, M313. .... B-16

Tables

Table 1-1. Selected characteristics of U.S. mortars and ammunition. 1-4
Table 1-2. Mortar ammunition color codes. 1-15
Table 2-1. Selected characteristics of the aiming circles, M2 and M2A2. 2-5
Table 2-2. Set-up distance from objects. 2-11
Table 2-3. Sightunit, M67, equipment data. 2-19
Table 2-4. Sightunit, M64-series, tabulated data. 2-21
Table 2-5. Boresight, M45-series, tabulated data. 2-24
Table 2-6. M115 boresight tabulated data. 2-26
Table 2-7. Sequence for transmission of fire commands. 2-43
Table 3-1. Tabulated data for the 60-mm mortar, M224. 3-2
Table 3-2. High-explosive ammunition for the 60-mm mortar, M224. 3-17
Table 3-3. Illumination ammunition for the 60-mm mortar, M224. 3-17
Table 3-4. Smoke, white phosphorus ammunition for the 60-mm mortar, M224. 3-18
Table 3-5. Training practice ammunition for the 60-mm mortar, M224. 3-18
Table 4-1. Tabulated data for the 81-mm mortar, M252. 4-4
Table 4-2. High-explosive ammunition for the 81-mm mortar, M252. 4-18
Table 4-3. Illumination ammunition for the 81-mm mortar, M252. 4-19
Table 4-4. Smoke, white phosphorus ammunition for the 81-mm mortar, M252. 4-20
Table 4-5. Training practice ammunition for the 81-mm mortar, M252. 4-20
Table 5-1. Tabulated data for the 120-mm mortar. 5-4
Table 5-2. Tabulated data for the mortar carrier, M1064A3. 5-25
Table 5-3. High-explosive ammunition for 120-mm mortars, M120 and M121. 5-39
Table 5-4. Illumination ammunition for 120-mm mortars, M120 and M121. 5-39
Table 5-5. Smoke, white phosphorus ammunition for 120-mm mortars, M120 and M121. 5-40

Table of Contents

Table 5-6. Training practice ammunition for 120-mm mortars, M120 and M121 5-40
Table 6-1. Function keys. 6-3
Table 8-1. Initial range change. 8-12
Table 9-1. Organization for conducting gunner's examination (ground-mounted). 9-4
Table 9-2. Organization for conducting gunner's examination (carrier-mounted). 9-15
Table A-1. Institution courses. A-3
Table B-1. Supply data for short-range training round, M880 .... B-9

Preface

This publication prescribes guidance for leaders and crewmen of mortar squads. It concerns mortar crew training, and it is used with the applicable technical manuals (TMs) and Army Training and Evaluation Programs (ARTEPs). It presents practical solutions to assist in the timely delivery of accurate mortar fires, but does not discuss all possible situations. Local requirements may dictate minor variations from the methods and techniques described herein. However, principles should not be violated by modification of techniques and methods.

The scope of this publication includes mortar crew training at the squad level. The 60-mm mortar, M224; 81-mm mortar, M252; and 120-mm mortars, M120/M121 are discussed, to include nomenclature, sighting, equipment, characteristics, capabilities, and ammunition. (For information on the tactics, techniques, and procedures that mortar sections and platoons use to execute the combat mission, refer to FM 7.90.)

This manual was revised to delete references to obsolete material and systems. In addition to various editorial corrections, this revision—

- Removes all references to M2 and M19 mortar systems, as they are now obsolete.
- Removes all references to M29 and M29A1 mortar systems, as they are now obsolete (except for M29A1 use with the M303 subcaliber insert).
- Removes all references to the sabot, as this round is now obsolete.
- Replaces all references to the five-man mortar squad with the term "four-man mortar squad" to reflect the new structure.
- Removes all references to the first and second ammunition bearers to reflect the new four-man mortar squad. All references now read "ammunition bearer."
- Replaces references to common terms with their accepted modifications. These modifications include—
  - Replacing the term "nuclear, biological, chemical (NBC)" with "chemical, biological, radiological, nuclear (CBRN).'
  - Replacing the term "battle dress uniforms (BDUs)" with "Army combat uniforms (ACUs)."
  - Replacing the term "light infantry" with "infantry brigade combat team (IBCT)."
  - Replacing the terms "mechanized infantry" and "armored infantry" with "heavy brigade combat team (HBCT)."

The provisions of this publication are the subject of STANAG 2321, NATO Code of Colors for the Identification of Ammunition (Except Ammunition of a Caliber Below 22 Millimeters).

This publication prescribes DA Form 5964-R (Gunner's Examination Scorecard–Mortars).

Uniforms depicted in this manual were drawn without camouflage for clarity of the illustration. Unless this publication states otherwise, masculine nouns and pronouns refer to both men and women.

Terms that have joint or Army definitions are identified in both the glossary and the text. Terms for which FM 3-22.90 is the proponent FM are indicated with an asterisk in the glossary. This publication applies to the Active Army, the Army National Guard/Army National Guard of the United States, and the United States Army Reserve, unless otherwise stated.

The proponent for this publication is the U.S. Army Training and Doctrine Command. The preparing agency is the U.S. Army Infantry School. Send comments and recommendations to benn.229-S3-doc-lit@conus.army.mil or, using DA Form 2028 (Recommended Changes to Publications and Blank Forms) or its format, write directly to:

> Commandant, U.S. Army Infantry School
> ATTN: ATSH-INB
> 6650 Wilkin Drive, Building 74, Room 102
> Fort Benning, Georgia 31905-5593
> Telephone: 706-545-8623 or DSN 835-8623
> Fax: 706-545-8600 or DSN 835-8600

This page intentionally left blank.

Chapter 1
Introduction

Mortars are suppressive indirect fire weapons. They can be employed to neutralize or destroy area or point targets, screen large areas with smoke, and provide illumination or coordinated high-explosive/illumination. The mortar platoon's mission is to provide close and immediate indirect fire support for maneuver battalions and companies.

SECTION I. GENERAL DOCTRINE

Doctrine demands the timely and accurate delivery of indirect, high-angle fire to meet the needs of supported units. All members of the indirect fire team must be trained to quickly execute an effective fire mission.

EFFECTIVE MORTAR FIRE

1-1. For mortar fire to be effective, it must be dense and must hit the target at the *right* time with the *right* projectile and fuze. Good observation is necessary for effective mortar fire. Limited observation results in a greater expenditure of ammunition and less effective fire. Some type of observation is desirable for every target to ensure that fire is placed on the target. Observation of close battle areas is usually visual. When targets are hidden by terrain features or when great distance or limited visibility is involved, observation can be achieved by radar or sound. When observation is possible, corrections can be made to place mortar fire on the target by adjustment procedures; however, lack of observation must not preclude firing on targets that can be located by other means.

1-2. Mortar fire must be delivered by the most accurate means that time and the tactical situation permit. When possible, survey data or systems, such as the Mortar Fire Control System (MFCS), are used to accurately locate the mortar position and target. Under some conditions, only a rapid estimate of the location of weapons and targets may be possible. To achieve the most effective massed fires, the MFCS should be used or a survey using accurate maps should be made of each mortar position, registration point, and target.

1-3. The immediate objective is to deliver a large volume of accurate and timely fire to inflict as many enemy casualties as possible. The number of casualties inflicted in a target area can usually be increased by surprise fire. If surprise massed fires cannot be achieved, the time required to bring effective fires on the target should be kept to a minimum. The greatest demoralizing effect on the enemy can be achieved by delivering the maximum number of effective rounds from all the mortars in the shortest possible time.

1-4. Mortar units must be prepared to accomplish multiple fire missions. They can provide an immediate, heavy volume of accurate fire for sustained periods.

1-5. In heavy brigade combat team (HBCT) companies, mortars are normally fired from mortar carriers; however, they maintain their capability to be ground-mounted. Firing from carriers permits rapid displacement and quick reaction. Infantry brigade combat team (IBCT) companies must fire their mortars from the ground.

MORTAR POSITIONS

1-6. Mortars should be employed in defilade to protect them from enemy direct fire and observation, and to take the greatest advantage of their indirect fire role. Although the use of defilade precludes sighting the weapons directly at the target (direct lay), it is necessary for survivability. Because mortars

Chapter 1

are indirect fire weapons, special procedures ensure that the weapon and ammunition settings used will cause the projectile to burst on or above the target. A coordinated effort by the indirect fire team ensures the timely and accurate engagement of targets.

SECTION II. INDIRECT FIRE TEAM

Indirect fire procedure is a team effort (Figure 1-1). Since the mortar is normally fired from defilade, the indirect fire team gathers and applies the required data. The team consists of a forward observer (FO), a fire direction center (FDC), and the gun squad.

APPLICATIONS

1-7. To successfully accomplish missions from a defilade position, certain steps must be followed in applying essential information and engaging targets. These steps include—
- Locate targets and mortar positions.
- Determine chart data (direction, range, and vertical interval from mortars to targets).
- Convert chart data to firing data.
- Apply firing data to the mortar and ammunition.

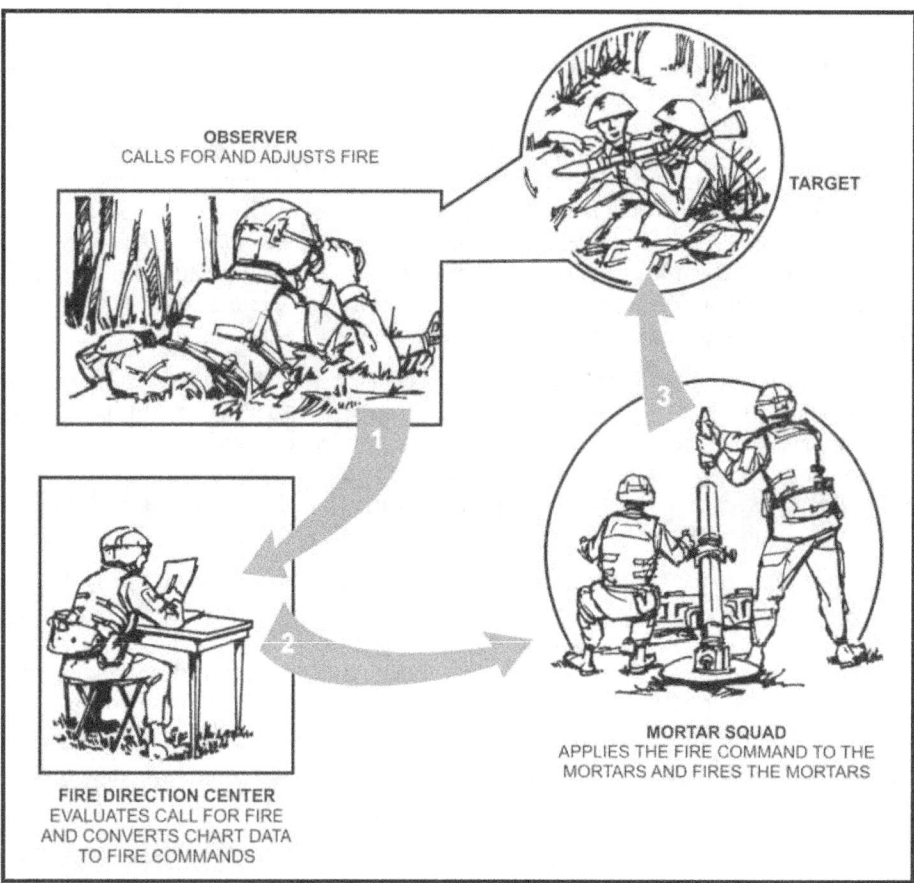

Figure 1-1. Indirect fire team.

TEAM MISSION

1-8. The team mission is to provide accurate and timely response to the unit it supports. Effective communication is vital to the successful coordination of the efforts of the indirect fire team.

Introduction

1-9. The battalion headquarters and headquarters company (HHC) fire support platoon provides the fire support teams (FISTs) to the battalion's maneuver companies upon deployment. The FISTs typically move to and remain with their supported companies and platoons. Each rifle company is supported by a 10-man FIST consisting of a lieutenant, staff sergeant, radio-telephone operator (RTO), and a driver with an armored personnel carrier (in HBCTs) or a high mobility multipurpose wheeled vehicle (HMMWV) (in IBCTs) at company headquarters, and six FOs (one 2-man team for each infantry platoon in the company). Each armor company is supported by a 4-man FIST consisting of a lieutenant, staff sergeant, RTO, and a driver with an armored personnel carrier at company headquarters. The FO's job is to find and report the location of targets, and to request and adjust fire.

1-10. The FDC has two computer personnel who control the mortar firing. They convert the data in a call for fire (CFF) from the FO into firing data that can be applied to the mortars and ammunition. The FDCs for medium and heavy mortars also have a driver.

1-11. A mortar squad consists of three to four mortarmen, depending on the system. The squad lays the mortar and prepares the ammunition using the data from the FDC fire command. When those data have been applied, the squad fires the mortar—it must also be able to fire without an FDC. Medium and heavy mortar squads also have a driver.

UNITED STATES MORTARS

1-12. U.S. mortars are smooth-bore, muzzle-loaded, and high-angle-of-fire weapons. Table 1-1 displays some selected characteristics of U.S. mortars and their ammunition. The mortar squad consists of three to four men, and the mortar components include the cannon, mount, and baseplate. Mortar ammunition consists of a fuze, cartridge, and propellant charges.

CREW

1-13. The mortar crew consists of three to four men—
- The squad leader is in charge of the squad and supervises the emplacement, laying, and firing of the mortar.
- The gunner manipulates the sight, elevating gear handle, and traversing assembly wheel. He also places firing data on the sight and lays the mortar for deflection and elevation.
- The assistant gunner loads the mortar and assists the gunner in shifting the mortar.
- The ammunition bearer prepares the ammunition.

1-14. The number of men assigned to a squad, however, may vary depending on the unit's mission.

EQUIPMENT

1-15. The mortar can be broken down into three parts—
(1) The cannon assembly consists of the barrel, which is sealed at the lower end with a breech plug. The muzzle end of the 81-mm and the 120-mm mortar has a cone-shaped blast attenuator device (BAD) fitted to reduce noise and blast overpressure. The 60-mm has a combination carrying handle and firing mechanism attached to the breech cap.
(2) The bipod provides front support for the barrel and carries the gears necessary to lay the mortar. It also has a slot to receive the sightunit. The 60-mm can fire without the bipod.
(3) The baseplate supports and aligns the mortar during firing. It has a large surface area to prevent the mortar from sinking into the ground during firing and has a socket that enables a full 360-degree traverse without moving the baseplate.

AMMUNITION

1-16. Mortar ammunition consists of a fuze, a cartridge, and propellant charges.

Chapter 1

Fuzes

1-17. U.S. mortars use a variety of fuzes.
- Point-detonating (PD), impact (IMP), or superquick (SQ) fuzes detonate the cartridge on impact with the ground.
- Near-surface burst (NSB) fuzes explode on or near the ground.
- Proximity (PROX) fuzes explode above the ground. 60/81 PRX detonates two meters above the ground, and 120 PRX detonates four meters above the ground.
- Delay (DLY) fuzes explode 0.05 seconds after impact.
- Time fuzes explode after a preselected time has elapsed from the round being fired (usually used with illumination).
- Multioption fuzes combine two or more of the other modes into one fuze.

Cartridge

1-18. The cartridge can contain high-explosive (HE), illumination (ILLUM), or smoke.
- HE is used against personnel and light materiel targets.
- ILLUM is used in night missions requiring illumination for assistance in observation.
- Smoke, with white or red phosphorus (WP, RP) filler, is used as a screening, signaling, casualty-producing, or incendiary agent.

Propellant Charges

1-19. Propellant charges provide the explosive force to propel the cartridge through the air. Propellant charges consist of a base charge (Charge 0) and from four to ten incremental charges. The number of incremental charges used is determined by the range and time of flight.

Table 1-1. Selected characteristics of U.S. mortars and ammunition.

CHARACTERISTIC	60-mm M224	81-mm M252	120-mm M120/121
Weight (pounds): Handheld:	47 18	121 NA	319 NA
Range (meters) HE Minimum: HE Maximum:	 70 3,490	 83 5,608	 200 7,200
Rate of Fire Maximum: Sustained:	 30 per min for 4 min 20 per min	 30 per min for 2 min 15 per min	 16 per min for 1 min 4 per min
Crew:	3	4	4
Weight of HE (pounds):	3.7 (M720A1/M888)	9.1 (M821/M889)	29.2 (M933/M934)
Illumination Candlepower: Duration: IR:	 300,000 55 secs Available	 600,000 50 to 60 secs Available	 1,000,000 46 to 60 secs Available
Smoke:	Available	Available	Available

Introduction

SECTION III. SAFETY PROCEDURES

Although safety is a command responsibility, each member of the mortar fire team must know safety procedures and enforce them. Safety is enhanced with MFCS features, such as the ability to view safety data and situational awareness information, as well as to input fire support coordination measures (FSCMs) and view them using the digital plot feature. No matter how sophisticated the system, it must be checked during live-fire exercises (LFXs).

DUTIES OF THE SAFETY OFFICER AND SUPERVISORY PERSONNEL

1-20. Safety officers must help commanders meet the responsibility of enforcing safety procedures. The safety officer has two principal duties: first, ensure that the section is properly laid so that, when rounds are fired, they land in the impact area; second, ensure that all safety precautions are observed at the firing point (FP).

DUTIES BEFORE DEPARTING FOR RANGE

1-21. The safety officer must read and understand—
- AR 385-63, Range Safety. 19 May 2003.
- Post range and terrain regulations.
- The terrain request of the firing area to know safety limits and coordinates of firing positions.
- Appropriate FMs and TMs pertaining to weapons and ammunition to be fired.

DUTIES OF SUPERVISORY PERSONNEL

1-22. Supervisory personnel must know the immediate action to be taken for firing accidents. The following is a list of *minimum* actions that must be taken if an accident occurs.
(1) Administer first aid to injured personnel, and then call for medical assistance.
(2) If the ammunition or equipment presents further danger, move all personnel and equipment out of the area.
(3) Do not change any settings on, or modify, the position of the mortar until an investigation has been completed.
(4) Record the ammunition lot number involved in the accident or malfunction and report it to the battalion ammunition officer. If a certain lot number is suspect, its use should be suspended by the platoon leader.

MORTAR RANGE SAFETY CHECKLIST

1-23. A mortar range safety checklist can be written for local use. The following is a suggested checklist (it can also include three columns on the right, titled "Yes," "No," and "Remarks").

Items to Check Before Firing

1-24. Is a range log or journal maintained by the officer in charge (OIC)?

1-25. Is radio or telephone communication maintained with—
- Range control?
- Unit S3?
- Firing crews?
- FOs?
- Road or barrier guards?

Chapter 1

1-26. Are the required emergency personnel and equipment present on the range?
- Properly briefed and qualified medical personnel.
- A wheeled or tracked ambulance.
- Firefighting equipment.

1-27. Are the following range controls and warning devices available, readily visible, and in use during the firing exercise?
- Barrier/road guards briefed and in position.
- Road barriers in position.
- Red range flag in position.
- Blinking red lights for night firing.
- Signs warning trespassers to beware of explosive hazards and not to remove duds or ammunition components from ranges.
- Noise hazard warning signs.

1-28. Are current copies of the following documents available and complied with?
- AR 385-63, Range Safety.
- Technical and field manuals pertinent to the mortar in use.
- Appropriate firing tables.
- Installation range regulations.

1-29. Are the following personal safety devices and equipment available and in use?
- Helmets.
- Protective earplugs.
- Protective earmuffs (for double hearing protection).

1-30. Is the ammunition the correct caliber, type, and quantity required for the day's firing? Are the rounds, fuzes, and charges—
- Stored in a location to minimize possible ignition or detonation?
- Covered to protect them from moisture and direct sunlight?
- Stacked on dunnage to keep them clear of the ground?
- Strictly accounted for by lot number?
- Exposed only immediately before firing?
- Stored separately from ammunition and protected from ignition?

1-31. Has the range safety officer verified the following?
- The mortar safety card applies to the unit and exercise.
- The firing position is correct and applies to the safety card, and the base mortar is within 100 meters of the surveyed FP.
- Boresighting and aiming circle declination are correct.
- The plotting board, mortar ballistic computer (MBC), MFCS, or lightweight handheld mortar ballistic computer (LHMBC) is correct.
- The FO has been briefed on the firing exercise and knows the limits of the safety fan.
- The lay of each mortar is correct.
- The safety stakes (if used) are placed along the right and left limits.
- Each safety noncommissioned officer (NCO) and gunner has been informed in writing of—
  - Right and left limits (deflection).
  - Maximum elevation and charge.
  - Minimum elevation and charge.
  - Minimum time setting for fuzes.
- All personnel at the firing position have been briefed on safety misfire procedures.
- If the safety card specified overhead fire, firing is in accordance with AR 385-63.

Introduction

> **NOTE:** Firing mortars over the heads of unprotected troops by Marine Corps units is not authorized or recommended for Army units. This restriction applies only during peacetime; it does not apply during combat.

- The mortars are safe to fire by checking—
    - Prefire safety checks for the specific mortar.
    - Mask and overhead clearance.
    - Weapons and ammunition.
    - Properly seated sights on weapons.
    - The lights on the sights and aiming stakes for night firing.
- The OIC is informed that the range is cleared to fire and that range control has placed it in a "wet" status.

Items to Check During Firing

1-32. Are the unit personnel adhering to the safety regulations?

1-33. Is each charge, elevation, and deflection setting checked before firing?

1-34. Does the safety NCO declare the mortar safe to fire before the squad leader announces, "Hang it, fire"?

1-35. Do all gun settings remain at last data announced until a subsequent fire command is issued by the FDC?

1-36. Are ammunition lots kept separate to avoid the firing of mixed lots?

Items to Check After Firing

1-37. Have the gunners and safety NCO verified that no loose propellants are mixed with the empty containers?

1-38. Has the safety NCO disposed of the unused propellants?

1-39. Has the unused ammunition been inventoried and repacked properly?

1-40. Have the proper entries been made in the equipment logbook (DA Form 2408-4 [Weapon Record Date])?

1-41. Has the OIC or safety officer notified range control of range status and other required information?

1-42. Has a thorough range police been conducted?

SAFETY CARD

1-43. The safety officer receives a copy of the safety card from the OIC or range control—depending on local regulation—before allowing fire to begin. He constructs a safety diagram based on the information on the safety card. A safety card should be prepared and approved for each firing position and type of ammunition used. The form of the card depends upon local regulations (training list, overlay, range bulletin). Even without a prescribed format, it should contain the following—

- Unit firing or problem number.
- Type of weapon and fire.
- Authorized projectile, fuze, and charge zone.
- Grid of the platoon center.
- Azimuth of left and right limits.
- Minimum and maximum ranges and elevations.
- Any special instructions to allow for varying limits on special ammunition or situations.

Chapter 1

SAFETY DIAGRAM AND SAFETY "T"

1-44. On receipt of the safety card or safety record, the safety officer constructs a safety diagram. The safety diagram is a graphic portrayal of the data on the safety card and does not need to be drawn to scale, but must accurately list the sight settings that delineate the impact area. The safety "T" serves as a convenient means of checking the commands announced to the gun crews against those commands that represent the safety limits. The construction of the safety diagram and safety "T" is the same for all mortars.

1-45. The diagram shows the right and left limits, and deflections corresponding to those limits; the maximum and minimum elevations; and the minimum fuze settings (when applicable) for each charge to be fired. The diagram also shows the minimum and maximum range lines, the left and right azimuth limits, the deflections corresponding to the azimuth limits, and the direction and mounting azimuth on which the guns are laid. The safety diagram must show only necessary information.

1-46. To accurately complete a safety diagram, the safety officer must use the information supplied by range control or, in the example in Figure 1-2, the safety card/record to—

    (1) Enter the known data (supplied from the safety card) on the safety diagram.
    (2) Determine the direction of fire or the center of the sector.
    (3) Determine the mounting azimuth.
    (4) Determine mils left and right deviations of the mounting azimuth.
    (5) Enter the referred deflection.
    (6) Determine deflections to left and right limits.
    (7) Determine minimum and maximum charges and elevations.
    (8) If illumination is to be used, determine, from the appropriate firing tables, the minimum and maximum charges and ranges to burst and impact for the canister. The minimum range is used to determine the minimum charge and range to burst. The maximum range is used to determine the maximum charge and range to impact.

1-47. An example of the preparation of a safety diagram and safety "T" follows.

Introduction

EXAMPLE

An 81-mm mortar section is firing at FP 52 with M821 HE and M853A1 ILLUM. The safety officer receives the safety record/card (Figure 1-2) from either range control or the OIC (in accordance with local range regulations).

ARTILLERY/MORTAR SAFETY RECORD

DATE: 10 December 2003

FIRING POINT: FP 52				WEAPONS: M252, M224, M120/121			
COORDINATES: GL12345678				ELEV		FUZE	
Weapon Projectile	Left Limit Mils	Right Limit Mils	Minimum Range Meters	Maximum Range Meters	Minimum Charge	Maximum Charge	Maximum Ordnance Meters or Feet
M821 HE	0500	0920	300	4000	0	4	NA
M853A1 IL	0500	0920	300	4000	0	4	NA

Figure 1-2. Example completed safety record or card.

The safety officer calculates the data for the safety diagram and then places the data for each type of ammunition on separate safety diagrams (Figure 1-3).

Calculate the direction of fire. Left limit is smaller than right limit.
Subtract the left limit azimuth from the right limit azimuth. Divide the result by two, and add that number to the left limit azimuth.

```
RIGHT LIMIT          + 0920
MINUS LEFT LIMIT     - 0500
SUM                    420
SUM ÷ 2 =              210
LEFT LIMIT           + 500
TOTAL                  710
```

The answer, 0710, is the direction of fire or the azimuth center of sector.
(To calculate the direction of fire if the left limit is larger than the right limit, add 6400 to the right limit, and use the same calculations as above. If the final answer is more than 6400, subtract 6400 to get the direction of fire.)

Round off the direction of fire or mounting azimuth.
For all mortars using the M16 plotting board, round off to nearest 50 mils. In this example, the safety officer rounds 0710 to 0700.

Enter the referred deflection.
The section sergeant provides the referred deflection. It can be any number, but 2800 is normally used.

Chapter 1

Determine deflection for the left and right limits.
- Determine the number of mils from the mounting azimuth to the left limit.

MOUNTING AZIMUTH	0700
LEFT LIMIT	-0500
MILS TO LEFT LIMIT	0200

- Using the LARS rule for referred deflection, calculate the left limit deflection.

CENTER OF SECTOR REFERRED DEFLECTION	2800
MILS TO LEFT LIMIT	+ 0200
LEFT LIMIT DEFLECTION	3000

- Determine the number of mils from the mounting azimuth to the right limit.

RIGHT LIMIT	0920
MOUNTING AZIMUTH	- 0700
MILS TO RIGHT LIMIT	0220

- Using the LARS rule for referred deflection, calculate the right limit deflection.

CENTER OF SECTOR REFERRED DEFLECTION	2800
MILS TO RIGHT LIMIT	+ 0220
RIGHT LIMIT DEFLECTION	2580

Determine minimum and maximum charges and elevations.

Use the firing tables (FTs) for the mortar being fired to determine minimum and maximum charges and elevations. For the example, the maximum range is 4,000 meters and the minimum range is 300 meters.

- M821 HE. Using FT-81-AR-2 for the M821 HE cartridge, the maximum and minimum charges and elevations are—

Max or Min	Charge	Elevation (mils)
Maximum	4	1185
Maximum	3	1055
Minimum	0	1256

- M853A1 ILLUM. Using FT-81-AR-2 for the M853A1 ILLUM cartridge, the maximum and minimum charges, elevations, and time settings are—

Max or Min	Charge	Elevation (mils)	Time Setting (seconds)
Maximum	4	1142	45.5
Maximum	3	0925	35.1
Minimum	1	1507	25.7

Introduction

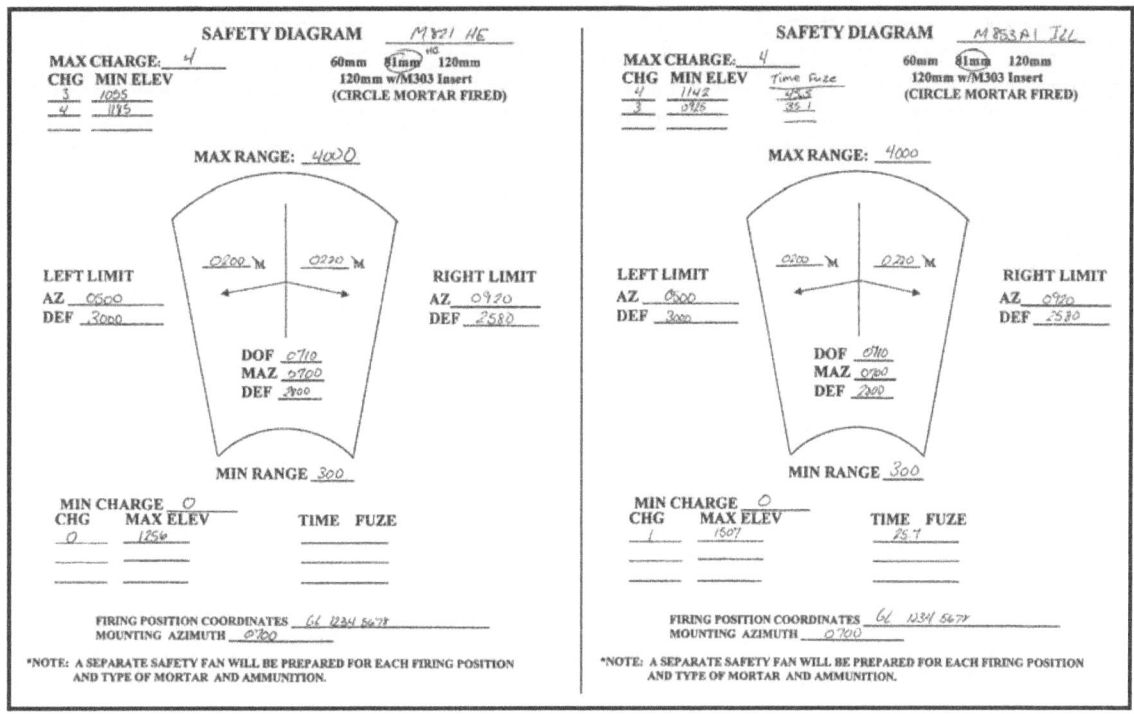

Figure 1-3. Safety diagrams for M821 HE and M853A1 ILLUM.

A safety "T" is made for each type of cartridge being fired. The requisite data from the safety diagram is transcribed onto the safety "T" (Figure 1-4 and Figure 1-5), and each gun has a copy.

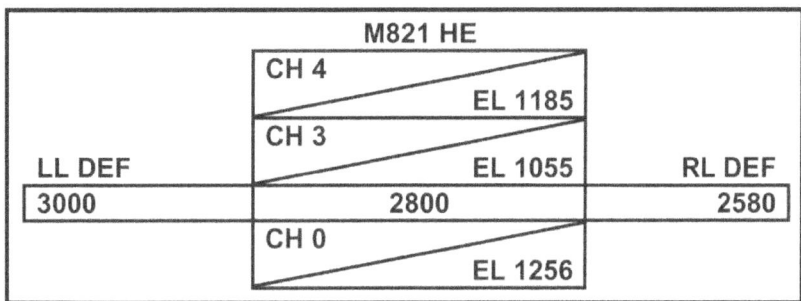

Figure 1-4. Safety "T" for M821 HE.

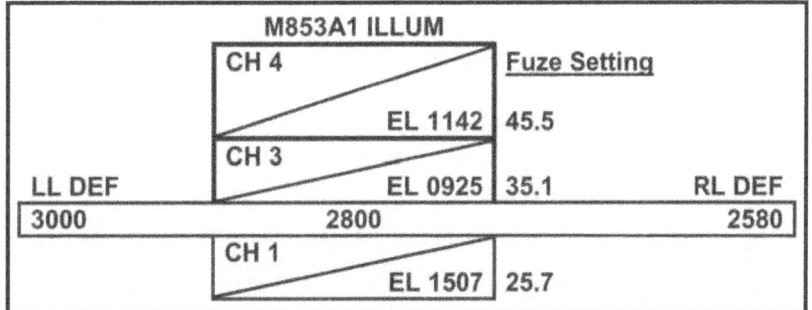

Figure 1-5. Safety "T" for M853A1 ILLUM.

7 December 2007   FM 3-22.90   1-11

SURFACE DANGER ZONES

1-48. The surface danger zone (SDZ) is the ground and airspace designated within the training complex for vertical and lateral containment of projectiles, fragments, debris, and components resulting from the firing, launching, or detonation of weapon systems. It is composed of an impact area and buffer zones.

COMPONENTS

1-49. SDZs are usually pie-shaped, with the target area in the center surrounded by concentric zones. Components include the firing position, impact areas, and buffer zones.

Firing Position

1-50. The firing position is the point or location at which the mortar is placed for firing.

Impact Area

1-51. The ground and associated airspace within the training complex used to contain fired or launched ammunition and explosives, and the resulting fragments, debris, and components from various weapon systems. Indirect fire weapon system impact areas include probable error for range and deflection.

Target Area

1-52. The target area contains the targets.

Dispersion Area

1-53. The dispersion area, by keeping the rounds inside the impact area, accounts for human error, gun or cannon tube wear, propellant temperature, and so on. Its width is based on the range and deviation probable errors for the maximum range of the charge permitted.
- The left and right dimensions are eight deflection probable errors (PE_D) from the left and right limits of the target area.
- The far dimensions are eight range probable errors (PE_R) from the far edge of the target area.

Buffer Zones

1-54. Buffer zones, or secondary danger areas, contain the fragments, debris, and components from frangible or explosive projectiles and warheads functioning on the edge of the target area. It has two parts: Area A and Area B.

Area A

1-55. Area A is the secondary danger area (buffer zone) that laterally parallels the impact area and dispersion area and contains fragments, debris, and components from frangible or explosive projectiles and warheads functioning on the right or left edge of the impact area or ricochet area. It starts at a 25-degree angle from the impact area (increased to 70 degrees for ranges at and beyond 600 meters for 60-mm, 940 meters for 81-mm, and 1,415 meters for 120-mm mortars). The width of area A depends on the mortar:
- 60-mm: 250 meters
- 81-mm: 400 meters
- 120-mm: 600 meters

Introduction

Area B

1-56. Area B is the secondary danger area (buffer zone) on the downrange (far) side of the impact area and contains fragments, debris, and components from frangible or exploding projectiles and warheads functioning on the far edge of the impact area and area A. The width of area B depends on the mortar:
- 60-mm: 300 meters
- 81-mm: 400 meters
- 120-mm: 600 meters

CONSTRUCTION

1-57. SDZs usually exist for approved FPs and can be found at range control or in the range book for the FP. If the SDZ has to be constructed, personnel should refer to DA Pam 385-63.

SECTION IV. AMMUNITION

A complete round of mortar ammunition contains all of the components needed to get the round out of the tube and to burst at the desired place and time. This section discusses the proper care and handling, color codes, and field storage of ammunition.

AMMUNITION CARE AND HANDLING

1-58. The key to proper ammunition functioning is protection. Rounds prepared but not fired should be returned to their containers, fin end first. Safety is always a matter of concern for all personnel and requires special attention where ammunition is concerned. Supervision is critical—improper care and handling can cause serious accidents, as well as inaccurate fire. Some of the principles of proper ammunition handling are—
- Never tumble, drag, throw, or drop individual cartridges or boxes of cartridges.
- Do not allow smoking, open flames, or other fire hazards around ammunition storage areas.
- Inspect each cartridge before it is loaded for firing. Dirty ammunition can damage the weapon or affect the accuracy of the round.
- Keep the ammunition dry and cool.
- Never make unauthorized alterations or mix components of one lot with another.

NOTE: For care and handling of specific mortar rounds, see corresponding chapters in this manual.

PROJECTILES/CARTRIDGES

1-59. Each projectile must be inspected to ensure that there is no leakage of the contents and that the projectile is correctly assembled.

BURNING OF UNUSED PROPELLING CHARGES

1-60. Mortar increments and propelling charges are highly flammable, and they must be handled with extreme care to prevent exposure to heat, flame, or any spark-producing source, such as the hot residue from burning increments or propelling charges that float downward after a cartridge leaves the cannon. Like other types of ammunition, increments and propelling charges must be kept cool and dry. Storing these items inside metal ammunition boxes until needed is an effective way to prevent premature combustion.

1-61. Unused charges must not be saved, but should be removed to a storage area until they can be burned or otherwise disposed of in accordance with local range or installation regulations or standing operating procedures (SOPs).

1-62. Burning increments create a large flash and lots of smoke. In a tactical environment, the platoon leader must ensure that burning increments do not compromise camouflage and concealment. The

Chapter 1

burning of increments in a dummy position, if established, can aid in the deception effort. The safety officer, in a range environment, supervises the disposal of unused propellant increments.

FUZES

1-63. Never fire a round with a fuze that is not authorized for that round. Specific fuzes available for each weapon system are discussed in this manual.

1-64. Fuzes are sensitive to shock and must be handled with care. Before fuzing a round, inspect the threads of the fuze and fuze well for cleanliness and crossed threads. The fuze should be screwed into the fuze well slowly until resistance is met and then firmly seated with a sharp twist of the M25 or M18 fuze wrench, as appropriate.

> **WARNING**
>
> **Premature detonation may occur if a fuze is not properly seated.**

1-65. To prevent accidental functioning of the point-detonating elements of M524-series fuzes, the fuzes must not be dropped, rolled, or struck under any circumstances. Any mechanical time (MT) fuze that is modified after it is set must be reset to SAFE, and the safety wires (if applicable) must be replaced before the fuze is repacked in the original carton.

1-66. All primers must be inspected before use for signs of corrosion. If a seal has been broken, it is likely that the primer has been affected by moisture, and it should be turned in.

SEGREGATION OF AMMUNITION LOTS

1-67. Different lots of propellant burn at different rates and give slightly different effects in the target area; therefore, the registration corrections derived from one lot do not always apply to another. Ammunition MUST be segregated by lot and weight zone. In the field storage area, on vehicles, or in a dump, ammunition lots should be roped off with communications wire or twine and conspicuously marked with a cardboard sign or other marker.

AMMUNITION COLOR CODES

1-68. Mortar ammunition is painted and marked with a color code for quick, accurate identification. A chart (Table 1-2) identifies rounds using NATO and U.S. color codes.

Table 1-2. Mortar ammunition color codes.

TYPE OF ROUND	NATO COLOR CODE			U.S. COLOR CODE		
	ROUND	MARKINGS	BAND	ROUND	MARKINGS	BAND
HIGH-EXPLOSIVE (Causes troop casualties and damage to light material)	Olive Drab	Yellow Drab	NA	Olive	Yellow	Yellow
WHITE PHOSPHORUS (To screen or signal, acts as an incendiary)	Light Green	Red	Yellow	Light Green	Red	Yellow
RED PHOSPHORUS (To screen or signal, acts as an incendiary)	Light Green	Black	Brown	Light Green	Black	Brown
ILLUMINATION (To illuminate, signal, and mark)	White	Black	NA	NA	NA	NA
TRAINING PRACTICE (For training and practice)	Blue	White	Brown or None	Blue	White	Brown

FIELD STORAGE OF AMMUNITION

1-69. Most ammunition components can be stored at temperatures as low as -80 degrees Fahrenheit for no longer than three days and as high as 160 degrees Fahrenheit for no longer than four hours.

1-70. Regardless of the method of storage, ammunition in the storage area faces certain hazards. The greatest hazards are weather; enemy fire; chemical, biological, radiological, nuclear (CBRN) contamination; improper handling; and accidental fires. Adhere to the following guidelines to properly store ammunition.

- Stack ammunition by type, lot number, and weight zone (Figure 1-6).

NOTE: WP ammunition must be stacked fuze-end up so that melted filler will settle at the bottom, reducing the chance of voids forming off the long axis of the cartridge. Voids off the long axis of the cartridge will cause the cartridge to fly erratically.

Chapter 1

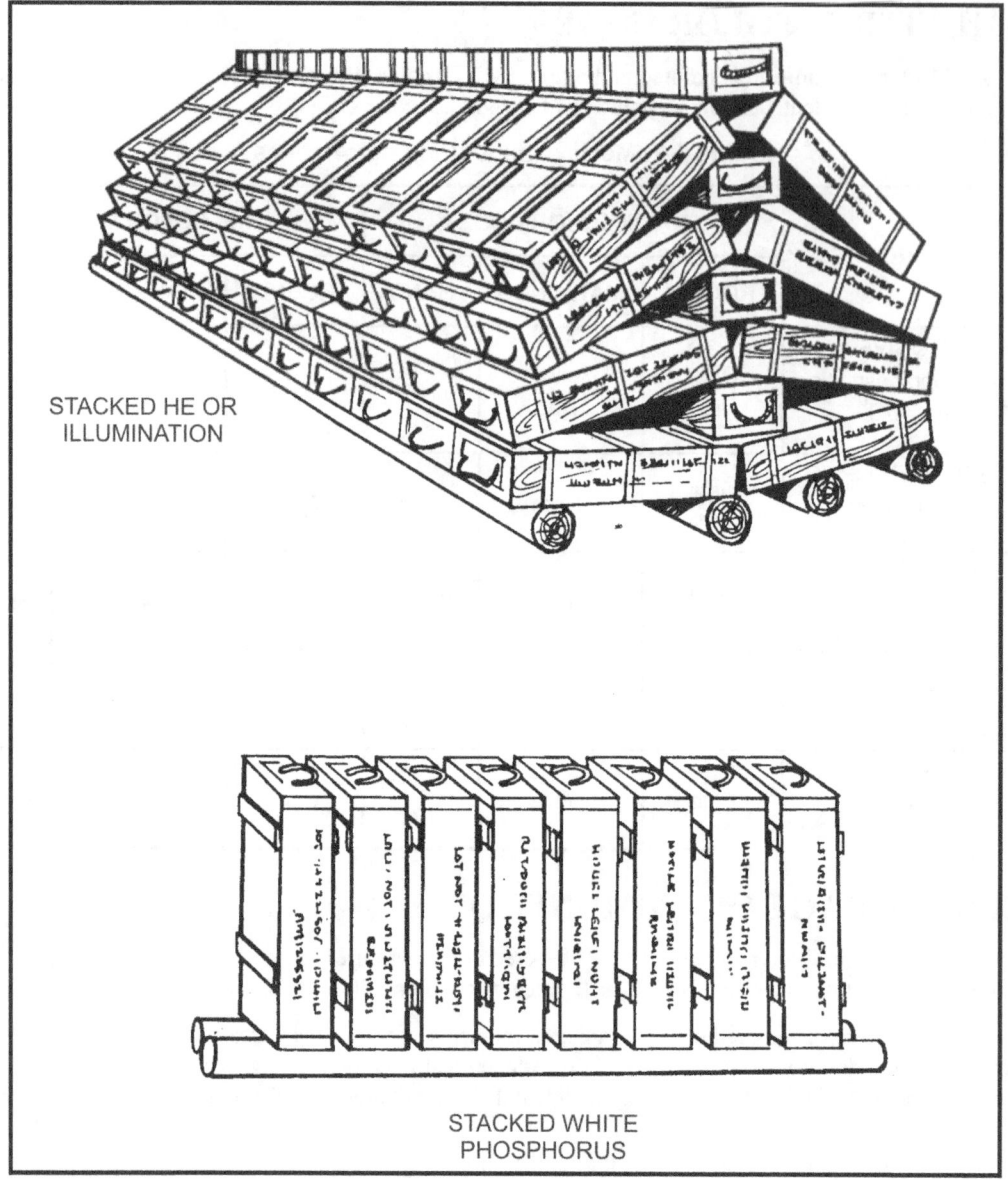

Figure 1-6. Stacked ammunition.

- If ammunition is stored on the ground, use at least 6 inches of strong dunnage under each stack.
- Keep the ammunition dry and out of direct sunlight by storing it in a vehicle or covering it with a tarpaulin. Be sure to provide adequate ventilation around the ammunition and between the covering material and the ammunition.
- Protect ammunition (as much as possible) from enemy indirect fires. If sandbags are used for protection, keep the walls at least 6 inches from the stacks and the roof at least 18 inches from the stacks to ensure proper ventilation.

1-71. With some PROX fuzes, the number of malfunctions can increase if fired in temperatures below 0 degrees Fahrenheit or above 120 degrees Fahrenheit. Powder temperature affects the muzzle velocity of a projectile and is of frequent concern to the FDC.

Chapter 2
Sighting and Fire Control Equipment

Proper employment of sighting and fire control equipment ensures effective fire against the enemy. This chapter describes this equipment and its applications.

SECTION I. COMPASS, M2

The M2 compass (Figures 2-1 through 2-3) is a multipurpose instrument used primarily to obtain azimuths and angles of sight. It also measures grid azimuths after the instrument has been declinated for the locality.

> **NOTE:** For detailed information, see TM 9-1290-333-15.

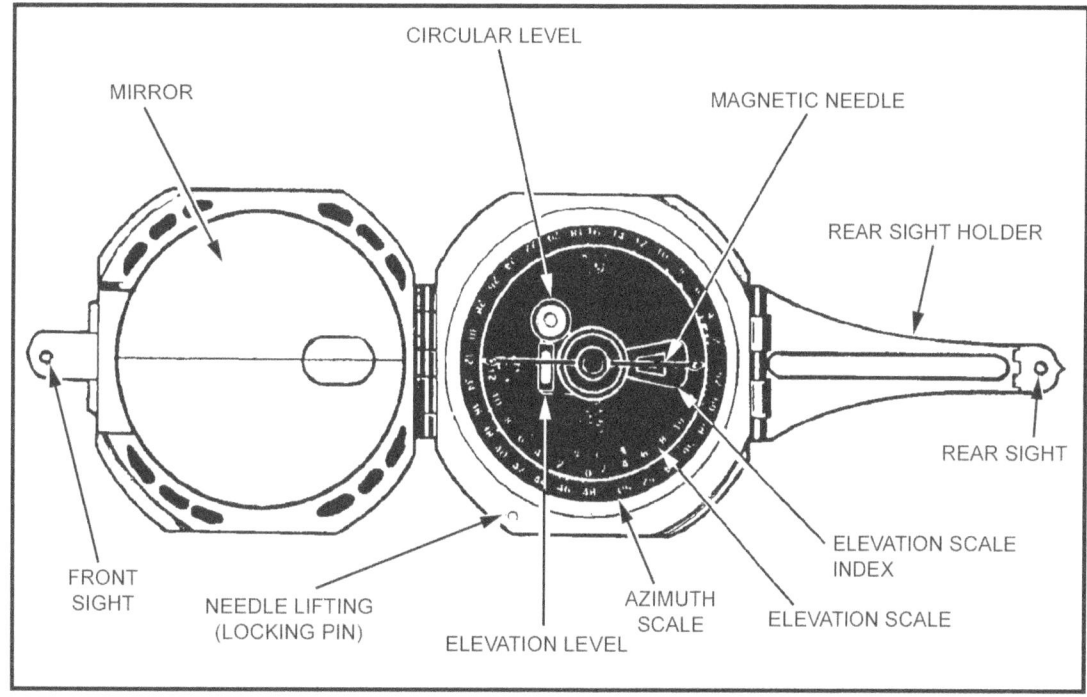

Figure 2-1. Compass, M2 (top view).

CHARACTERISTICS

2-1. The main characteristics of the M2 compass include the following:
- Angle-of-sight scale: 1200-0-1200 mils.
- Azimuth scale: 0 to 6400 mils.
- Dimensions closed: 2 3/4 inches by 1 1/8 inches.
- Weight: 8 ounces.

Chapter 2

DESCRIPTION

2-2. The principal parts of the compass include—
- Compass body assembly.
- Angle-of-sight mechanism.
- Magnetic needle and lifting mechanism.
- Azimuth scale and adjuster.
- Front and rear sight.

COMPASS BODY ASSEMBLY

2-3. This assembly consists of a nonmagnetic body and a circular glass window that covers the instrument and keeps dust and moisture from its interior, protecting the compass needle and angle-of-sight mechanism. A hinge assembly holds the compass cover in the position in which it is placed. A hole in the cover coincides with a small oval window in the mirror on the inside of the cover. A sighting line is etched across the face of the mirror.

ANGLE-OF-SIGHT MECHANISM

2-4. The angle-of-sight mechanism is attached to the bottom of the compass body. It consists of an actuating (leveling) lever located on the back of the compass, a leveling assembly with a tubular elevation level, and a circular level. The instrument is leveled with the circular level to read azimuths and with the elevation level to read angles of sight. The elevation (angle -of -sight) scale and the four points of the compass, represented by three letters and a star, are engraved on the inside bottom of the compass body. The elevation scale is graduated in two directions; in each direction, it is graduated from 0 to 1200 mils in 20-mil increments and numbered every 200 mils.

MAGNETIC NEEDLE AND LIFTING MECHANISM

2-5. The magnetic needle assembly consists of a magnetized needle and a jewel housing that serves as a pivot. The north-seeking end of the needle is white. (The newer compasses have the north and south ends of the needle marked "N" and "S" in raised, white lettering.) On some compasses, a thin piece of copper wire is wrapped around the needle for counterbalance. A lifting pin projects above the top rim of the compass body. The lower end of the pin engages the needle-lifting lever. When the cover is closed, the magnetic needle is automatically lifted from its pivot and held firmly against the window of the compass.

AZIMUTH SCALE AND ADJUSTER

2-6. The azimuth scale is a circular dial geared to the azimuth scale adjuster. This permits rotation of the azimuth scale about 900 mils in either direction. The azimuth index provides a means of orienting the azimuth scale at 0 or the declination constant of the locality. The azimuth scale is graduated from 0 to 6400 in 20-mil increments and numbered at 200-mil intervals.

FRONT AND REAR SIGHT

2-7. The front sight is hinged to the compass cover. It can be folded across the compass body, and the cover closed. The rear sight is made in two parts—a rear sight and a holder. When the compass is not being used, the rear sight and holder are folded across the compass body and the cover is closed.

USE

2-8. The compass should be held as steadily as possible to obtain accurate readings. The use of a sitting or prone position, a rest for the hand or elbows, or a solid nonmetallic support helps eliminate unintentional movement of the instrument. When being used to measure azimuths, the compass must not be near metallic objects.

Sighting and Fire Control Equipment

2-9. To measure a magnetic azimuth—

(1) Zero the azimuth scale by turning the scale adjuster.

(2) Place the cover at an angle of about 45 degrees to the face of the compass so that the scale reflection is viewed in the mirror.

(3) Adjust the front and rear sights to the desired position. Sight the compass using any of these methods:

- Raise the front sight and the extended rear sight assembly perpendicular to the face of the compass (Figures 2-2 and 2-3). Sight over the tips of the front and rear sights. If the object is above the line of sighting, fold the rear sight toward the eye, as needed. The instrument is correctly aligned when, with the level centered, the operator sees the tips of the sights and the center of the object at the same time.

- Raise the rear sight approximately perpendicular to the face of the compass. Sight on the object through the opening in the rear sight holder and through the window in the cover. Keep the compass level, and raise or lower the eye along the opening in the rear sight holder until the black center line of the window bisects the object and the opening in the rear sight holder.

- Fold the rear sight holder out parallel with the face of the compass with the rear sight perpendicular to its holder. Sight through or over the rear sight and view the object through the window in the cover. If the object sighted is at a lower elevation than the compass, raise the rear sight holder as needed. The compass is correctly sighted when the compass is level and the operator sees the black center line of the window bisecting the rear sight and the object sighted.

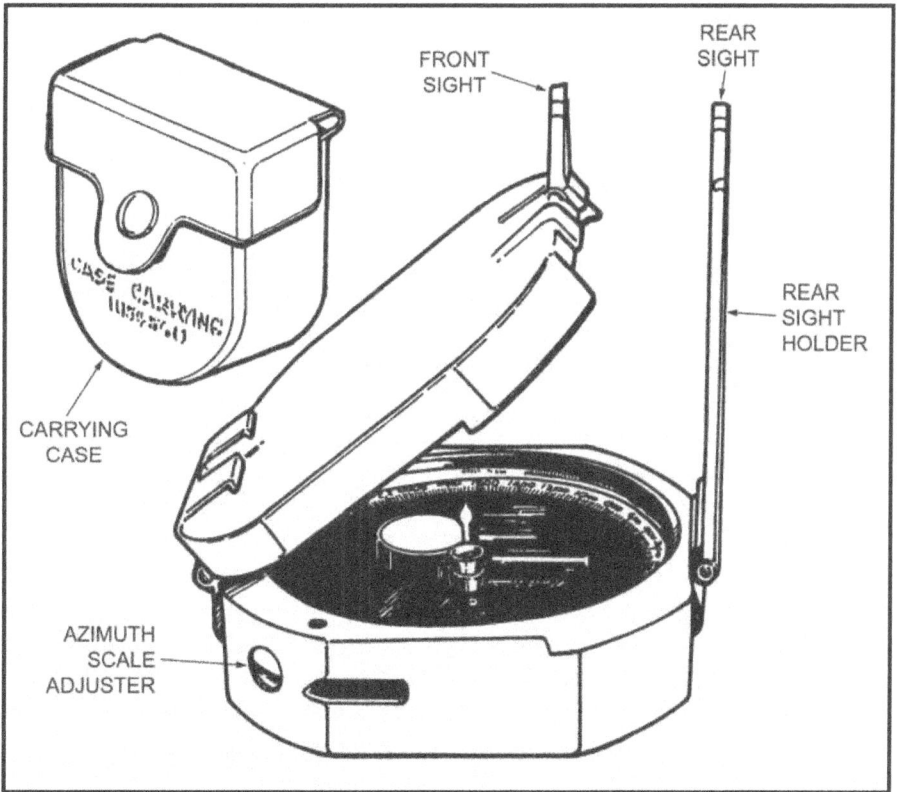

Figure 2-2. Compass, M2 (side view).

Chapter 2

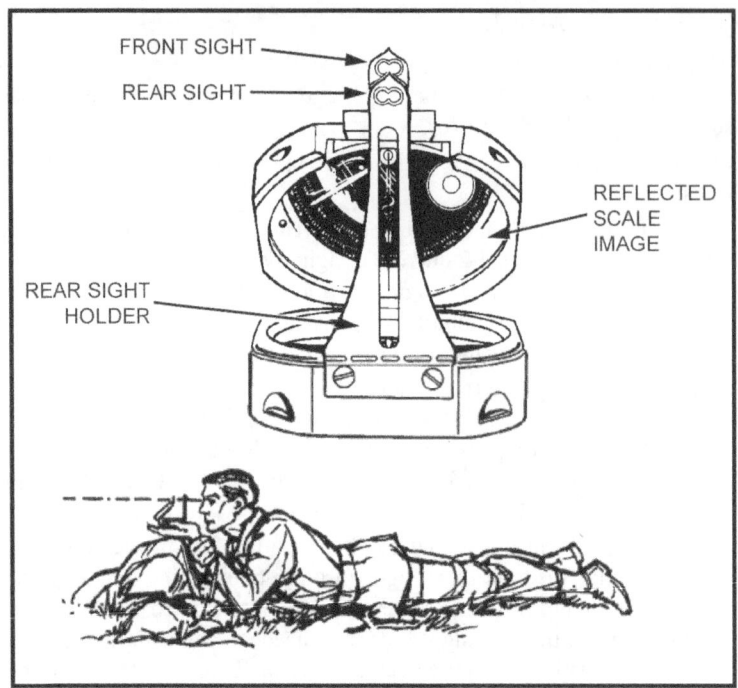

Figure 2-3. Compass, M2 (user's view).

(4) Hold the compass in both hands, at eye level, with the arms braced against the body and the rear sight near the eyes. For precise measurements, rest the compass on a nonmetallic stake or object.

(5) Level the instrument by viewing the circular level in the mirror and moving the compass until the bubble is centered. Sight on the object, look in the mirror, and read the azimuth indicated by the black (south) end of the magnetic needle.

2-10. To measure a grid azimuth—

(1) Index the known declination constant on the azimuth scale by turning the azimuth scale adjuster. Be sure to loosen the locking screw on the bottom of the compass. (The lightweight plastic M2 compass has no locking screw.)

(2) Measure the azimuth as described above. The azimuth measured is a grid azimuth.

2-11. To measure an angle of sight or vertical angle from the horizontal—

(1) Hold the compass with the left side down (cover to the left), and fold the rear sight holder out parallel to the face of the compass, with the rear sight perpendicular to the holder. Position the cover so that, when looking through the rear sight and the aperture in the cover, the elevation vial is reflected in the mirror.

(2) Sight on the point to be measured.

(3) Center the bubble in the elevation level vial (reflected in the mirror) with the level lever.

(4) Read the angle on the elevation scale opposite of the index mark. The section of the scale graduated counterclockwise from 0 to 1200 mils measures plus angles of sight. The section of the scale graduated clockwise from 0 to 1200 mils measures minus angles of sight.

2-12. To declinate the M2 compass from a surveyed declination station free from magnetic attractions—

(1) Set the M2 compass on an aiming circle tripod over the orienting station, and center the circular level.

(2) Sight in on the known, surveyed azimuth marker.

(3) Using the azimuth adjuster scale, rotate the azimuth scale until it indicates the same as the known, surveyed azimuth.

(4) Recheck the sight picture and azimuth to the known point. Once the sight picture is correct and the azimuth reading is the same as the surveyed data, the M2 is declinated.

2-13. To use the field-expedient method to declinate the M2 compass—
(1) Using the azimuth adjuster scale, set off the grid-magnetic (G-M) angle shown at the bottom of all military maps.
(2) Once the G-M angle has been set off on the azimuth scale, the M2 compass is declinated.

SECTION II. AIMING CIRCLES, M2 AND M2A2

The aiming circle is used to measure azimuth and elevation angles with respect to a preselected baseline. It is a low-power telescope that is mounted on a composite body and contains a magnetic compass, adjusting mechanisms, and leveling screws for establishing a horizontal plane. The instrument is supported by a baseplate for mounting on a tripod. Angular measurements in azimuth are indicated on graduated scales and associated micrometers. (For more detailed information, see TM 9-1290-262-10.)

CHARACTERISTICS

2-14. Selected characteristics of the aiming circles are shown in Table 2-1.

Table 2-1. Selected characteristics of the aiming circles, M2 and M2A2.

CHARACTERISTIC	M2	M2A2
Weight (w/o equipment)	9 pounds	9 pounds
Weight (w/equipment less batteries)	21 pounds	21 pounds
Azimuth rotation	6400 mils	6400 mils
Elevation (maximum)	800 mils	1100 mils
Depression (maximum)	400 mils	400 mils
Magnification	4 power	4 power
Field of view	10 degrees	10 degrees

DESCRIPTION

2-15. M2 and M2A2 aiming circles consist of an elbow telescope mounted on orienting and elevating mechanisms, which are contained within a main housing. The main housing, in turn, is supported by adjusting screws through the baseplate.

USE

2-16. The M2 or M2A2 aiming circle (Figure 2-4) is used for the precise measurement of the azimuth and elevation angles of a ground or aerial target with the respect to a preselected baseline as required for the orientation of indirect fire weapons. It can also be used for general topographical surveying.

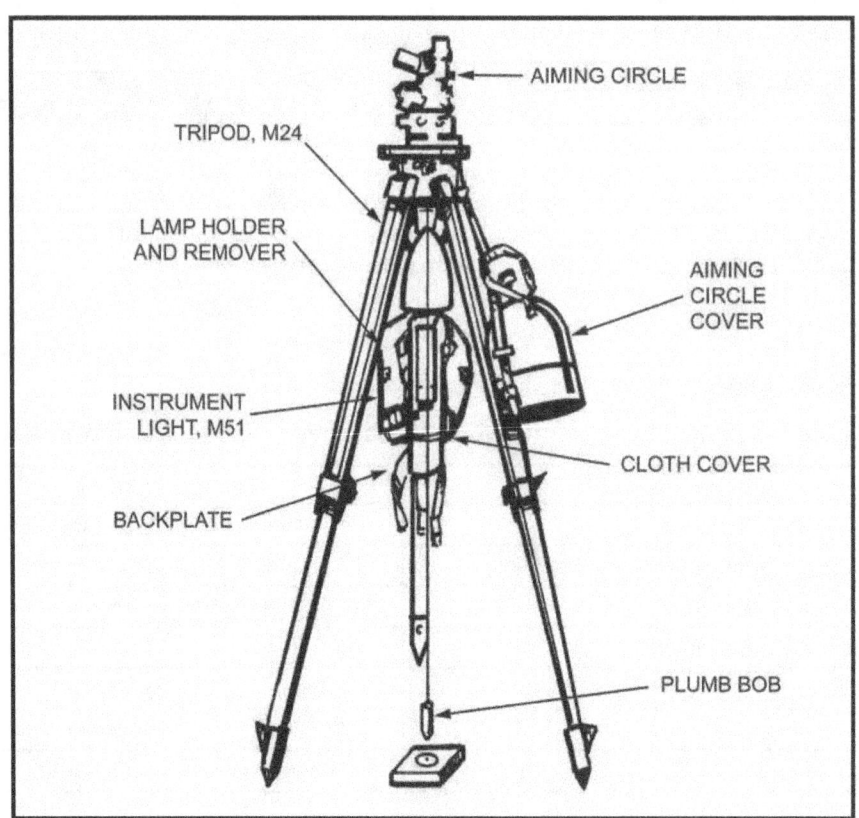

Figure 2-4. Aiming circles, M2 and M2A2, and accessory equipment.

Sighting and Fire Control Equipment

ORIENTING AND ELEVATING MECHANISMS

2-17. The orienting and elevating mechanisms permit unlimited azimuth orienting movement (360 degrees [6400 mils]) and limited elevation and depression (M2, 1200 mils; M2A2, 1500 mils). Two orienting knobs (Figure 2-5) control the azimuth orienting rotation. The azimuth micrometer knob controls azimuth measurement. The elevation micrometer knob controls elevation and depression movement. The azimuth micrometer worm can be disengaged to provide rapid azimuth measurement of movement by exerting pressure on the azimuth micrometer knob against the pressure of an internal spring-loaded plunger. Releasing the pressure on the azimuth micrometer knob allows the mechanism to reengage. A similar throw-out mechanism permits the azimuth orienting worm to also be disengaged to provide rapid azimuth orienting movement.

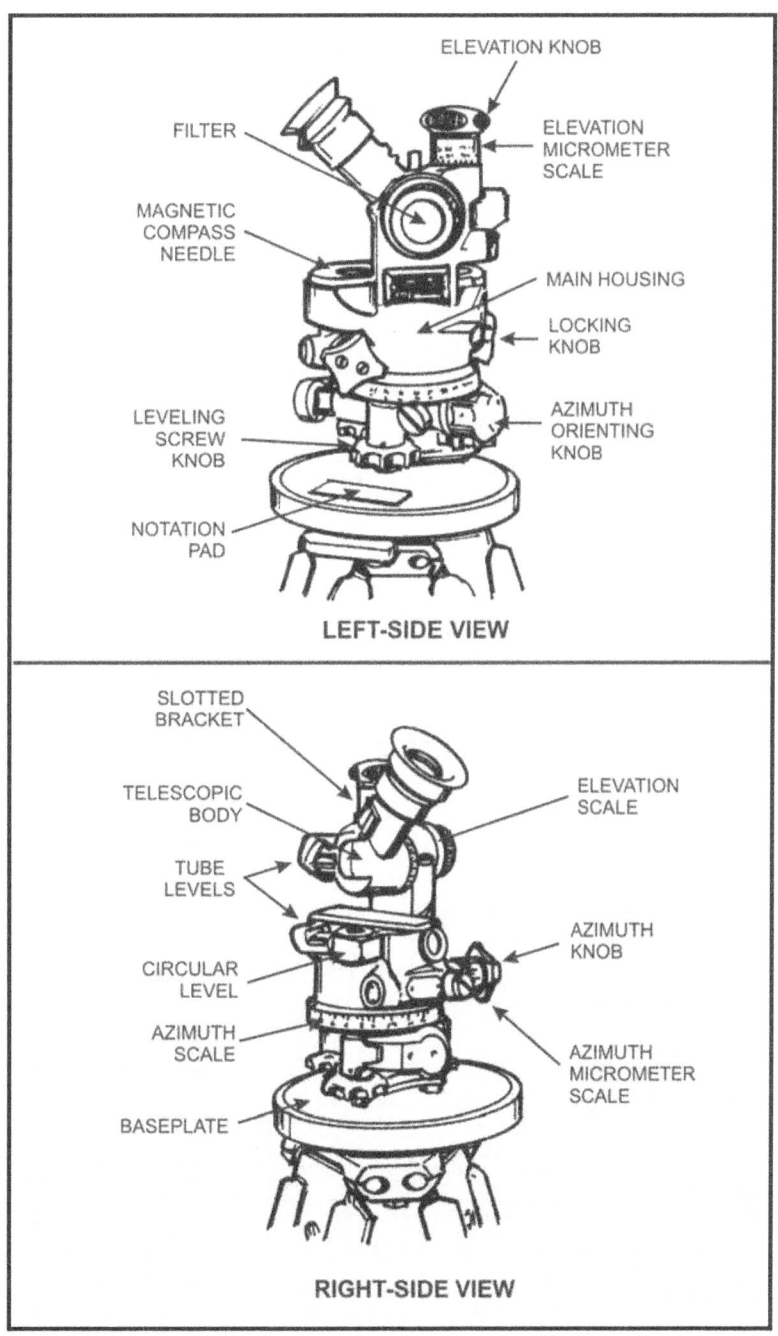

Figure 2-5. Aiming circle, M2.

TELESCOPE

2-18. The telescope of the aiming circle is a four-power, fixed-focus, elbow-type instrument. The reticle of the telescope contains cross lines graduated to give azimuth and elevation angular readings from 0 to 85 mils in 5-mil increments. Thus, the FO can read small angular values directly from the reticle without referring to the azimuth and elevation micrometer scales. An externally stowed filter is provided for protection against the rays of the sun. A slotted bracket provides the means of securing the lamp bracket on one lead wire of the M51 instrument light so that illumination of the reticle during night operation can be accomplished. The reflector can be illuminated and used in conjunction with the sightunits on the mortars during night operations to backlight the vertical centerline of the aiming circle.

LEVELS

2-19. Three levels are contained within the telescope body and main housing of the aiming circle. One tubular level, held between two bosses on the telescope body, is used to establish a true horizontal line-of-sight. The two bosses supporting this level are machined to form an open sight for approximate alignment of the telescope and target, and for quick or emergency sighting. One circular level and one tubular level are held within bosses on the main housing. The circular level is used for rough leveling of the aiming circle, and the tubular level is used for fine leveling adjustments. The three leveling screws on the baseplate are used to level the instrument and each is controlled by a leveling screw knob.

MAGNETIC COMPASS NEEDLE

2-20. A magnetic compass needle is located in a recess in the top of the housing. A magnifier and rectangular reticle located at one end of the recess enable the FO to observe the end of the compass needle and to align the line of sight of the telescope with the needle. The compass needle can be locked in position by actuating the locking lever on the side of the housing.

AZIMUTH AND ELEVATION SCALES

2-21. Azimuth and elevation scales are employed to measure azimuth or elevation angles. The scales provide coarse readings and the micrometer provides fine readings. The two readings added together give the angle. Graduation intervals and numeral scales are graduated into relatively large round number intervals for convenience in reading. The scale intervals are in graduations of 100 mils.

- The azimuth scale is graduated from 0 to 6400 mils (zero equals 6400). The upper series forms the main azimuth scale, colored black and numbered at 200-mil intervals. The lower series, colored red, is numbered from 0 to 3200 mils (the large zero in the main scale equals 3200). The red scale should only be used when verifying the lay of the aiming circle with another aiming circle.
- The azimuth micrometer scale is graduated at 1-mil intervals and numbered from 0 to 100 at ten 10-mil intervals.
- The elevation scale is graduated and numbered on both sides of 0. Minus (red) readings represent depression and plus (black) readings represent elevations at 100-mil intervals from minus 400 to plus 800 mils.
- The elevation micrometer scale is graduated at 1-mil intervals from 0 to 99 mils—large zero is designated 0 and 100. Red numerals represent depression and black numerals represent elevation.

NOTATION STRIP

2-22. A notation strip is provided on the baseplate. This strip is a raised and machined surface on which scale readings, settings, or other data can be recorded for reference.

ACCESSORY EQUIPMENT

2-23. The accessory equipment for the M2 aiming circle includes the aiming circle cover, M24 tripod, and the accessory kit that includes the M51 instrument light, backplate, cloth cover, plumb bob, and a lamp holder and remover. This equipment is mounted on the M24 tripod when the instrument is set up for use.

AIMING CIRCLE COVER

2-24. The aiming circle cover is a metal cover that protects and houses the aiming circle when not in use. It attaches to the baseplate of the aiming circle and can be carried by means of its strap. When the aiming circle is in use, the cover is placed on the tripod.

TRIPOD, M24

2-25. The M24 tripod comprises three telescoping wooden legs hinged to a metal head, which contains a captive screw for attaching the aiming circle. When not in use, the tripod cover should be fitted on the head and the legs retracted and held by a strap. The aiming circle cover and cloth cover with attached accessory equipment can be mounted on its legs when the aiming circle is set up for use. A hook is also provided to suspend the plumb bob and attached thread.

INSTRUMENT LIGHT, M51

2-26. The M51 instrument light is used with the M2 aiming circle during night operations and for certain test and adjustment procedures. The light is powered by two flashlight batteries and contains two attaching lead wires. A lamp bracket attached to one lead wire can be inserted into the slotted bracket of the aiming circle telescope for illumination of the telescope reticle, while a hand light is attached to the other lead wire and used for general-purpose illumination. Rotation of the rheostat knob turns the two lamps on and off and increases or decreases the intensity of illumination.

BACKPLATE

2-27. The backplate secures and protects the instrument light and lamp bracket, hand light, and lead wires of the light. The plate with the attached instrument light is stored within the cloth cover.

CLOTH COVER

2-28. The cloth cover is used to store the backplate, its attached instrument light, the plumb bob, and the lamp holder/remover. When the aiming circle is set up for use, the cloth cover is mounted on one of the legs of the M24 tripod.

PLUMB BOB

2-29. The plumb bob is used to aid in orienting the aiming circle over an exact point, such as a declination station marker, on the ground. It is composed of a pointed weight attached to a nylon thread that can be suspended from the hook under the tripod head when in use. The effective length of the thread can be adjusted by means of the slide.

LAMP HOLDER AND REMOVER

2-30. A lamp holder and remover are used to hold and remove incandescent lamps for the M51 instrument light.

Chapter 2

SETUP AND LEVELING OF THE AIMING CIRCLE

NOTE: The aiming circle must always be level during operation.

2-31. To set up and level the aiming circle—

(1) Remove the strap from the tripod legs, loosen the leg clamp thumbscrews, extend the legs so that the tripod is about chest high, and tighten the leg clamp thumbscrews. Spread the legs about 18 inches apart, adjust the legs so the tripod head is about level, and plant the feet firmly in the ground.

(2) Remove the tripod head cover. Open the baseplate cover of the aiming circle head. Keeping the baseplate cover pointed toward you, thread the tripod guide screw assembly into the aiming circle until it is firmly seated. The base of the aiming circle should not protrude over the machine surface of the tripod head. Pull out and down on the strap latch assembly. Remove the cover and hang it on the tripod head cover.

(3) If the instrument is to be set up over an orienting point, attach the plumb bob to the hook. Adjust the tripod legs and aiming circle head until it is over the point.

(4) Loosen the leveling screws to expose sufficient threads (3/8 to 1/2 inch) on the three screws to permit the instrument to be leveled. Number the leveling screws clockwise, 1, 2, and 3 (Figure 2-6). Now place the tubular level vial over the notation pad. Grasp leveling screw number 1 between the thumb and forefinger of the right hand, and grasp leveling screw number 2 between the thumb and forefinger of the left hand. Turn the screws so that the thumbs move toward or away from each other. Using these two leveling screws, center the bubble. The bubble moves in the same direction as the left thumb.

(5) Rotate the aiming circle head until the magnifier is over the notation pad. Level the tubular level by turning only level screw number 3. The bubble should now remain level in any direction that the aiming circle is rotated. A variation of one graduation from the center of the vial is acceptable. If the bubble does not remain level, repeat this procedure.

NOTE: If the spring plate is bent, the aiming circle cannot be leveled and must be turned in to the maintenance support activity.

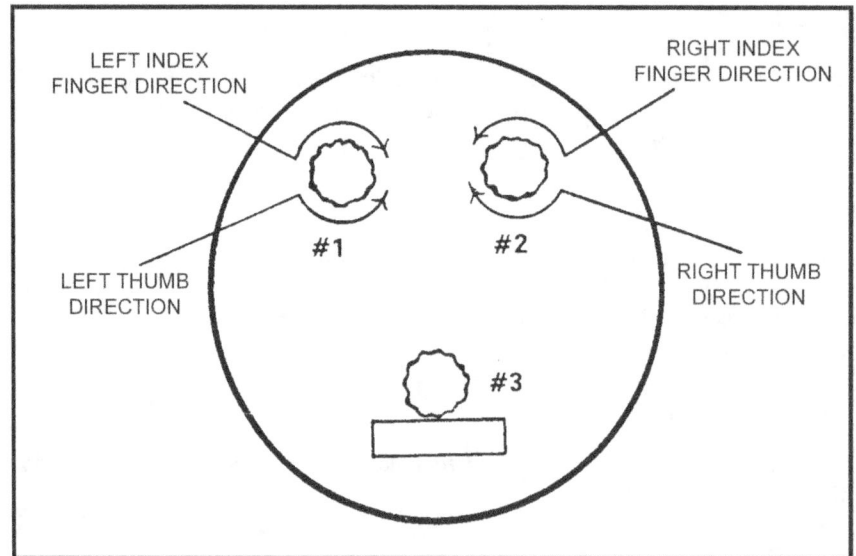

Figure 2-6. Leveling screws.

(6) The aiming circle should be set up at the distance indicated by Table 2-2.

Table 2-2. Set-up distance from objects.

OBJECT	DISTANCE (METERS)
High-tension power lines	150
Electronic equipment	150
Railroad tracks	75
Tanks and trucks	75
Vehicles	50
Barbed wire	30
Mortars or telegraph wire	25
Helmets and so forth	10

DECLINATION CONSTANT

2-32. Since the magnetic needle of an aiming circle points toward magnetic north and not the grid north on a map, it is necessary to correct for this difference by using the declination constant. The declination constant of an instrument is the clockwise angle between grid north and magnetic north; that is, the grid azimuth of magnetic north. This constant differs slightly for different instruments and must be recorded on each instrument. The constant also varies for the same instrument in different localities. Use the following to determine the declination constant.

DECLINATION STATION

2-33. Declination stations are established by fires battalion survey teams to determine the declination constants of instruments and to correct for local attractions, annual variations, and instrument errors. When a unit moves from one locality to another, a station should be established where all instruments are declinated. If the declination constants for all instruments of a unit are determined at the same station, grid azimuths measured with each instrument will agree with the map grid, and all instruments will agree with each other. The point chosen for the declination station must have a view of at least two distant, well-defined points with a known grid azimuth. Two additional points are desirable, one in each quadrant, as a check.

PROCEDURE FOR DECLINATING THE AIMING CIRCLE AT A DECLINATION STATION

2-34. Where a declination station is available, the procedure for declinating the aiming circle is as follows:
  (1) Set up and fine level the aiming circle directly over the declination station marker using the plumb bob.
  (2) Place the grid azimuth of the first azimuth marker on the scales using the recording motion. Place the vertical cross line of the telescope on the azimuth marker using the nonrecording (orienting) motion. The aiming circle is now oriented on grid north.
  (3) With the recording motion, rotate the instrument to zero. Release the magnetic needle and look through the magnifier. Center the north-seeking needle using the recording motion, and then relock the magnetic needle.
  (4) Notice the new azimuth on the scale, which is the declination constant—record it.
  (5) Recheck the aiming circle level and repeat steps 2 through 4 using the remaining azimuth markers until three readings have been taken. If there is only one marker, repeat the entire procedure twice using the same marker.
  (6) Find the average declination constant using these three readings.

Chapter 2

EXAMPLE 1

1st point reading   =   6399 mils
2nd point reading   =   6398 mils
3rd point reading   =   6398 mils

Total   =   19195 mils

19195 ÷ 3 =   6398.3 (rounded off to the nearest whole number) =
              6398 mils (average declination constant)

EXAMPLE 2

1st point reading   =   0030 mils
2nd point reading   =   0031 mils
3rd point reading   =   0029 mils

Total   =   0090 mils

0090 ÷ 3 = 0030 mils (average declination constant)

---

(7) Record the average declination constant in pencil on the notation (strip) pad of the aiming circle as its declination constant. All readings should be within 2 mils of each other; if not, repeat steps 2 through 4. Ensure the aiming circle is directly over the station marker to obtain the 2-mil tolerance. If the desired 2-mil accuracy is not gained after two tries, the aiming circle is defective and should be turned in for repair.

USE OF THE GRID-MAGNETIC ANGLE

2-35. If an aiming circle is used in a new area without a declination station, a declination constant can be determined by using the G-M angle from a map. When the G-M angle (converted to mils) is westerly, it is subtracted from 6400 mils. The remainder is the declination constant. When the G-M angle is easterly, the angle (in mils) is the declination constant.

PROCEDURE FOR DECLINATING AN AIMING CIRCLE WHEN A DECLINATION STATION IS NOT AVAILABLE

> **NOTE:** This procedure is the least desirable and should be used only when no other means are available. It does not compensate for the error that could be inherent in the aiming circle.

2-36. To declinate an aiming circle when a declination station isn't available—

(1) Determine the G-M angle from the map of the area in which the aiming circle is to be used. This G-M angle is used as indicated below.

(2) In 1, Figure 2-7, the difference between grid north and magnetic north is 200 mils (westerly). This total is then subtracted from 6400 mils. The declination constant that can be used is 6200 mils.

(3) In 2, Figure 2-7, the difference between grid north and magnetic north in a clockwise direction is 120 mils. This can be used as the declination constant.

Sighting and Fire Control Equipment

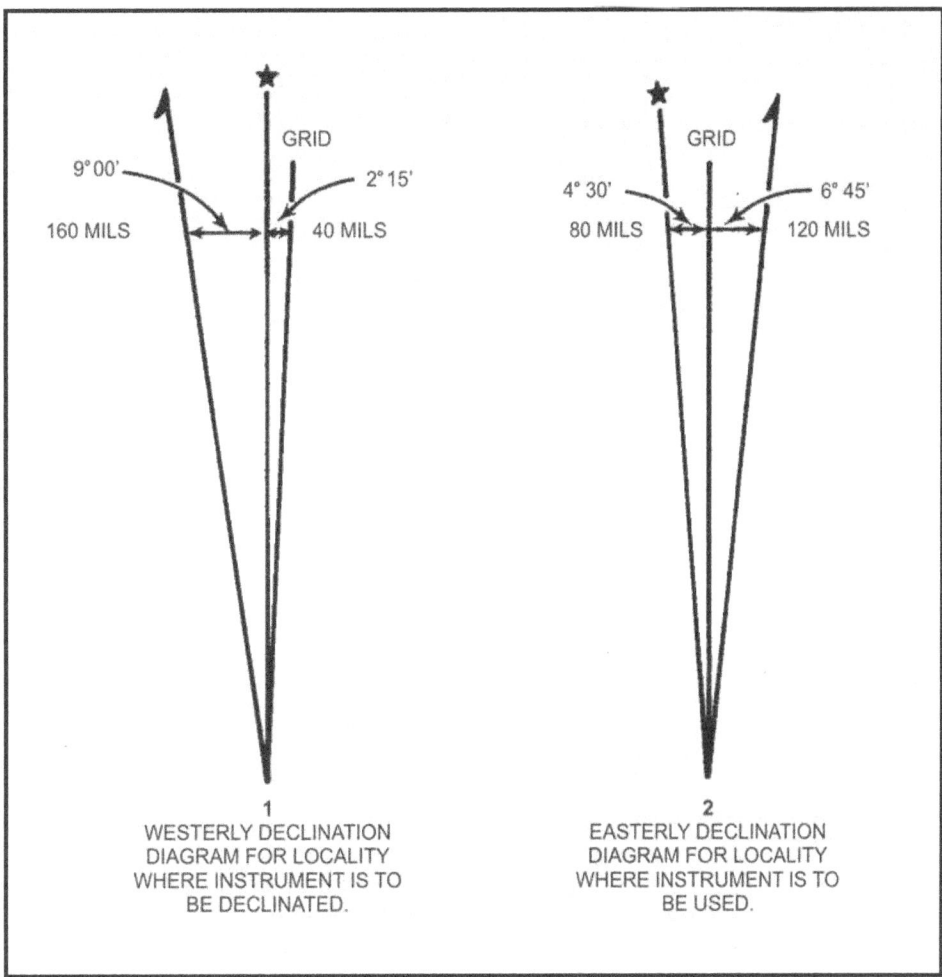

Figure 2-7. Marginal data from a map.

WHEN TO DECLINATE THE AIMING CIRCLE

2-37. Certain rules prescribe how often and under what circumstances the aiming circle should be declinated to determine and keep the declination constant current. These rules are as follows:

- The aiming circle should be declinated when it is moved 25 miles or more from the area in which it was last declinated. In some locations, a move of less than 25 miles could require a redeclination of the aiming circle.
- The aiming circle must be declinated after an electrical storm or after receiving a severe shock, such as a drop from the bed of a truck to the ground. The magnetic needle is a delicately balanced mechanism, and any shock can cause a significant change in the declination constant.
- The aiming circle should be declinated every 30 days to guard against changes that may have occurred due to unreported accidents to the instrument. If a radical change is observed, the instrument should be declinated again within a few days to determine if the observed change was due to a magnetic storm or is a real change in the characteristics of the instrument.
- The aiming circle should be declinated when it is initially received and when it is returned from support maintenance repair.

ORIENTING OF THE INSTRUMENT ON GRID NORTH TO MEASURE GRID AZIMUTH TO OBJECTS

2-38. The procedure to orient the aiming circle on grid north to measure grid azimuth to objects is as follows:

(1) Level the instrument.
(2) Set the azimuth micrometer and the azimuth scale on the declination constant of the instrument.
(3) Release the magnetic needle.
(4) With the orienting knob, align the south end of the needle accurately with the center etched line by using the magnetic needle magnifier.
(5) Lock the magnetic needle and close the orienting knob covers.
(6) Using the throw-out mechanism (azimuth knob), turn the telescope until the vertical line of the reticle is about on the object.
(7) By rotating the azimuth knob, bring the vertical line exactly on the object.
(8) Read the azimuth to the object on the azimuth and micrometer scales.

MEASURING OF THE HORIZONTAL ANGLE BETWEEN TWO POINTS

NOTE: To measure the horizontal angle between two points, at least two measurements should be made.

2-39. To measure the horizontal angle between two points—

(1) Set the azimuth micrometer and the azimuth scale at zero.
(2) Rotate the instrument using the orienting knob throw-out mechanism until the vertical line of the telescope is about on the left edge of the left-hand object.
(3) Lay the vertical line exactly on the right edge of the left-hand object by rotating the orienting knob.
(4) Using the throw-out mechanism (azimuth knob), turn the telescope clockwise until the vertical line is about on the left edge of the right-hand object.
(5) Lay the vertical line exactly on the left edge of the right-hand object by turning the azimuth knob.
(6) Read the horizontal angle on the scales and record the value to the nearest 0.5 mil. This completes the first repetition.
(7) Rotate the aiming circle, using the lower motion, until the vertical cross line is again on the rear station.

NOTE: The value obtained from the first repetition is still on the scales.

(8) Rotate the aiming circle body, using the upper motion, until the vertical cross line is again on the forward station.
(9) Read and record the accumulated value of the two measurements of the angle to the nearest 0.5 mil. This completes the second repetition.
(10) Divide the second reading by 2 to obtain the mean angle to the nearest 0.1 mil. This mean angle must be within 0.5 mil of the first reading. If it is not, the measurement is void and the angle is measured again.

ORIENTING OF THE 0-3200 LINE ON A GIVEN GRID AZIMUTH

2-40. The procedure for orienting the 0-3200 line of the aiming circle on a given grid azimuth is illustrated below. In this example, the mounting azimuth is 5550 mils and the aiming circle is assumed to have a declination constant of 6380 mils.

   (1) Set up and level the aiming circle.
   (2) Subtract the announced mounting azimuth from the declination constant of the aiming circle (adding 6400 to the declination constant of the aiming circle if the mounting azimuth is larger). In this case, subtract the mounting azimuth 5550 from the declination constant 6380.

SOLUTION

Declination constant	6380 mils
Announced mounting azimuth	-5550 mils
Remainder	830 mils

   (3) Set the remainder on the azimuth and micrometer scales of the aiming circle. In this case, the remainder is 830 mils (recording motion).
   (4) Release the compass needle. Look through the window in the cover housing and rotate the instrument until the needle floats freely using the orienting knob throw-out mechanism. For fine adjustments, use orienting knobs until the magnetic needle is exactly centered on the etched marks on the magnifier. Relock the compass needle to orient the 0-3200 line of the aiming circle on the mounting azimuth. In this case, the grid azimuth is 5550 mils (Figure 2-8).
   (5) Do not disturb the lower motion of the aiming circle once it is oriented.

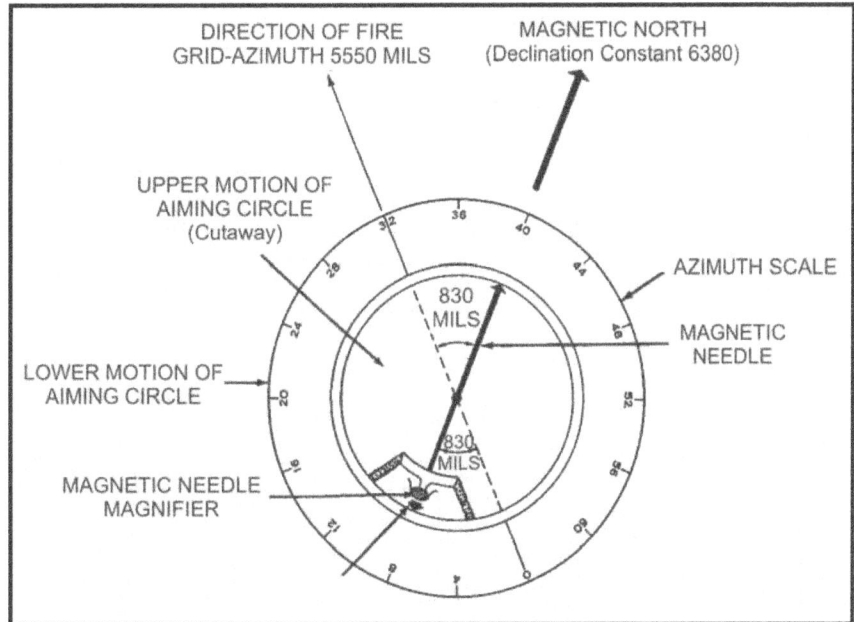

Figure 2-8. Aiming circle oriented in desired direction of fire.

ORIENTING OF THE 0-3200 LINE ON A GIVEN MAGNETIC AZIMUTH

2-41. The procedure for orienting the 0-3200 line of the aiming circle on a given magnetic azimuth is as follows:

   (1) Subtract the announced magnetic azimuth from 6400.

(2) Set the remainder on the azimuth and micrometer scales of the aiming circle.

(3) Release the compass needle and rotate the orienting knob until the magnetic needle is exactly centered in the magnetic needle magnifier. Lock the compass needle.

(4) The 0-3200 line of the aiming circle is now oriented on a given magnetic azimuth.

VERIFICATION OF THE LAY OF THE PLATOON

2-42. To verify the lay of the platoon—

(1) After the section or platoon is laid, the leader verifies the lay by using another M2 aiming circle, which is referred to as the safety circle.

(2) The leader or designated safety officer sets up and orients an M2 aiming circle by using the method that was used with the lay circle (Figure 2-9). The aiming circle must be located where it can be seen by all mortars and should not be closer than 10 meters to the lay circle.

(3) After picking up a line of sight on the lay circle, the safety circle operator commands, LAY CIRCLE REFER, AIMING POINT THIS INSTRUMENT. The lay circle operator sights his instrument onto the safety circle by use of the recording motion.

(4) When the aiming circle is used to orient another aiming circle for direction, the reading between the two circles will be 3200 mils apart, because both circles measure horizontal clockwise angles from the line of fire. To prevent confusion, remember that if you see red, read red. One half of the aiming circle azimuth scale has a second red scale that goes in the opposite direction of the black scale.

(5) There should be no more than 10 mils difference between the circles.

(6) If the lay circle and the safety circle deflection are within the 10-mil tolerance, the instrument operator on the safety circle places the deflection reading by the lay circle on the upper motion of the safety circle. With the lower motion, the instrument sights back on the lay circle. This serves to align the 0-3200 line of the safety circle parallel to the 0-3200 line of the lay circle.

(7) The instrument operator on the safety circle commands, PLATOON, REFER AIMING POINT THIS INSTRUMENT. All gunners refer and announce the deflection to the safety circle. If the deflection referred by the mortar is within 10 mils, the operator on the safety circle announces that the mortar is safe. Once the mortars are safe, the operator announces, "THE PLATOON IS SAFE."

(8) The platoon leader walks the gun line and visually checks the guns to ensure they are parallel. An M2 compass should also be used to ensure the guns are on the azimuth of fire.

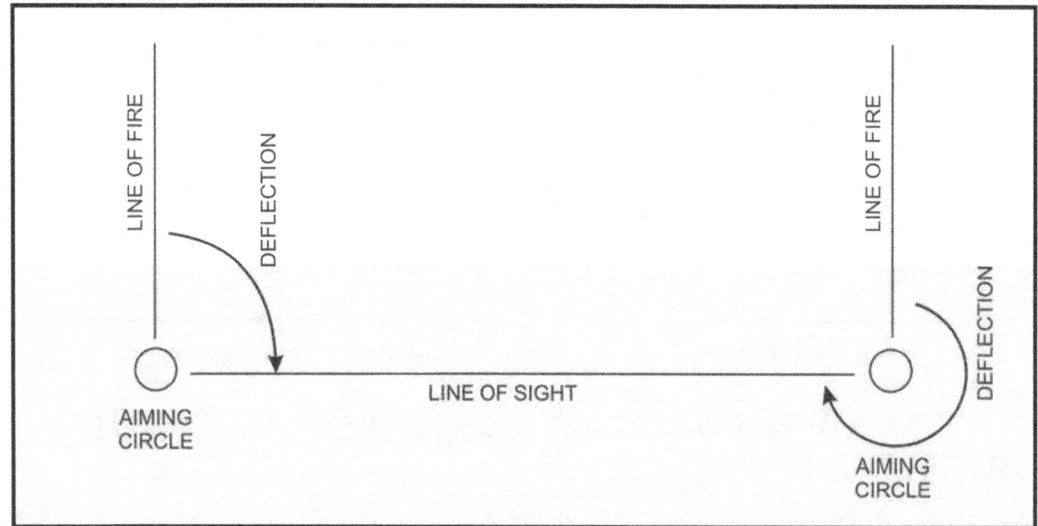

Figure 2-9. Method used to orient an aiming circle, M2.

ORIENTING BY ORIENTING ANGLE

2-43. Orienting by orienting angle eliminates errors that result from the use of the magnetic needle of the aiming circle.

NOTE: The magnetic needle is not used to orient the aiming circle.

2-44. An orienting angle is the horizontal clockwise angle from the mounting azimuth to the orienting line, the vertex being at the orienting station. It is a line of known direction established on the ground near the firing section, which serves as a basis for laying for direction. This line is established by a survey team.

2-45. To orient the aiming circle using the orienting angle—
   (1) Set the aiming circle over the orienting station and level it.
   (2) Place the orienting angle on the azimuth scale.
   (3) Sight on the far end of the orienting line, using the lower motion.

NOTE: Before performing the next step, close the orienting knob cover to prevent unintentional movement of the lower motion.

   (4) The 0-3200 line of the aiming circle is now oriented parallel to the mounting azimuth. Example: the azimuth of the orienting line is 3200 mils. The azimuth on which the section leader wishes to lay the section is 1600 mils. The orienting angle is 1600 mils (Figure 2-10).

Azimuth of orienting line	3200 mils
Minus mounting azimuth	− 1600 mils
Orienting angle	1600 mils

   (5) The aiming circle is set up over the orienting station using the plumb bob. The upper motion is used to set off 1600 mils on the aiming circle. The section leader sights on the end of the orienting line using the lower motion. The 0-3200 line of the aiming circle is now oriented.

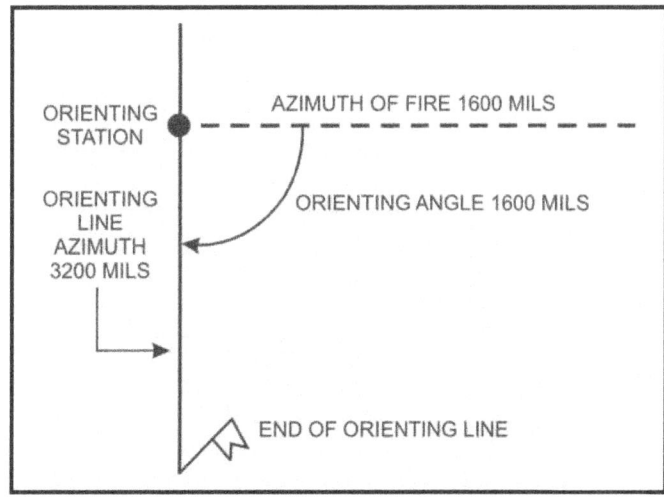

Figure 2-10. Orienting by orienting angle.

DISASSEMBLY OF THE AIMING CIRCLE

2-46. Instructions for disassembling the aiming circle are located inside the circle cover.

Chapter 2

CARE AND MAINTENANCE

2-47. The aiming circle cannot withstand rough handling or abuse. Proper care prolongs its life and ensures better results for the user. Inaccuracies or malfunctions result from mistreatment. The following precautions must be observed:

- Since stops are provided on instruments to limit the travel of the moving parts, do not attempt to force the rotation of any knob beyond its stop limit.
- Keep the instrument as clean and dry as possible. If the aiming circle is wet, dry it carefully.
- When not in use, keep the equipment covered and protected from dust and moisture.
- Do not point the telescope directly at the sun unless a filter is used; the heat of the focused rays can damage optical elements.
- Keep all exposed surfaces clean and dry to prevent corrosion and etching of the optical elements.
- To prevent excessive wear of threads and other damage to the instrument, do not tighten leveling, adjusting, and clamping screws beyond a snug contact.

NOTE: Only maintenance personnel are authorized to lubricate the aiming circle.

SECTION III. SIGHTUNITS

Sightunits are used to lay mortars for elevation and deflection. M67 and M64-series sightunits are the standard sighting devices used with mortars.

SIGHTUNIT, M67

2-48. The M67 sightunit (Figure 2-11) is used to lay all mortars for deflection and elevation. Lighting for night operations using the sightunit is provided by radioactive tritium gas contained in phosphor-coated glass vials. The sightunit is lightweight and portable. It is attached to the bipod mount by means of a dovetail. Coarse elevation and deflection scales and fine elevation and deflection scales are used in conjunction with elevation and deflection knob assemblies to sight the mortar system.

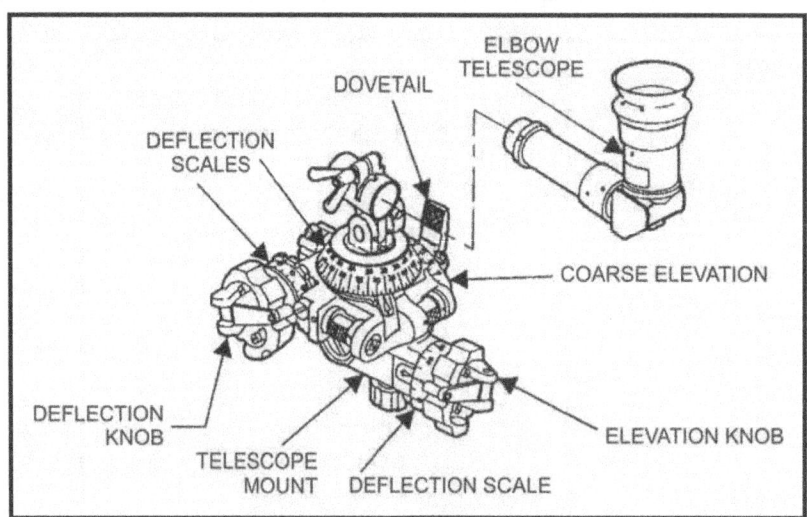

Figure 2-11. Sightunit, M67.

MAJOR COMPONENTS

2-49. The M67 sightunit consists of two major components: the elbow telescope and the telescope mount.

Sighting and Fire Control Equipment

Elbow Telescope

2-50. The elbow telescope is 4.0-power, hermetically sealed with a tritium illuminated crosshair reticle.

Telescope Mount

2-51. The telescope mount, provided with tritium back-lighted level vials, indexes, and translucent plastic scales, is used to orient the elbow telescope in azimuth and elevation.

> **CAUTION**
> When not in use, store the sightunit in its carrying case.

EQUIPMENT DATA

2-52. The equipment data for the M67 sightunit are shown in Table 2-3.

Table 2-3. Sightunit, M67, equipment data.

Weight	2.9 pounds (1.3 kilograms)
Field of view	10 degrees
Magnification	4.0 X nominal, 3.5 effective
Length	5 3/8 (13.7 centimeters)
Width	4 3/8 (11.1 centimeters)
Height	8 1/2 inches (21.6 centimeters)
Light source	Self-contained, radioactive tritium gas (H^3)

SIGHTUNIT, M64-SERIES

2-53. The sightunit (Figure 2-12) is the device on which the gunner sets deflection and elevation to hit targets by using the elevation level vial and the cross-level vial. After the sight has been set for deflection and elevation, the mortar is elevated or depressed until the elevation bubble on the sight is level. The mortar is then traversed until a proper sight picture is seen (using the aiming posts as the aiming point) and cross-level bubble is level. The mortar is laid for deflection and elevation when all bubbles are level. After the ammunition has been prepared, it is ready to be fired.

MAJOR COMPONENTS

2-54. The two major components are the elbow telescope and sight mount. The elbow telescope has an illuminated cross line. The sight mount has a dovetail, locking knobs, control knobs, scales, cranks, and locking latch.

Dovetail

2-55. The dovetail is compatible with standard U.S. mortars. When the dovetail is properly seated in the dovetail slot, the locking latch clicks. The locking latch is pushed toward the barrel to release the sight from the dovetail slot for removal.

Locking Knobs

2-56. The red locking knobs lock the deflection and elevation mechanisms of the sight during firing.

Micrometer Knobs

2-57. The elevation and deflection micrometer knobs are large for easy handling. Each knob has a crank for large deflection and elevation changes.

Chapter 2

Scales

2-58. All scales can be adjusted to any position. Micrometer scales are white. The elevation micrometer scale and fixed boresight references (red lines) above the coarse deflection scale and adjacent to the micrometer deflection scale are slipped by loosening slot-headed screws. Coarse deflection scales and micrometer deflection scales are slipped by depression and rotation. The coarse elevation scale is factory set and should not be adjusted at crew level. (If the index does not align with the coarse elevation scale within ±20 mils when boresighting at 800 mils, field-level maintenance should be notified.) The screws that maintain the coarse elevation scale are held in place with locking compound. If the screws are loosened and then tightened without reapplying the locking compound, the coarse elevation scale can shift during firing.

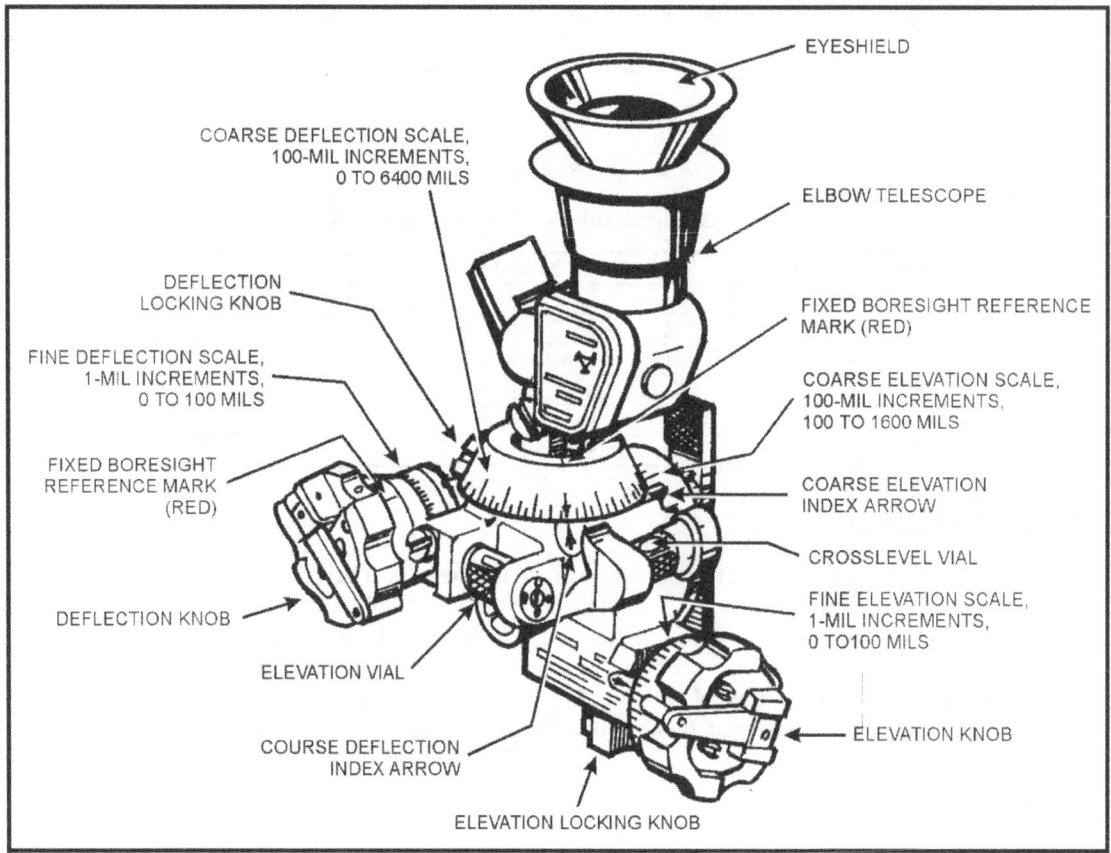

Figure 2-12. Sightunit, M64-series.

ILLUMINATION

2-59. Instrument lights are not needed when using the sightunit at night. Nine parts of the sight are illuminated by tritium gas.

- Telescope.
- Coarse elevation scale.
- Coarse elevation index arrow.
- Elevation vial.
- Fine elevation scale.
- Coarse deflection index arrow.
- Cross-leveling vial.
- Fine deflection scale.
- Coarse deflection scale.

TABULATED DATA

2-60. The tabulated data of the M64-series sightunit are shown in Table 2-4.

Table 2-4. Sightunit, M64-series, tabulated data.

Weight	2.5 pounds (1.1 kilograms).
Field of view	17 degrees (302 mils).
Magnification	1.5 unity power
Length	4 3/8 inches (11 centimeters).
Width	4 3/4 inches (12 centimeters).
Height	7 3/8 inches (19 centimeters).
Light source	Self-contained, radioactive tritium gas (H^3).

CAUTION
When not in use, store the sightunit in its carrying case.

OPERATION OF SIGHTUNITS

2-61. The operation of all the U.S. sightunits is similar.

ATTACHING THE SIGHTUNIT

2-62. Insert the dovetail of the telescope mount into the sight socket. Press the locking lever inward, seat the mount firmly, and release the locking lever.

NOTE: Until the baseplate is firmly seated, remove the sight from the mortar before firing each round.

PLACING THE SIGHTUNIT INTO OPERATION

2-63. To place the sightunit into operation, use the following procedures.

Setting the Deflection

2-64. Setting the correct deflection on the sight places the mortar, once the sight is leveled, in the direction commanded by the FDC.

2-65. To place a deflection setting on the sight, turn the deflection knob. Before attempting to place a deflection setting on the sightunit, ensure that the deflection locking knob is released. After placing a setting on the sight, lock the locking knob to lock the data onto the sight and to ensure that the scale does not slip during firing.

2-66. When setting the deflection, it is necessary to use the red fixed coarse scale and the red fixed micrometer scale to obtain the desired setting. Set the first two digits of the deflection on the coarse scale and the last two on the micrometer scale.

NOTE: The black coarse scale and the black micrometer scale are slip scales.

2-67. Setting a deflection on the deflection scale does not change the direction in which the barrel is pointing (the lay of the mortar). It only moves the vertical line off (to the left or right) the aiming line. The deflection placed on the sight is the deflection announced in the fire command. Place a deflection on the sight before elevation.

Setting the Elevation

2-68. Setting the correct elevation on the sight places the mortar, once the sight is leveled, on the angle of fire commanded by the FDC.

2-69. To set for elevation, turn the elevation knob. This operates both the elevation micrometer and coarse elevation scales. Both scales must be set properly to obtain the desired elevation. To place an elevation of 1065 mils for example, turn the elevation knob until the fixed index opposite the moving coarse elevation scale is between the black 1000- and 1100-mil graduations on the scale (the graduations are numbered every 200 mils), and then use the micrometer knob to set the last two digits (65).

2-70. Setting an elevation on the elevation scale does not change the elevation of the mortar barrel.

2-71. Before setting elevations on the sight, unlock the elevation locking knob. Once the elevation is placed on the sight, lock the elevation locking knob. This ensures the data placed on the sight do not accidentally change.

CORRECT SIGHT PICTURE

2-72. Sight the mortar on the aiming posts with the vertical reticle line aligned on the left side of the aiming post. Check to ensure the bubbles are level. Repeat adjustments until a correct sight picture is obtained with the bubbles level.

LAYING FOR DIRECTION

2-73. In laying the mortar for direction, the two aiming posts do not always appear as one when viewed through the sight(s). This separation is caused by one of two things: either a large deflection shift of the barrel or a rearward displacement of the baseplate assembly caused by the shock of firing. (See Chapter 3, Figure 3-7 for an example of a compensated sight picture.)

2-74. When the aiming posts appear separated, the gunner cannot use either one as his aiming point. To lay the mortar correctly, he takes a compensated sight picture. To take a compensated sight picture, he traverses the mortar until the sight picture appears with the left edge of the far aiming post placed exactly midway between the left edge of the near aiming post and the vertical line of the sight. This corrects for the displacement. A memory trick for correcting displacement is: *Hey diddle diddle, far post in the middle.*

2-75. The gunner determines if the displacement is caused by traversing the mortar or by displacement of the baseplate assembly. To do so, he places the referred deflection on the sight and lays on the aiming posts. If both aiming posts appear as one, the separation is caused by traversing. Therefore, the gunner continues to lay the mortar as described and does not realign the aiming posts. When the posts appear separated, the separation is caused by displacement of the baseplate assembly. The gunner notifies his squad leader, who requests permission from the section/platoon leader to realign the aiming posts. (For more information, see Chapter 3.)

REPLACING THE SIGHTUNIT IN THE CARRYING CASE

2-76. Before returning the sightunit to the carrying case, close the covers on the level vials and set an elevation of 800 mils and deflection of 3800 mils on the scales. Place the elbow telescope in the left horizontal position. All crank handles should be folded into the inoperative position.

CARE AND MAINTENANCE OF SIGHTUNITS

2-77. Although sightunits are rugged, the units could become inaccurate or could malfunction if abused or handled roughly. To properly care for and maintain the sightunit—

- Avoid striking or otherwise damaging any part of the sight. Be particularly careful not to burr or dent the dovetail bracket. Avoid bumping the micrometer knobs, telescope adapter, and level vials. Except when using the sight, keep the metal vial covers closed.
- Keep the sight in the carrying case when not in use. Keep it as dry as possible, and do not place it in the carrying case while it is damp.
- When the sight fails to function correctly, return it to the field-level maintenance unit for repair. Members of the mortar crew are not authorized to disassemble the sight.
- Keep the optical parts of the telescope clean and dry. Remove dust from the lens with a clean camel's-hair brush. Use only lens cleaning tissue to wipe these parts. Do not use ordinary polishing liquids, pastes, or abrasives on optical parts. Use only authorized lens cleaning compound for removing grease or oil from the lens.
- Occasionally oil only the sight locking devices by using a small quantity of light preservative lubricating oil. To prevent accumulation of dust and grit, wipe off excess lubricant that seeps from moving parts. Ensure that no oil gets on the deflection and elevation scales. (Oil removes the paint from the deflection scale.) No maintenance is authorized.

RADIOACTIVE TRITIUM GAS

2-78. Both the M67 and M64 use radioactive tritium gas (H^3) for illumination during night operations. The gas is sealed in glass tubes and is not hazardous when the glass tubes are intact. If there is no illumination, the range patrol officer (RPO) or CBRN officer should be notified. (For information on first aid, see FM 4-25.11.)

> **WARNING**
>
> Do not try to repair or replace the radioactive material. If skin contact is made with tritium, wash the area immediately with nonabrasive soap and water.

IDENTIFICATION

2-79. Radioactive self-luminous sources are identified by means of warning labels (Figure 2-13), which should not be defaced or removed. If necessary, they must be replaced immediately.

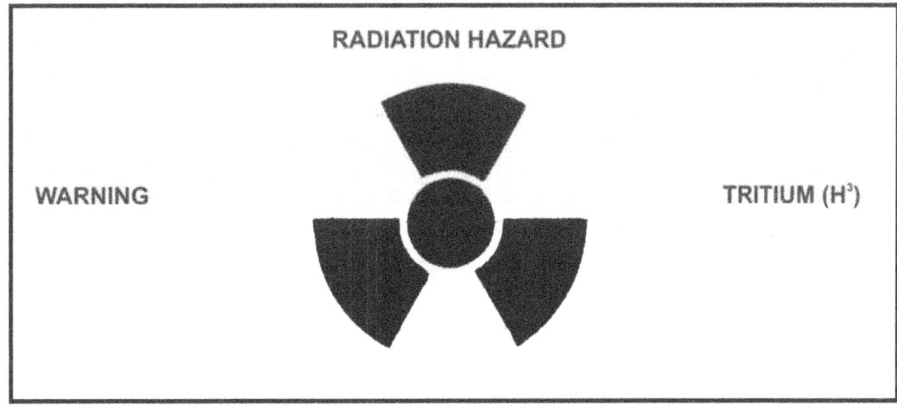

Figure 2-13. Warning label for tritium gas (H^3).

STORAGE AND SHIPPING

2-80. All radioactively illuminated instruments or modules that are defective must be evacuated to a sustainment-level maintenance activity. Defective items must be placed in a plastic bag and packed in the shipping container from which the replacement was taken. Spare equipment must be stored in the shipping container as received until installed on the weapon. Such items should be stored in an outdoor shed or unoccupied building.

SECTION IV. BORESIGHTS

Boresights are adjusted by the manufacturer and should not require readjustment as a result of normal field handling.

BORESIGHT, M45-SERIES

2-81. The boresight, M45-series, detects deflection and elevation errors in the sight.

COMPONENTS

2-82. The boresight, M45, (Figure 2-14) consists of an elbow telescope, telescope clamp, body, two strap assemblies, and clamp assembly.

Elbow Telescope

2-83. The elbow telescope establishes a definite line of sight.

Telescope Clamp

2-84. The telescope clamp maintains the line of sight in the plane established by the centerline of the V-slides.

Body

2-85. The body incorporates two perpendicular V-slides. It contains level vials (preset at 800 mils elevation) that are used to determine the angle of elevation of 800 mils and whether the V-slides are in perpendicular positions. It also provides the hardware to which the straps are attached.

Strap Assemblies

2-86. Two strap assemblies are supplied with each boresight and marked for cutting in the field to the size required for any mortar.

Clamp Assembly

2-87. The clamp assembly applies tension to the strap assemblies to secure the boresight against the mortar barrel.

TABULATED DATA

2-88. The tabulated data of the M45-series boresight are shown in Table 2-5.

Table 2-5. Boresight, M45-series, tabulated data.

Weight	2.5 pounds
Field of view	12 degrees
Magnification	3 power

Sighting and Fire Control Equipment

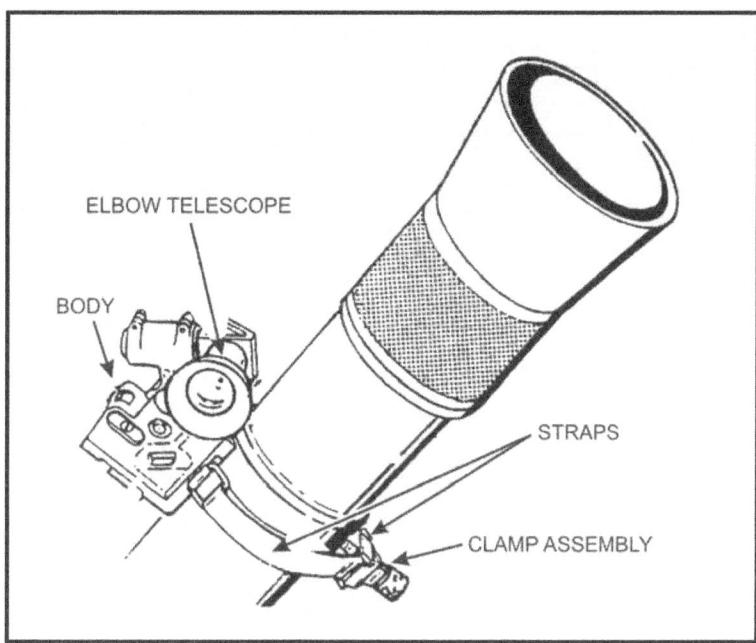

Figure 2-14. Boresight, M45.

INSTALLATION

2-89. Installation procedures for the M45 boresight are as follows:
   (1) Remove the boresight, clamp assembly, and straps from the carrying case. Grasp the boresight by the body to prevent damaging the telescope.
   (2) Place the ring over the hook and attach the strap snap to the eye provided on the strap shaft.
   (3) If necessary, release the catches and reset the straps to the proper length.
   (4) Remove any burrs or projecting imperfections from the seating area of the mortar barrel to ensure proper seating of the boresight. Attach the boresight to the barrel below and touching the upper stop band on the M252 mortar. However, attach the boresight about 1 inch from the muzzle of the barrel on all other mortars.

PRINCIPLES OF OPERATION

2-90. Except for the M115, the boresight is constructed so that the telescope line of sight lies in the plane established by the center lines of the V-slides. When properly secured to a mortar barrel, the centerline of the contacting V-slide is parallel to the centerline of the barrel. Further, the cross-level vial, when centered, indicates that the center lines of both slides, the elbow telescope, and the barrel lie in the same vertical plane. Therefore, the line of sight of the telescope coincides with the axis of the barrel, regardless of which V-slide of the boresight is contacting the barrel. The elevation vial is constructed with a fixed elevation of 800 mils.

BORESIGHT, M115

2-91. The boresight, M115, (Figure 2-15) detects deflection and elevation errors in the sight. The boresight has three plungers that keep it in place when mounted in the muzzle of the barrel. The telescope has the same field of view and magnification as the M64-series sightunit. The elevation bubble levels only at 800 mils.

COMPONENTS

2-92. The components of the M115 boresight are the body, telescope, and leveling bubbles (one for cross-leveling and one for elevation).

Second Cross-Level Bubble

2-93. A second cross-level bubble is used as a self-check of the M115. After leveling and cross-leveling, the M115 can be rotated 180 degrees in the muzzle until the second cross-level bubble is centered. The image of the boresight target should not vary in deflection. A large deviation indicates misalignment between the cross-level bubble and lenses.

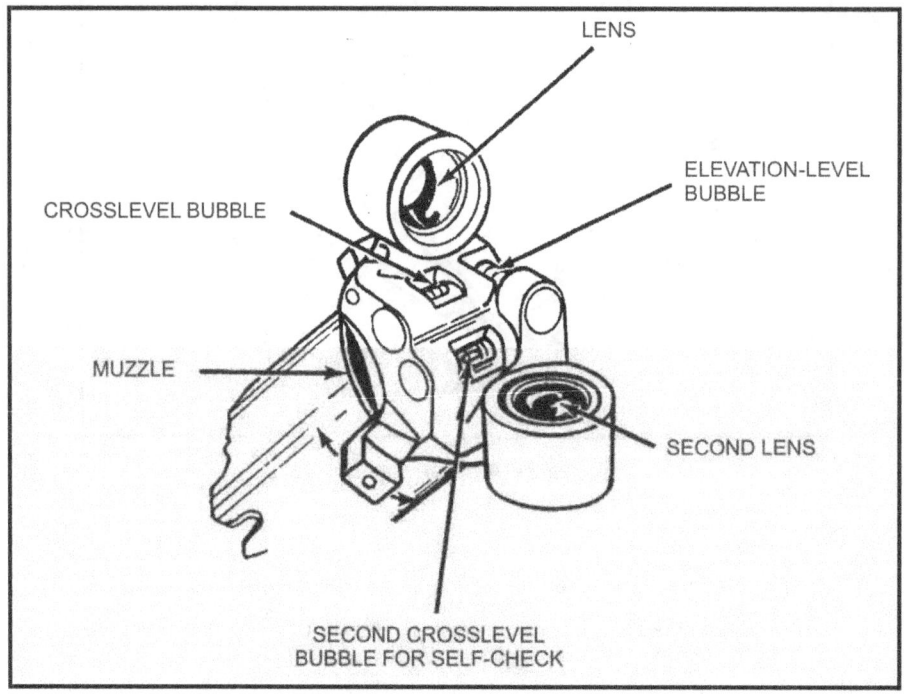

Figure 2-15. Boresight, M115.

TABULATED DATA

2-94. The tabulated data of the M115 boresight are shown in Table 2-6.

Table 2-6. M115 boresight tabulated data.

Weight	5 ounces
Field of view	17 degrees
Magnification	1.5 power

SIGHT CALIBRATION

2-95. Always calibrate the mortar sight to the mortar on which it will be mounted. This procedure is necessary since the sight socket that receives the sightunit is a machined part and varies in accuracy with each mortar. There is no set rule for frequency of calibration. The sight should be calibrated each time the mortar is mounted in a new location, since the movement might disturb the setting of the elevation and deflection scales. Time available and accuracy dictate the frequency of calibration.

BORESIGHT METHOD OF CALIBRATION

2-96. Once the mortar has been mounted, place the M67 or the M64-series sightunit into position in the sight socket. Place a deflection of 3200 mils and an elevation of 0800 mils on the scales. For the M252 mortar, place a deflection of 0 mils on the sight. Align the vertical cross line of the sight on an aiming point (at least 200 meters distant) by shifting the bipod. If necessary, use the traversing mechanism; however, keep the mortar within two turns of center of traverse (four turns of center of

Sighting and Fire Control Equipment

traverse for the 120-mm mortar). Make a visual check of the mortar for cant; if cant exists, remove the cant and re-lay, if necessary.

ELEVATION SETTING

2-97. Use the following procedures to set the elevation.

(1) Install the boresight on the mortar barrel. Center the cross-level vial by rotating the boresight slightly around the outside diameter of the mortar barrel. Slight movements are made by loosening the clamp screw and lightly tapping the boresight body. When the bubble centers, tighten the clamp screw.

(2) Elevate or depress the mortar barrel until the boresight elevation level vial is centered. The mortar is now set at 800 mils (45 degrees) elevation.

(3) Using the elevation micrometer knob, elevate or lower the sightunit until the elevation level bubble is centered. If necessary, cross-level the sightunit.

(4) Recheck all level bubbles.

(5) The reading on the coarse elevation scale of the sightunit should be 800 mils and the reading on the elevation micrometer scale should be 0. If adjustment is necessary, proceed as indicated below. Loosen the two screws on the elevation micrometer knob and slip the elevation micrometer scale until the 0 mark on the micrometer scale coincides with the reference mark on the housing. Tighten the two screws to secure the micrometer scale.

NOTE: Do not adjust the M64-series sightunit coarse elevation scale. If it does not line up with the 0800-mil mark, turn it in to field-level maintenance.

(6) Recheck all level bubbles.

DEFLECTION SETTING

2-98. Use the following procedures to set the deflection.

(1) Check again to ensure that the sight setting reads 3200 on the fixed deflection (red) scale and elevation 800 mils. Set zero deflection for the M252 mortar.

(2) Traverse the mortar no more than two turns of center of traverse (four turns for the 120-mm mortar) and align the vertical cross line of the boresight on the original aiming point. Adjust the boresight to keep the cross-level bubble centered since the mortar could cant during traversing. (If the mortar is initially mounted on the aiming point, it decreases the amount of traverse needed to align the cross line on the aiming point.) Also, the elevation level bubble may need to be leveled.

(3) After the boresight is aligned on the aiming point, level the sight by centering the cross-level bubble. Rotate the deflection micrometer knob until the sight is aligned on the aiming point. The coarse deflection scale should read 3200 mils and the micrometer scales should read 0. If adjustment is necessary, loosen the two screws on the deflection micrometer knob and slip the micrometer deflection scale until the arrow on the index is aligned with the zero mark on the micrometer scale.

(4) To verify proper alignment, remove the boresight and place it in position on the under side of the cannon as shown in Figure 2-16. Center the boresight cross-level bubble and check the vertical cross line to see if it is still on the aiming point. If cant exists, the vertical cross line of the boresight is not on the aiming point. This indicates that the true axis of the bore lies halfway between the aiming point and where the boresight is now pointing.

Chapter 2

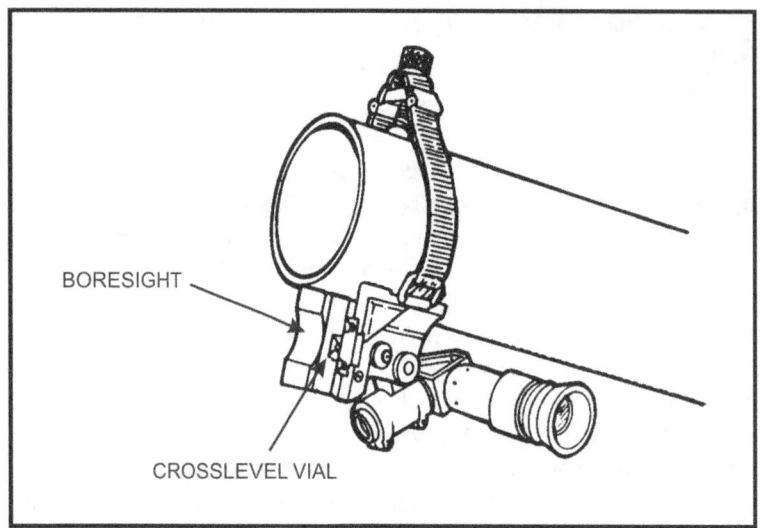

Figure 2-16. Verifying proper alignment of the boresight device.

(5) To correct this error, look through the boresight, traverse the mortar onto the aiming point. If bubbles are level, use the deflection micrometer knob and place the vertical cross line of the sight back onto the aiming point. With the sight in this position, index one-half of the mil variation between the sight and boresight. Slip zero on the micrometer scale to the index mark—for example, the mil variation is 10 mils and one-half of this value is 5 mils. Loosen the two screws on the deflection micrometer and index zero.

(6) Check all level bubbles, sightunit, and boresight.

(7) With a deflection on the micrometer scale of half the value of the original mil variation, both the sightunit and boresight are on the aiming point. If an error exists, repeat the procedure outlined above.

(8) Using the M64-series sightunit, adjust the deflection scale and micrometer scale of the sightunit to zero. To do this, loosen the deflection knob screws and slip the scale to zero. Adjust the deflection micrometer scale to zero by pushing in on the micrometer knob retaining button and slipping the scale to zero.

(9) Check again all level bubbles, and the lay of the sightunit and the boresight on the aiming point.

REMOVAL

2-99. Use the following procedures to remove the boresight.
   (1) Loosen the clamp screw, releasing the boresight from the barrel.
   (2) Swing the elbow telescope until it is about parallel with the elevation level bubble.
   (3) Release the clamp assembly and straps by removing the ring from the hook and strap shaft.
   (4) Stow the clamp assembly and straps in the corner compartment. Put the boresight in the center compartment of the carrying case.

CALIBRATION FOR DEFLECTION USING THE M2 AIMING CIRCLE

2-100. Two methods can be used to calibrate the sight for deflection using the M2 aiming circle: the angle method and the distant aiming point (DAP) method.

CALIBRATION FOR DEFLECTION USING THE ANGLE METHOD

2-101. This method (Figure 2-17) uses the geometric property that the alternate interior angles formed by a line intersecting two parallel lines are equal.

(1) Set up the aiming circle 25 meters to the rear of the mounted mortar. The mortar is mounted at 800 mils elevation and at a deflection of 3200 mils.

(2) With the aiming circle fine leveled, index 0 on the azimuth scale and azimuth micrometer scale. Using the orienting motion (nonrecording motion), align the vertical line of the reticle of the telescope so that it bisects the baseplate.

(3) Traverse and cross-level the mortar until the center axis of the barrel from the baseplate to the muzzle is aligned with the vertical line of the aiming circle telescope reticle.

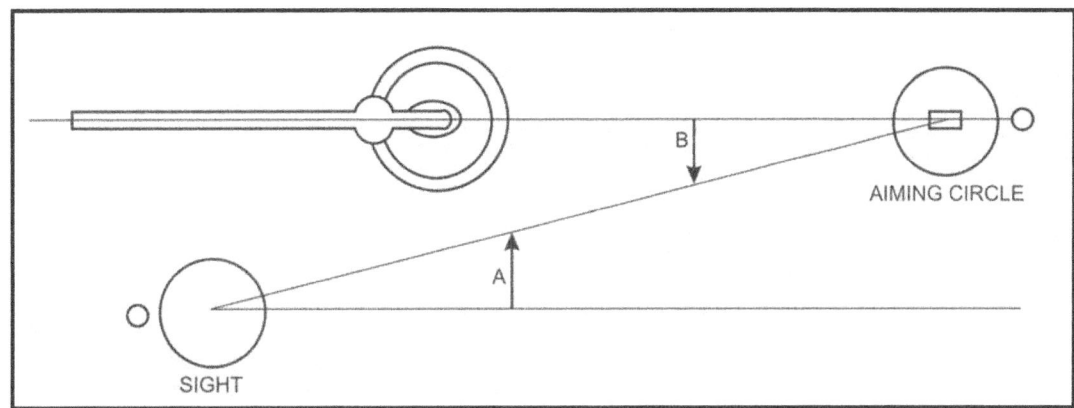

Figure 2-17. Calibration for deflection using the angle method.

(4) Turn the deflection knob of the sight until the vertical line is centered on the lens of the aiming circle and read angle A, opposite the fixed index.

(5) Turn the azimuth micrometer knob of the aiming circle until the vertical line of the telescope is laid on the center of the sight lens and read angle B, opposite the azimuth scale index. If the sight is in calibration, angles A and B will be equal. If they are not equal the sight is adjusted by loosening the two screws in the face of the deflection knob of the sight and slipping the micrometer deflection scale until the scale is indexed at the same reading as angle B of the aiming circle.

CALIBRATION FOR DEFLECTION USING THE AIMING POINT METHOD

2-102. This method (Figure 2-18) places the aiming circle and the sightunit on essentially the same line of sight.

(1) Set up the aiming circle and fine level. Align the vertical line of the telescope on a DAP (a sharp, distinct object not less than 200 meters in distance).

(2) Move the mortar baseplate until the baseplate is bisected by the vertical line of the telescope of the aiming circle. Mount the mortar at an elevation of 800 mils. Traverse and cross-level the mortar until the axis of the barrel from the baseplate to the muzzle is bisected by the vertical line of the aiming circle (the mortar should be mounted about 25 meters from the aiming circle).

NOTE: Indexing the aiming circle at 0 is not necessary; only the vertical line is used to align the mortar with the DAP.

(3) The aiming circle operator moves to the mortar and lays the vertical line of the sight on the same DAP. If the sight is calibrated, the deflection scales of the sight are slipped to a reading of 3200.

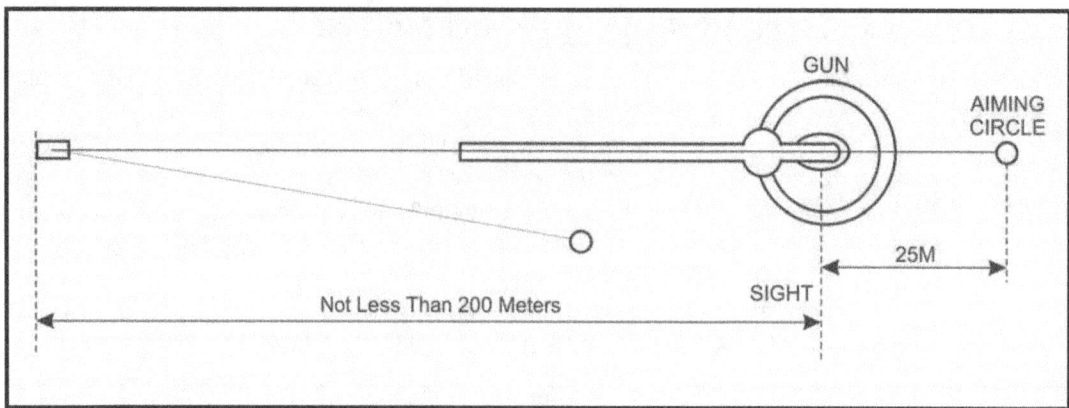

Figure 2-18. Calibration for deflection using the distant aiming point method.

SECTION V. OTHER EQUIPMENT

Other equipment required to operate and employ mortars is discussed in this section.

AIMING POSTS, M14 AND M1A2

2-103. The M14 aiming posts (Figure 2-19) are used to establish an aiming point (reference line) when laying the mortar for deflection. They are made of aluminum tubing and have a pointed tip on one end. Aiming posts have red and white stripes so they can be easily seen through the sight. The M14 aiming post comes in a set of eight segments, plus a weighted stake for every 16 segments to be used as a driver in hard soil. The stake has a point on each end and, after emplacement, it can be mounted with an aiming post. The segments can be stacked from tip to tail, and they are carried in a specially designed case with a compartment for each segment. Four M1A2 aiming posts are provided with the mortar. They may be stacked up to two high.

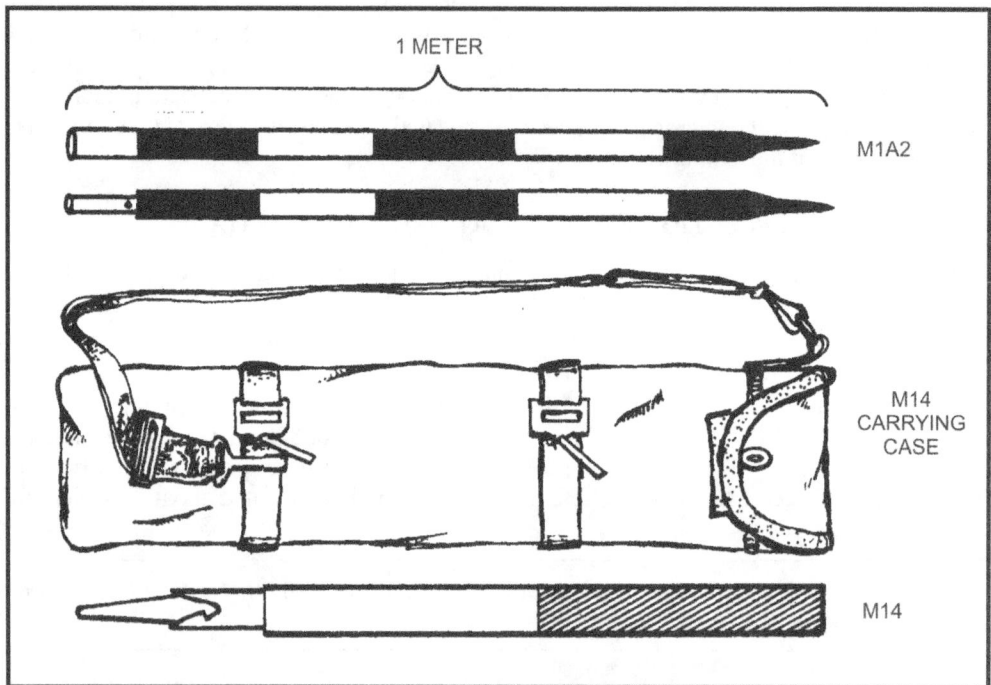

Figure 2-19. Aiming posts, M14 and M1A2.

AIMING POST LIGHTS, M58 AND M59

2-104. Aiming post lights (Figure 2-20) are attached to the aiming posts so they can be seen at night through the sight. The near post must have a different color light than that of the far post. Aiming post lights come in sets of three—two green (M58) and one orange (M59). An extra third light is issued for the alternate aiming post. Each light has a clamp, tightened with a wing nut, for attachment to the aiming post. The light does not have a cover for protection when not in use and does not need batteries.

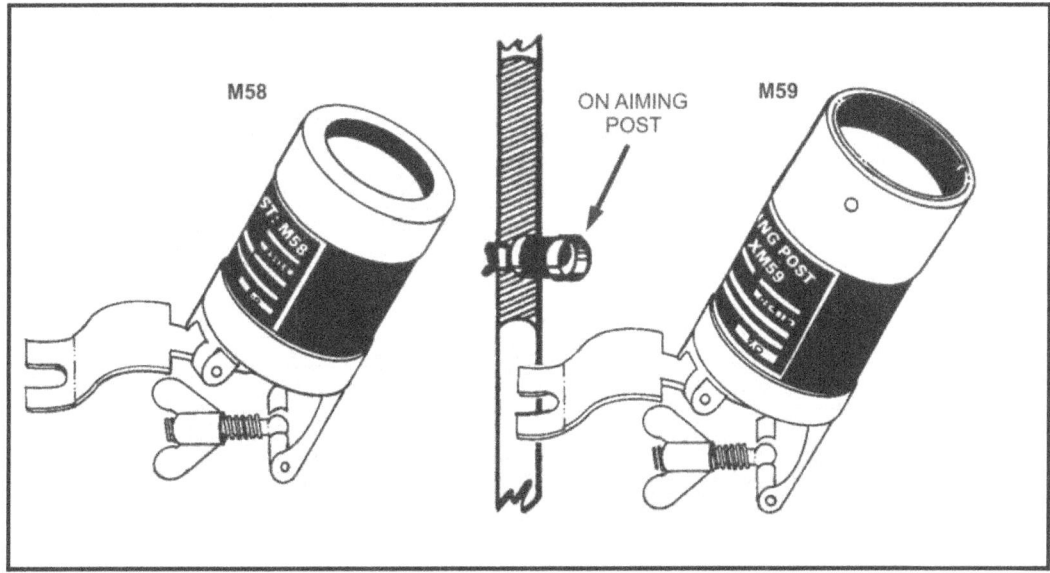

Figure 2-20. Aiming post lights, M58 and M59.

WARNING

Radioactive material (tritium gas [H^3]) is used in the M58 and M59 aiming post lights. Radioactive leakage may occur if the M58 and M59 aiming post lights are broken or damaged. If exposed to a broken or damaged M58 or M59 aiming post light or if skin contact is made with any area contaminated with tritium, immediately wash with nonabrasive soap and water, and notify the local RPO.

Chapter 2

SECTION VI. LAYING OF THE SECTION

When all mortars in the section are mounted, the section leader lays the section parallel on the prescribed azimuth with an aiming circle. The mortar section normally fires a parallel sheaf (Figure 2-21). To obtain this sheaf, it is necessary to lay the mortars parallel. When a section moves into a firing position, the FDC determines the azimuth on which the section is to be laid and notifies the platoon sergeant (section sergeant). Before laying the mortars parallel, the section leader must calibrate the mortar sights. All mortars are then laid parallel using the aiming circle, mortar sight, or compass. The section is normally laid parallel by the following two steps:

 (1) Establish the 0-3200 line of the aiming circle parallel to the mounting azimuth.
 (2) Lay the section parallel to the 0-3200 line of the aiming circle (reciprocal laying).

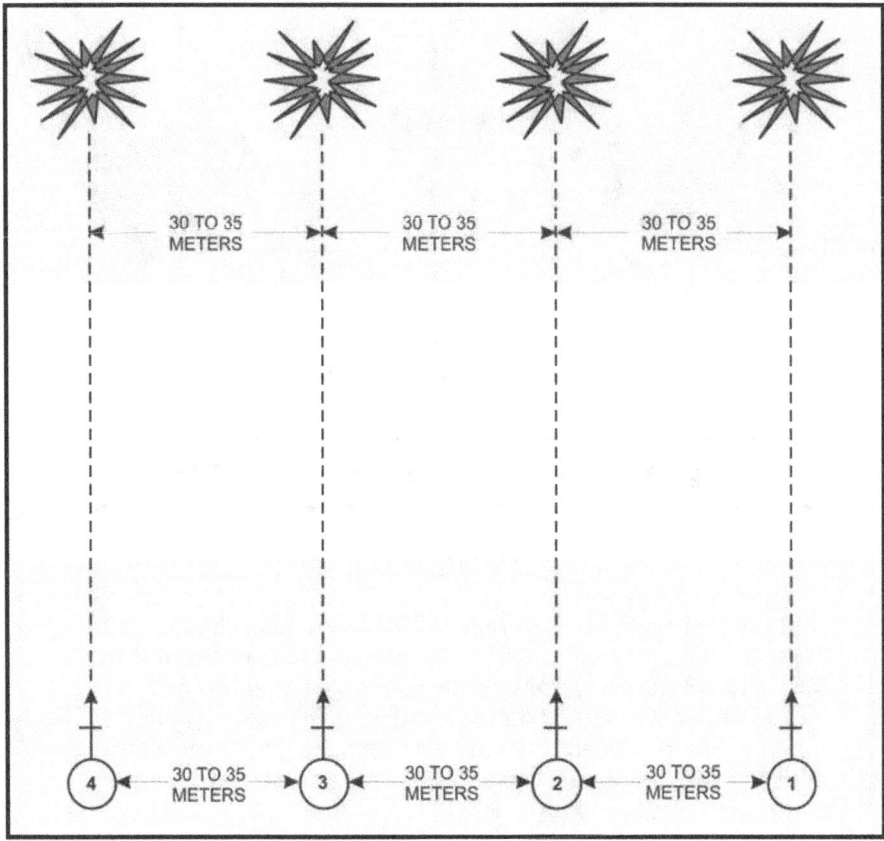

Figure 2-21. Parallel sheaf.

RECIPROCAL LAYING

2-105. Reciprocal laying is a procedure by which the 0-3200 line of one instrument (aiming circle) and the 0-3200 line of another instrument (sightunit) are laid parallel (Figure 2-22). When the 0-3200 lines of an aiming circle and the 0-3200 line of the sightunit are parallel, the barrel is parallel to both 0-3200 lines, if the sight has been properly calibrated. The principle of reciprocal laying is based on the geometric theorem that states if two parallel lines are cut by a transversal, the alternate interior angles are equal. The parallel lines are the 0-3200 lines of the instruments, and the transversal is the line of sight between the two instruments. The alternate interior angles are the equal deflections placed on the instruments.

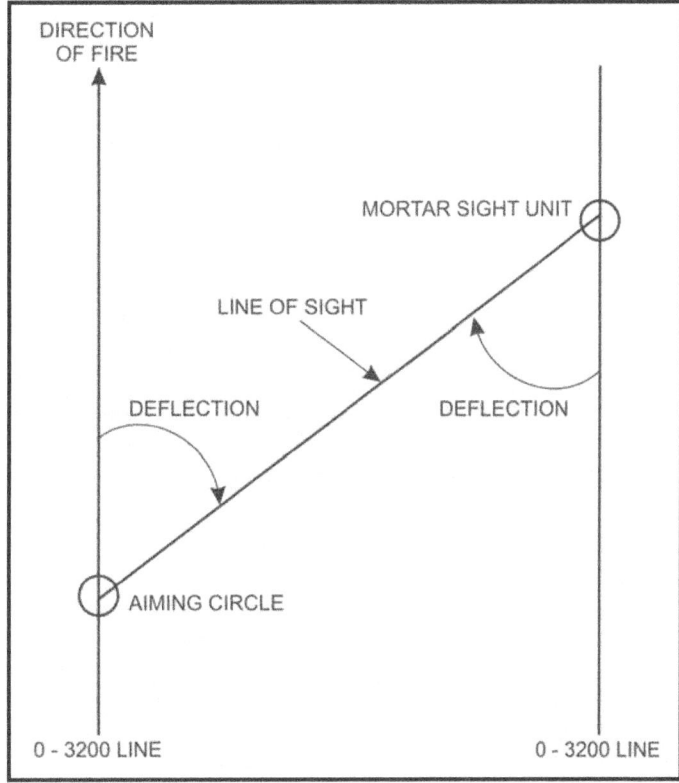

Figure 2-22. Principle of reciprocal laying.

2-106. Orient the aiming circle so that the 0-3200 line of the aiming circle is parallel to the mounting azimuth. The section leader announces to the mortar gunners (either by voice or visual signal), "Section, aiming point this instrument." The gunners turn their sights until the vertical cross line of the sight is sighted on the lens of the aiming circle and the mortar is level. The gunners announce (either by voice or visual signal), "Number (1, 2, and so on), aiming point identified." The section leader, using the upper motion, sights on the lens of the sightunit, reads the deflection on the azimuth micrometer scales, and announces the deflection to the gunner on the mortar. The gunner sets the deflection on the sightunit and causes the mortar to be moved until the vertical cross line of the sight is again sighted on the lens of the aiming circle and the mortar is level.

2-107. When the sight has been sighted on the aiming circle, the gunner reports, "Ready for recheck." The platoon sergeant (section sergeant) again sights on the lens of the sightunit, and reads and announces the deflection. This procedure is repeated until the gunner reports a difference of ZERO (or ONE) MIL between successive deflections. The mortar has then been laid.

Reciprocal Laying on a Grid Azimuth

2-108. This paragraph discusses the commands and procedures used in reciprocal laying of the mortar section on a given grid azimuth. The example below illustrates the commands and procedures. The example uses a mounting azimuth of 5550 mils and an aiming circle with a declination constant of 450 mils.

(1) The FDC directs the section to lay the mortar parallel on a mounting (grid) azimuth.

(2) The platoon sergeant (section sergeant) receives the command MOUNTING AZIMUTH FIVE FIVE FIVE ZERO (5550 mils) from the FDC.

(3) The section leader calculates the number of mils to set on the aiming circle:

Declination constant	450 mils
	+ 6400 mils
Sum	6850 mils
Minus the mounting (grid) azimuth	- 5550 mils
Remainder to set on aiming circle	1300 mils

(4) The platoon sergeant (section sergeant) mounts and levels the aiming circle at a point from which he can observe the sights of all the mortars in the section (normally the left front or left rear of the section).

(5) He places 1300 mils on the azimuth and micrometer scales of the aiming circle (recording motion).

(6) Using the orienting knob, he centers the magnetic needle in the magnetic needle magnifier. This orients the 0-3200 line of the aiming circle in the desired direction (mounting azimuth 5550 mils).

(7) The platoon sergeant (section sergeant) announces, "Section, aiming point this instrument."

(8) All gunners refer their sights to the aiming circle with the vertical cross line laid on the center of the aiming circle. The gunner then announces, "Number two (one or three), aiming point identified."

(9) To lay the mortar barrel parallel to the 0-3200 line of the aiming circle (Figure 2-23), the platoon sergeant (section sergeant) turns the upper motion of the aiming circle until the vertical cross line is laid on the center of the lens of the mortar sight. He reads the azimuth and micrometer scales and announces the deflection; for example, "Number two, deflection one nine nine eight (1998)."

(10) The gunner repeats the announced deflection, "Number two, deflection one nine nine eight," and places it on his sight. Assisted by the assistant gunner, he lays the mortar so that the vertical line is once again laid on the center of the aiming circle after the gunner announces, "Number two, ready for recheck."

(11) Using the upper motion, the platoon sergeant (section sergeant) again lays the vertical cross line of the aiming circle on the lens of the mortar sight. He reads the new deflection from the azimuth and micrometer scales and announces the reading; for example, "Number two, deflection two zero zero zero."

Sighting and Fire Control Equipment

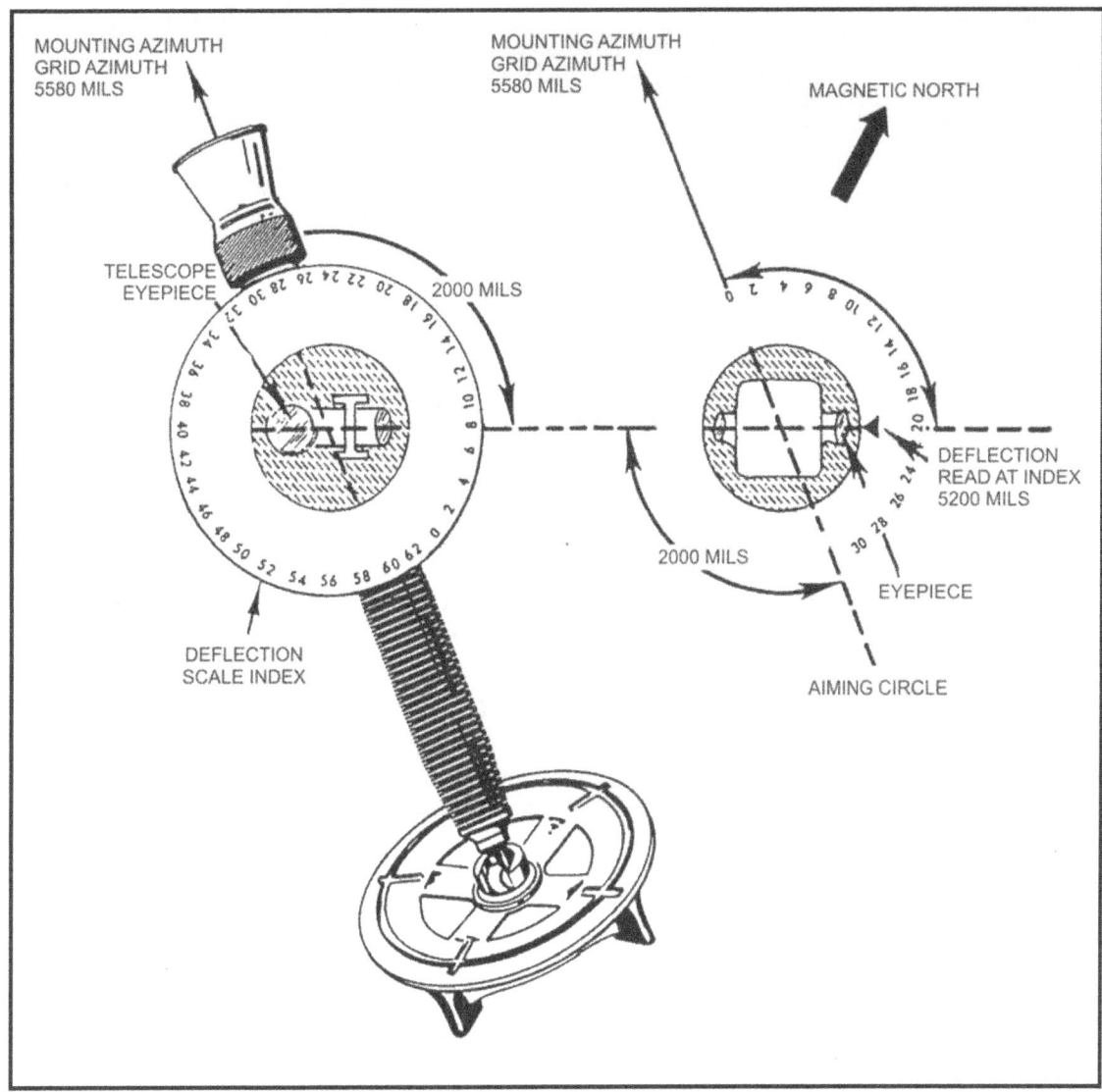

Figure 2-23. Mortar laid parallel with the aiming circle.

(12) The gunner repeats the new deflection (Number two, deflection two zero zero zero) and places it on his sight. Assisted by the assistant gunner, he lays the mortar with the vertical cross line of the sight on the center of the aiming circle and announces, "Number two, ready for recheck."

(13) The above procedure is repeated until the mortar sight and aiming circle are sighted on each other with a difference of not more than ONE mil between the deflection readings. When so laid, the gunner announces, "Number two (one or three), zero mils (one mil), mortar laid." The mortar barrel is now laid parallel to the 0-3200 line of the aiming circle.

(14) The section leader uses the same procedure to lay each of the other mortars in the section parallel. When all mortars are parallel to the 0-3200 line of the aiming circle, they are parallel to each other and laid in the desired azimuth (Figure 2-24).

Chapter 2

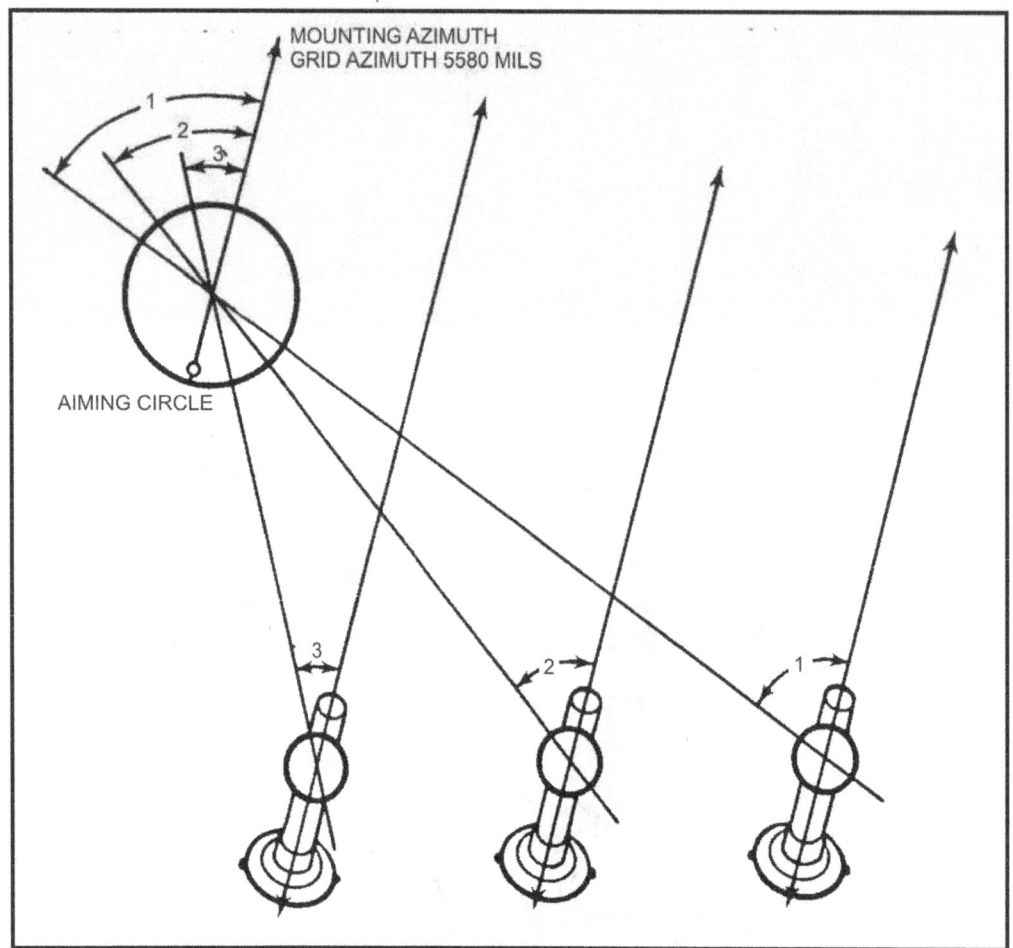

Figure 2-24. Mortars laid parallel in the desired azimuth.

> **NOTE:** The section can be laid parallel by laying all mortars at the same time. The instrument operator reads deflections to each of the mortars in turn. As soon as the gunner of any mortar announces, "Ready for recheck," the instrument operator reads the new deflection to that mortar. By laying all mortars at the same time, the section is quickly ready to fire.

(15) As soon as each mortar is laid, the platoon sergeant (section sergeant) commands DEFLECTION TWO EIGHT ZERO ZERO (2800), REFER, PLACE OUT AIMING POSTS. (The aiming posts are normally placed out on a referred deflection of 2800 mils.) The gunner, without disturbing the lay of the mortar, places the announced deflection on his sight and aligns the aiming posts with the vertical line of the mortar sight. He then announces, "Up."

(16) When all mortar gunners announce, "Up," the instrument operator covers the head of the aiming circle, but leaves the instrument in position to permit a rapid recheck of any mortar, if necessary.

RECIPROCAL LAYING ON A MAGNETIC AZIMUTH

2-109. Although the section is normally laid parallel on a grid azimuth, it can be laid parallel on a magnetic azimuth by subtracting the magnetic mounting azimuth from 6400 mils and by setting the remainder on the azimuth and micrometer scales of the aiming circle. The section leader orients the instrument and lays the section.

Sighting and Fire Control Equipment

RECIPROCAL LAYING USING THE ORIENTING ANGLE

2-110. The mortars of each section can be laid parallel more accurately if the instrument operator lays the section parallel by using the orienting angle. He sets up and levels the aiming circle; orients the aiming circle, and lays the section.

RECIPROCAL LAYING USING THE MORTAR SIGHTS

2-111. The mortar section can be laid parallel by using the mortar sights (Figure 2-25). For this method, it is best to have the mortars positioned so that all sights are visible from the base mortar. The base mortar (normally No. 2) is laid in the desired direction of fire by compass or by registration on a known point. After the base mortar is laid for direction, the remaining mortars are laid parallel to the base mortar as follows:

(1) The section leader moves to the mortar sight of the base mortar and commands SECTION, AIMING POINT THIS INSTRUMENT. The gunners of the other mortars refer their sights to the sight of the base mortar and announce, "Aiming point identified."

(2) The section leader reads the deflection from the red scale on the sight of the base mortar. He then determines the back azimuth of that deflection and announces it to the other gunners.

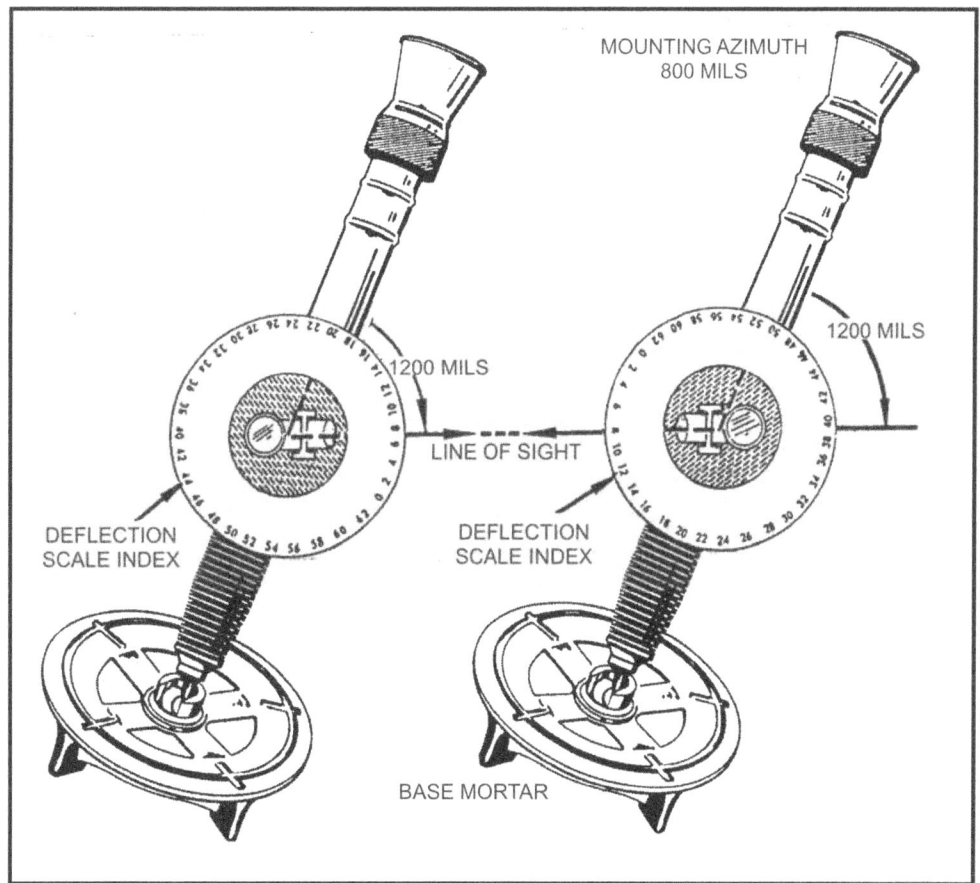

Figure 2-25. Mortar laid parallel with sights.

Chapter 2

EXAMPLE

Base mortar mounting azimuth: 0800 mils
Base deflection to # 3 gun: 1200 mils

Add or subtract 3200 mils and announce the reciprocal deflection to the gun to be laid: "#3 GUN, DEFLECTION 4400 MILS."

> **NOTE:** A back azimuth is determined by adding or subtracting 3200 to the initial deflection—for example, "Number three, deflection one two zero zero."

(3) Each gunner repeats the announced deflection for his mortar, places the deflection on his sight, and re-lays on the sight of the base mortar. When the lens of the base mortar sight is not visible, the gunner lays the vertical cross line of his sight on one of the other three mortar sights (Figure 2-26). He is laid in by this mortar once it is parallel to the base mortar sightunit. He then announces, "Number one (or three), ready for recheck."

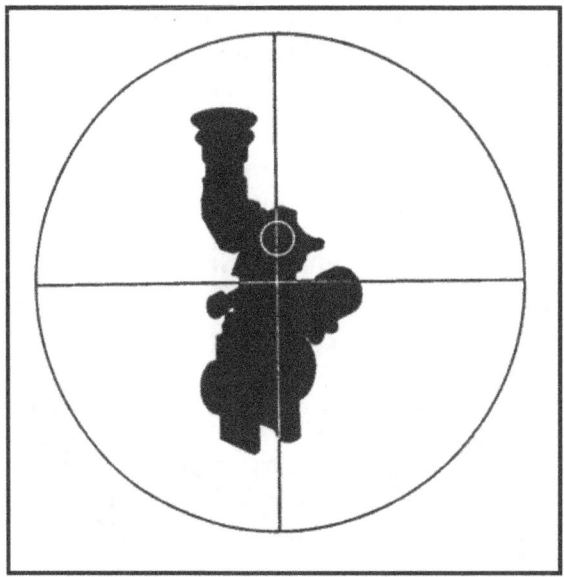

Figure 2-26. Sighting on the mortar sight.

(4) After each mortar has been laid parallel within zero (or one mil), the mortar barrels are parallel to the base mortar.
(5) As soon as each mortar is laid, the section leader commands NUMBER THREE, REFER DEFLECTION TWO EIGHT ZERO ZERO (2800), PLACE OUT AIMING POSTS.

RECIPROCAL LAYING USING THE M2 COMPASS

2-112. A rapid means of laying the section parallel is by using the compass. This is an alternate means and is used only when an aiming circle is not available or when time dictates. It is not as accurate as the methods previously described.

(1) Before mounting the mortars, each squad leader places a base stake in the ground to mark the approximate location of the mortar.
(2) The section leader announces the desired mounting azimuth; for example, "Mount mortars, magnetic azimuth two two zero zero."

(3) Each squad leader places his compass on the base stake marking the location of his mortar, and orients the compass on the desired mounting azimuth. By sighting through the compass, he directs the ammunition bearer in aligning the direction stakes along the mounting (magnetic) azimuth.

(4) Each mortar is then mounted and laid on the direction stakes with a deflection of 3200 placed on the sight. The mortar barrels are now laid parallel. Once laid for direction, the referred deflection is announced and the aiming posts are emplaced. The direction stakes may be used as the aiming posts but this limits the available amount of mils before the barrel blocks the line of sight.

NOTE: Recognizing the difference in individual compasses, the section leader can prescribe that all mortars be laid with one compass. This eliminates some mechanical error. It is also possible to lay only the base mortar as described above and then lay the remaining mortars parallel using the mortar sight method.

PLACING OUT AIMING POSTS

2-113. When a firing position is occupied, the section leader must determine in which direction the aiming posts are to be placed out. Factors to consider are terrain, sight blockage, traffic patterns in the section area, and the mission. After the section leader has laid the section for direction, he commands the section to place out aiming posts on a prescribed deflection. The gunner uses the arm-and-hand signals displayed in Figure 2-27 to direct the ammunition bearer to place out the aiming posts.

REFERRED DEFLECTION IS 2800 MILS

2-114. The section leader commands SECTION, REFER DEFLECTION TWO EIGHT ZERO ZERO, PLACE OUT AIMING POSTS. If possible, the aiming posts should be placed out to the left front on a referred deflection of 2800 mils. This direction gives a large latitude in deflection change before sight blockage occurs. Under normal conditions the front aiming post is placed out 50 meters and the far aiming post 100 meters.

REFERRED DEFLECTION IS NOT 2800 MILS

2-115. The section leader commands SECTION, REFER DEFLECTION XXX, PLACE OUT AIMING POSTS. When local terrain features do not permit placing out the aiming posts at a referred deflection of 2800 mils, the following procedure is used:

(1) The initial steps are the same as those given above. The gunner refers the sight to the back deflection of the referred deflection and directs the ammunition bearer to place out two aiming posts 50 and 100 meters from the mortar position.

(2) If the gunner receives a deflection that would be obscured by the barrel, he indexes the referred back deflection, and lays in on the rear aiming post.

ANY DIRECTION FIRE CAPABILITY

2-116. Two more aiming posts may be placed out to prevent a sight block if the mortar is used in a 6400-mil capability. The section leader must then select an area to the rear of each mortar where aiming posts can be placed on a common deflection for all mortars. A common deflection of 0700 mils is preferred; however, any common deflection to the rear may be selected when obstacles, traffic patterns, or terrain prevent use of 0700 mils.

PROCEDURES WHEN AIMING POSTS CANNOT BE LAID ON THE PRESCRIBED REFERRED DEFLECTION

2-117. If the gunner cannot lay his aiming posts on the prescribed referred deflection, he will determine which general direction allows a maximum traverse before encountering a sight block. He refers the sight to that direction and indexes any deflection to the nearest 100 mils. He then places out the aiming posts, assisted by the ammunition bearer, and informs the squad leader of his deflection. The squad leader then supervises the gunner in slipping his scale back to the referred deflection.

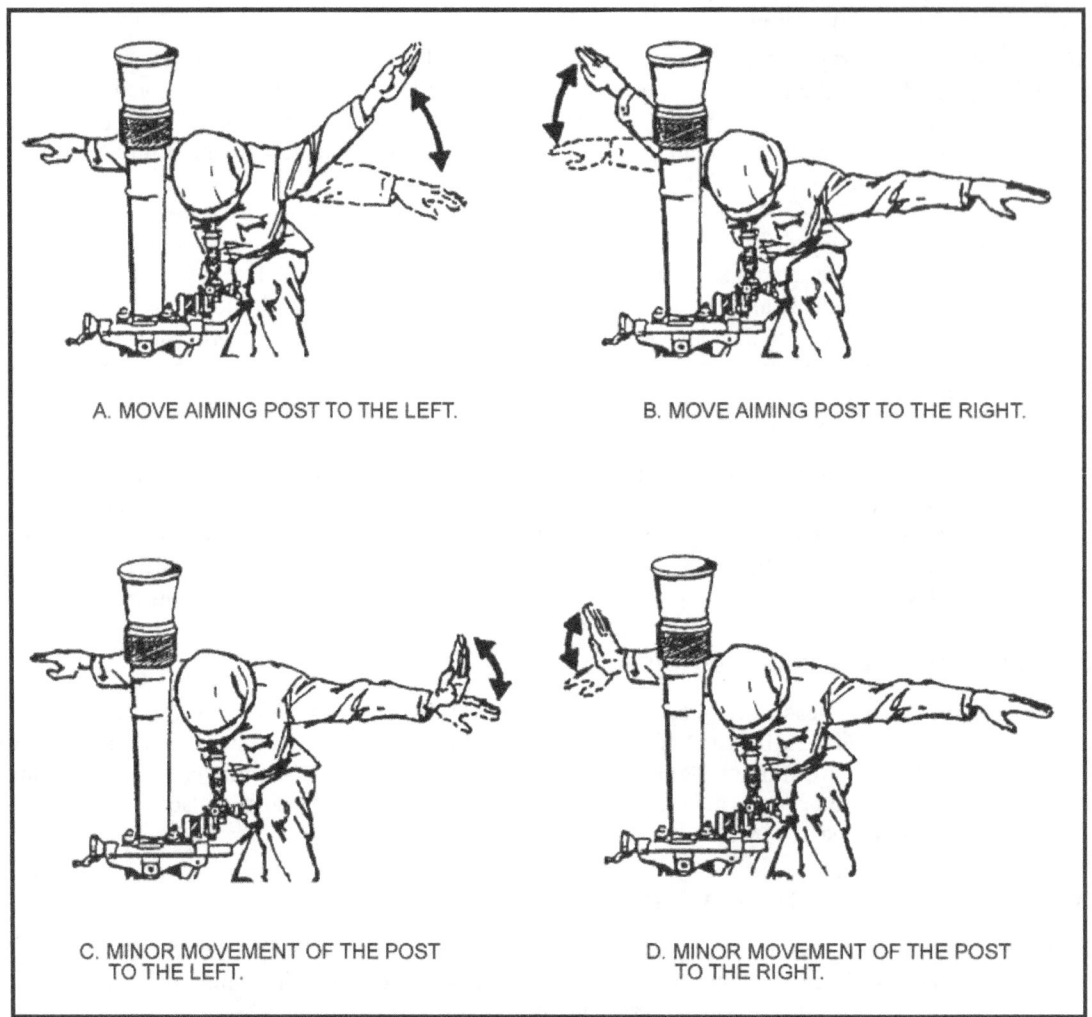

Figure 2-27. Arm-and-hand signals used in placing out aiming posts.

Sighting and Fire Control Equipment

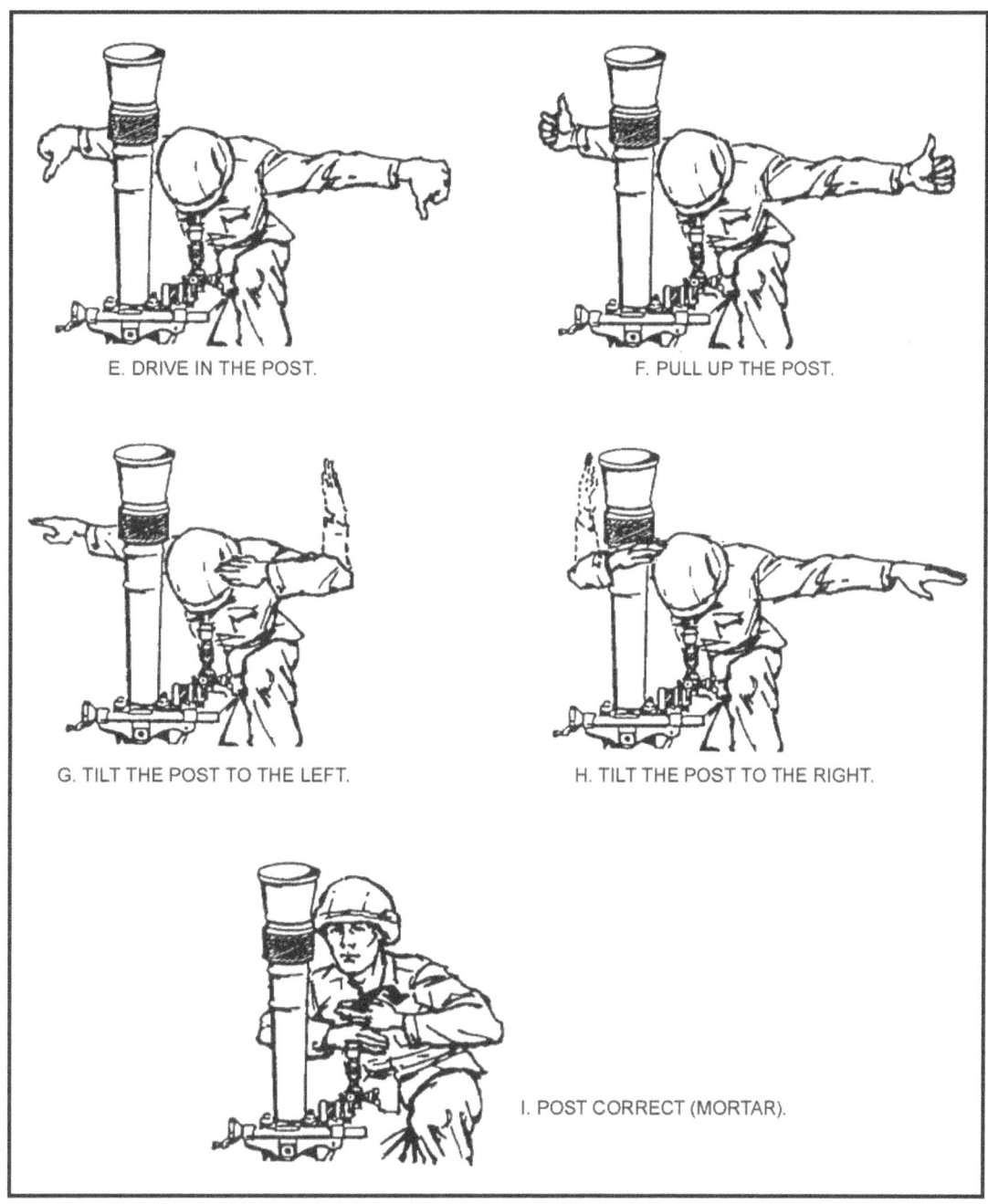

Figure 2-27. Arm-and-hand signals used in placing out aiming posts (continued).

Chapter 2

SECTION VII. LOADING AND FIRING

Upon receiving a fire command from the section leader, the entire gun crew repeats each element of it. The gunner places the firing data on the sight and, assisted by the assistant gunner, lays the mortar. The ammunition bearer repeats the charge element when announced by the gunner and prepares the round with that charge. (If a fuze setting is announced, the ammunition bearer also repeats the setting and places it on the fuze.) He completes his preparation of the cartridge to include safety checks, removing charges, and removing any safety wires. The squad leader, for safety purposes, spot-checks the ammunition and the data on the sight and the lay of the mortar. He then commands FIRE.

FIRING OF THE GROUND-MOUNTED MORTAR

2-118. The crew fires the mortar as follows:

(1) To settle the baseplate, the gunner removes the sight, being careful not to disturb the lay of the mortar. He continues to remove the sight until the baseplate assembly is settled and there is no danger of the sight becoming damaged from the recoil of the mortar. The bipod assembly can slide up the barrel when the gunner fires the 81-mm mortar, M252 or the 120-mm mortar, M120. The gunner must not try to place the sight back into position until the baseplate is settled.

(2) The ammunition bearer passes a round to the assistant gunner, holding the round with the palms of both hands up and near each end of the round so that the fuze is pointing in the general direction of the mortar.

(3) The assistant gunner takes the round from the ammunition bearer with his right hand, palm up and his left hand, palm down. He grasps the body of the round near the center, guides it into the barrel to a point beyond the narrow portion of the body of the shell, and releases the round. He cuts both hands sharply away and down along the barrel. At the same time, he pivots to the left and bends toward the ammunition bearer, extending his hands to receive the next round. He is careful not to disturb the lay of the mortar as he loads the round, which could cause considerable dispersion in the target area and create unsafe conditions due to erratic fire.

> **CAUTION**
> Do not load or fire the mortar while wearing gloves.

TARGET ENGAGEMENT

2-119. Target engagement is achieved through fire commands, which are the technical instructions issued to mortar crews. The basis for these commands is the data processed in the FDC. There are two types of commands: initial fire commands, issued to start a fire mission; and subsequent fire commands, issued to change firing data and to cease firing. The elements of both commands follow the same sequence. However, subsequent commands include only such elements that are changed, except for the elevation element, which is always announced. A correct fire command is brief and clear, and includes all the elements necessary for accomplishing the mission. The commands are sent to the section leader, or individual squad leaders if the section is dispersed, by the best available means. To limit errors in transmission, the person receiving the commands at the mortar position repeats each element as it is received. The sequence for the transmission of fire commands is shown in Table 2-7.

Table 2-7. Sequence for transmission of fire commands.

SEQUENCE	EXAMPLE
Mortars to follow	Section
Shell and fuze	HE quick
Mortars to fire	Number two
Method of fire	One round
Deflection	Two eight hundred
Charge	Charge 3
Time	
Elevation	One two seven zero

NOTE: All fire commands will follow this sequence. Elements not necessary for the proper conduct of fire are omitted.

EXECUTION OF FIRE COMMANDS

2-120. The various fire commands are explained in this paragraph.

MORTARS TO FOLLOW

2-121. This element serves two purposes: it alerts the section for a fire mission and it designates the mortars that are to follow the commands. The command for all mortars in the section to follow the fire command is SECTION. Commands for individual or pairs of mortars are given a number (ONE, TWO, and so forth).

SHELL AND FUZE

2-122. This element alerts the ammunition bearers as to what type of ammunition and fuze action to prepare for firing; for example, HE QUICK; HE DELAY; HE PROXIMITY; and so forth.

MORTAR(S) TO FIRE

2-123. This element designates the specific mortar(s) to fire. If the mortars to fire are the same as the mortars to follow, this element is omitted. The command to fire an individual mortar or any combination of mortars is NUMBER(s) ONE (THREE and so forth).

METHOD OF FIRE

2-124. In this element, the mortar(s) designated to fire in the preceding element is told how many rounds to fire, how to engage the target, and any special control desired. Also included are the number and type ammunition to be used in the fire-for-effect phase.

Volley Fire

2-125. A volley can be fired by one or more mortars. The command for volley fire is (so many) ROUNDS. Once all mortars are reported up, they fire on the platoon sergeant's (section sergeant's) or squad leader's command. If more than one round is being fired by each mortar, the squads fire the first round on command and the remaining as rapidly as possible consistent with accuracy and safety, and without regard to other mortars. If a specific time interval is desired, the command is (so many) ROUNDS AT (so many) SECONDS INTERVAL, or (so many) ROUNDS PER MINUTE. In this case, a single round for each mortar, at the time interval indicated, is fired at the platoon sergeant's (section sergeant's) or squad leader's command.

Chapter 2

Section Right (Left)

2-126. This is a method of fire in which mortars are discharged from the right (left) one after the other, normally at 10-second intervals. The command for section fire from the right (left) flank at intervals of 10 seconds is SECTION RIGHT (LEFT), ONE ROUND. Once all mortars are reported up, the platoon sergeant (section sergeant) gives the command FIRE; for example, SECTION RIGHT, ONE ROUND; the platoon sergeant (section sergeant) commands FIRE ONE; 10 seconds later FIRE TWO, and so forth.

2-127. If the section is firing a section left, the fire begins with the left-most gun and works to the right. The command LEFT (RIGHT) designates the flank from which the fire begins. The platoon sergeant (section sergeant) fires a section right (left) at 10-second intervals unless he is told differently by the FDC; for example, SECTION LEFT, ONE ROUND, TWENTY-SECOND INTERVALS.

2-128. When firing continuously at a target is desired, the command is CONTINUOUS FIRE. When maintaining a smoke screen is desired, firing a series of sections right (left) may be necessary. In this case, the command is CONTINUOUS FIRE FROM THE RIGHT (LEFT). The platoon sergeant (section sergeant) then fires the designated mortars consecutively at 10-second intervals unless a different time interval is specified in the command.

2-129. Changes in firing data (deflections and elevations) are applied to the mortars in turns of traverse or elevation so as not to stop or break the continuity of fire; for example, NUMBER ONE, RIGHT THREE TURNS; NUMBER TWO, UP ONE TURN. When continuous fire is given in the fire command, the platoon sergeant (section sergeant) continues to fire the section until the FDC changes the method of fire or until the command END OF MISSION is given.

Traversing Fire

2-130. In traversing fire, rounds are fired with a designated number of turns of traverse between each round. The command for traversing fire is (so many) ROUNDS, TRAVERSE RIGHT (LEFT) (so many) TURNS. At the platoon sergeant's (section sergeant's) or squad leader's command FIRE, all mortars fire one round, traverse the specified number of turns, fire another round, and continue this procedure until the number of rounds specified in the command have been fired.

Searching Fire

2-131. Searching fire is fired the same as volley fire except that each round normally has a different range. No specific order is followed in firing the rounds. For example, the assistant gunner does not start at the shortest range and progress to the highest charge or vice versa, unless instructed to do so. Firing the rounds in a definite sequence (high to low or low to high) establishes a pattern of fire that can be detected by the enemy.

At My Command

2-132. If the FDC wants to control the fire, the command AT MY COMMAND is placed in the method of fire element of the fire command. Once all mortars are reported up, the platoon sergeant (section sergeant) reports to the FDC: SECTION READY. The FDC then gives the command FIRE.

Do Not Fire

2-133. The FDC can command DO NOT FIRE immediately following the method of fire. DO NOT FIRE then becomes a part of the method of fire. This command is repeated by the platoon sergeant (section sergeant). As soon as the weapons are laid, the platoon sergeant (section sergeant) reports to FDC that the section is laid. The command for the section of fire is the command for a new method of fire not followed by DO NOT FIRE.

DEFLECTION

2-134. This element gives the exact deflection setting to be placed on the mortar sight. It is always announced in four digits, and the word DEFLECTION always precedes the sight setting; for example, DEFLECTION, TWO EIGHT FOUR SEVEN (2847). When the mortars are to be fired with different deflections, the number of the mortar is given and then the deflection for that mortar; for example, NUMBER THREE, DEFLECTION TWO FOUR ZERO ONE (2401).

CHARGE

2-135. This element gives the charge consistent with elevation and range as determined from the firing tables; for example, CHARGE FOUR (4). The word CHARGE always precedes the amount; for example, ONE ROUND, CHARGE FOUR (4).

TIME

2-136. The computer tells the ammunition bearer the exact time setting to place on the PROX, MTSQ, and MT fuze. The command for time setting is TIME (so much); for example, TIME TWO SEVEN. The command for a change in time setting is a new command for time.

ELEVATION

2-137. The elevation element is always given in the fire command. It serves two purposes: it gives the exact elevation setting to place on the mortar sight, and it gives permission to fire when the method of fire is "when ready." When the method of fire is "at my command," the section fires at the section leader's command.

ARM-AND-HAND SIGNALS

2-138. When giving the commands FIRE or CEASE FIRING, the section leader or squad leader uses both arm-and-hand signals and voice commands (Figure 2-28).

READY

2-139. The signal for "I am ready" or "Are you ready?" is to extend the arm toward the person being signaled. Then, the arm is raised slightly above the horizontal, palm outward.

FIRE

2-140. The signal to start fire is to drop the right arm sharply from a vertical position to the side. When the section leader wishes to fire a single mortar, he points with his arm extended at the mortar to be fired, then drops his arm sharply to his side.

CEASE FIRING

2-141. The signal for cease firing is to raise the hand in front of the forehead, palm to the front, and to move the hand and forearm up and down several times in front of the face.

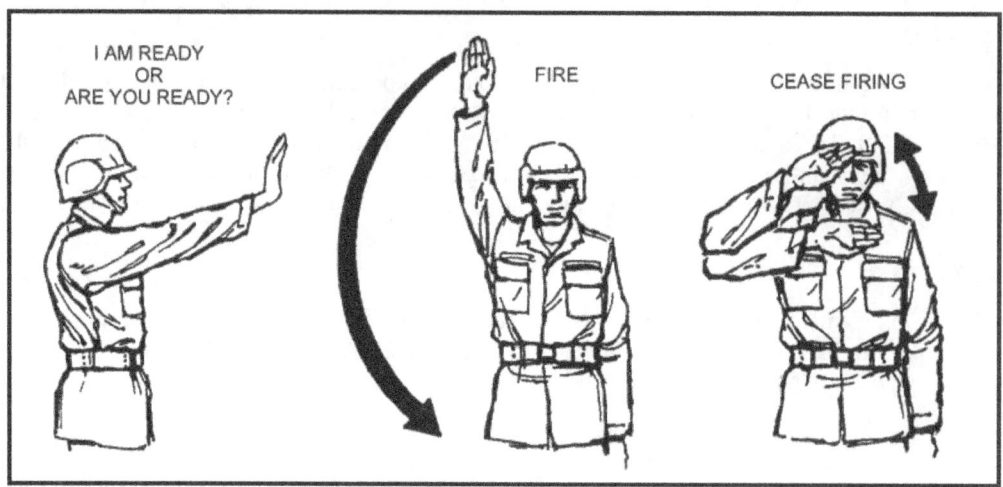

Figure 2-28. Arm-and-hand-signals for ready, fire, and cease firing.

SUBSEQUENT FIRE COMMANDS

2-142. Only the elements that change from the previous fire command are announced in subsequent fire commands except for the elevation element (command to fire), which is always announced in every fire command.

DEFLECTION

2-143. Changes in direction are given in total deflection to be placed on the sight; for example, DEFLECTION TWO EIGHT ONE TWO (2812).

ELEVATION

2-144. When a change is made in mortars to fire or in the method of fire, the subsequent command includes one or both of these elements and the elevation. When the elevation does not change, the command ELEVATION (so many mils) is given (same as that given in the previous command).

CEASE FIRING/CHECK FIRE

2-145. To interrupt firing, CEASE FIRING or CHECK FIRE is commanded.

2-146. CEASE FIRING indicates to the section the completion of a fire mission, but not necessarily the end of the alert. Firing is renewed by issuing a new initial fire command.

2-147. CHECK FIRE indicates a temporary cessation of firing and allows firing to be resumed with the same data by the command RESUME FIRING or by a subsequent fire command.

END OF MISSION

2-148. So that the mortar crews can relax between fire missions, the end of the alert is announced by the command END OF MISSION. All gunners then lay their mortars as directed by the FDC. Upon completion of a fire mission, all mortars normally lay on final protective fire (FPF) data unless otherwise directed. The platoon sergeant (section sergeant) is responsible for ensuring that the mortars are laid on FPF data and that the prescribed amount of ammunition for the FPF is prepared and on position.

REPEATING AND CORRECTING OF FIRE COMMANDS

2-149. If the platoon sergeant (section sergeant) or squad member fails to understand any elements of the fire command, he can request that element be repeated by stating, for example, "Say again deflection, elevation," and so forth. To avoid further misunderstanding, the repeated element is prefaced with "I say again deflection (repeats mils)."

INITIAL FIRE COMMAND

2-150. In an initial fire command, an incorrect element is corrected by stating, "Correction," and giving only the corrected element.

SUBSEQUENT FIRE COMMAND

2-151. In a subsequent command, an incorrect element is corrected by stating, "Correction," and then by repeating all of the subsequent commands. (The term "correction" cancels the entire command.)

REPORTING OF ERRORS IN FIRING

2-152. When any squad member discovers that an error has been made in firing, he immediately notifies his squad leader, who in turn notifies the FDC. Such errors include, but are not limited to, incorrect deflection or elevation settings, incorrect laying of the mortar, or ammunition improperly prepared for firing. Misfires are also reported this way. Errors should be promptly reported to the FDC to prevent loss of time in determining the cause and required corrective action.

NIGHT FIRING

2-153. When firing the mortar at night, the mission dictates whether noise and light discipline are to be sacrificed for speed. To counteract the loss of speed for night firing, the gunner must consider presetting both fuze and charge for ILLUM rounds with the presetting of charges for other rounds. The procedure for manipulating the mortar at night is the same as during daylight operations. To assist the gunner in these manipulations, the sight reticle is illuminated, and the aiming posts are provided with lights.

2-154. The instrument lights illuminate the reticle of the sights and make the vertical cross lines visible. The hand light on the flexible cord is used to illuminate the scales and bubbles.

2-155. An aiming post light is placed on each aiming post to enable the gunner to see the aiming posts. Aiming posts are placed out at night similar to the daylight procedure. The lights must be attached to the posts before they can be seen and positioned by the gunner. The gunner must issue commands such as NUMBER ONE, MOVE RIGHT, LEFT, HOLD, DRIVE IN, POST CORRECT. Tilt in the posts is corrected at daybreak. Some of the distance to the far post can be sacrificed if it cannot be easily seen at 100 meters. However, the near post should still be positioned about half the distance to the far post from the mortar. The far post light should be a different color from the one on the near post and be positioned so it appears slightly higher. Adjacent squads should alternate post lights to avoid laying on the wrong posts; for example, 1ST SQUAD, NEAR POST—GREEN LIGHT, FAR POST—RED LIGHT; 2D SQUAD, NEAR POST—RED LIGHT, FAR POST—GREEN LIGHT. (The M58 light is green and the M59 light is orange.)

2-156. The mortar is laid for deflection by placing the vertical cross line of the sight in the correct relation to the center of the lights attached to the aiming posts. The procedure for laying the mortar is the same as discussed in Section VI.

2-157. The night lights can be used to align the aiming posts without using voice commands.
   (1) The gunner directs the ammunition bearer to place out the aiming posts. The ammunition bearer moves out 100 meters and turns on the night light of the far aiming post. The gunner holds the instrument night light in his right (left) hand and, by moving the light to the right (left), directs the ammunition bearer to move to the right (left). To ensure that the ammunition bearer sees the light moving only in the desired direction, the gunner places his thumb over the light when returning it to the starting position. The gunner continues to direct the ammunition bearer to move the aiming post until it is properly aligned.

(2) The gunner moves the instrument light a shorter distance from the starting position when he wants the ammunition bearer to move the aiming post a short distance.

(3) The gunner holds the light over his head (starting position) and moves the light to waist level when he wants the ammunition bearer to place the aiming post into the ground. In returning the instrument light to the starting position, the gunner covers the light with his thumb to ensure that the ammunition bearer sees the light move only in the desired direction.

(4) The gunner uses the same procedure described above when he wants the ammunition bearer to move the aiming post light to a position corresponding to the vertical hairline in the sight after the aiming post has been placed into the ground.

(5) The gunner reverses the procedure described above when he wants the ammunition bearer to take the aiming post out of the ground. The gunner places the uncovered light at waist level and moves it to a position directly above his head. He then directs alignment as required.

(6) When the gunner is satisfied with the alignment of the aiming posts, he signals the ammunition bearer to return to the mortar positions by making a circular motion with the instrument light.

NOTE: When the night light is used to signal, the gunner directs the light toward the ammunition bearer.

Chapter 3
60-mm Mortar, M224

The M224 60-mm Lightweight Company Mortar System (LWCMS) provides ranger and IBCT companies with effective and lightweight fire support. Its relatively short range and the small explosive charge of its ammunition can be compensated for with careful planning and expert handling. The mortar can be fired accurately with or without an FDC.

SECTION I. SQUAD AND SECTION ORGANIZATION AND DUTIES

This section discusses the organization and duties of the 60-mm mortar squad and section.

ORGANIZATION

3-1. If the mortar section is to operate quickly and effectively in accomplishing its mission, mortar squad members must be proficient in individually assigned duties. Correctly applying and performing these duties enables the mortar section to perform as an effective fighting team. The section leader commands the section and supervises the training of the elements. He uses the chain of command to assist him in effecting his command and supervising duties.

DUTIES

3-2. The mortar squad consists of three Soldiers. Each squad member is cross-trained to perform all duties involved in firing the mortar. The positions and principal duties are as follows.

SQUAD LEADER

3-3. The squad leader is in position to best control the mortar squad. He is positioned to the right of the mortar, facing the cannon. He is also the FDC.

GUNNER

3-4. The gunner is on the left side of the mortar where he can manipulate the sight, elevating gear handle, and traversing assembly wheel. He places firing data on the sight and lays the mortar for deflection and elevation. Assisted by the squad leader (or ammunition bearer), he makes large deflection shifts by shifting the bipod assembly.

AMMUNITION BEARER

3-5. The ammunition bearer is to the right rear of the mortar. He prepares the ammunition and assists the gunner in shifting and loading the mortar. He swabs the cannon every 10 rounds or after each end of mission.

Chapter 3

SECTION II. COMPONENTS

The 60-mm mortar, M224, is a muzzle-loaded, smooth-bore, high-angle-of-fire weapon that can be fired in the conventional or handheld mode. It can be drop-fired or trigger-fired and has five major components (Figure 3-1).

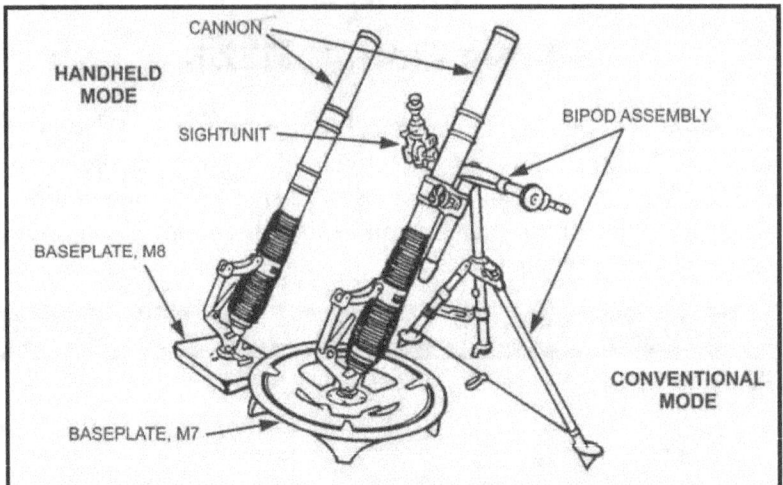

Figure 3-1. 60-mm mortar, M224, handheld and conventional mode.

TABULATED DATA

3-6. The tabulated data for the 60-mm mortar are as shown in Table 3-1.

Table 3-1. Tabulated data for the 60-mm mortar, M224.

	CONVENTIONAL MODE	HANDHELD MODE
WEIGHTS		
Complete mortar	46.5 pounds	18.0 pounds
Cannon, M225	14.4 pounds	14.4 pounds
Bipod Assembly, M170	15.2 pounds	
Sightunit, M64A1	2.5 pounds	
Baseplate, M7	14.4 pounds	
Baseplate, M8		3.6 pounds
RANGE		
Cannon, M225		
Minimum*	70 meters	75 meters
Maximum	3,490 meters	1,340 meters
RATES OF FIRE		
Maximum		
M720/M888	30 rounds for first 4 minutes	
M49A4	30 rounds for 1 minute; 18 rounds for next 4 minutes	No limit at charges 0 and 1
Sustained		
M720/M888	20 rounds per minute indefinitely	
M49A4	8 rounds per minute indefinitely	
TYPE OF FIRE	Drop-fire	Drop-fire (charges 0 and 1 only)
	Trigger-fire	Trigger-fire**
CARRYING OPTIONS	One-man carry	One-man carry
	Two-man carry	
	Three-man carry	
*Minimum SAFE range. The specific ammunition minimum ranges are design minimums.		
**Do not trigger-fire above charge 1.		

CANNON ASSEMBLY, M225

3-7. The cannon assembly (Figure 3-2) has one end closed by a breech cap. The breech cap end of the cannon has cooling fins on the outside that reduce heat generated during firing. Attached to the breech cap end is a combination carrying handle and firing mechanism. The carrying handle has a trigger, firing selector, range indicator, and auxiliary carrying handle. On the outside of the cannon is an upper and a lower firing saddle. The lower saddle is used when firing at elevations of 1100 to 1511 mils; the upper saddle is used when firing at elevations of 0800 to 1100 mils.

NOTE: When the bipod is positioned in the upper saddle, one turn of the traversing handwheel will move the cannon 10 mils. When the bipod is positioned in the lower saddle, one turn of the traversing handwheel will move the cannon 15 mils.

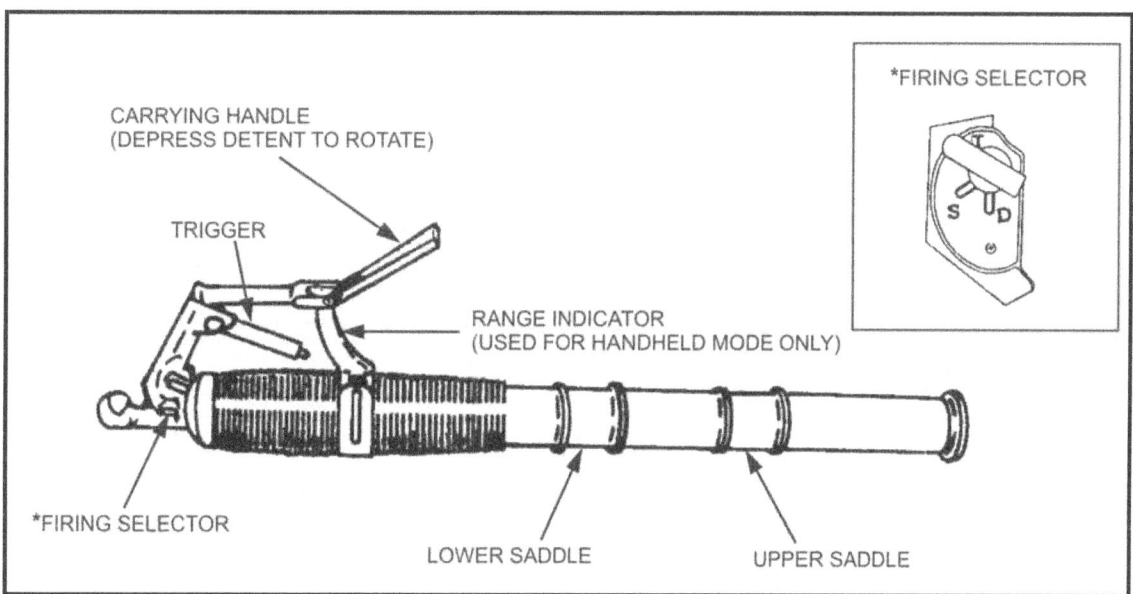

Figure 3-2. Cannon assembly, M225.

BIPOD ASSEMBLY, M170

3-8. The bipod assembly (Figure 3-3) can be attached to the cannon either before or after assembly of the cannon to the baseplate. It consists of seven subassemblies.

COLLAR ASSEMBLY

3-9. The collar assembly, with an upper and lower half, is hinged on the left and secured by a locking knob on the right. The collar fastens in one of the two firing saddles (depending on the elevation being fired), securing the bipod to the cannon.

SHOCK ABSORBERS

3-10. Two shock absorbers located on the underside of the collar assembly protect the bipod and sight from the shock of recoil during firing.

TRAVERSING MECHANISM

3-11. The traversing mechanism moves the collar assembly left or right when the traversing hand crank is pulled out and turned. The hand crank is turned clockwise to move the cannon to the right, and counterclockwise to move the cannon to the left. The left side of the traversing mechanism has a dovetail slot to attach the sight to the bipod.

ELEVATING MECHANISM

3-12. The elevating mechanism is used to elevate or depress the cannon by turning the hand crank at the base of the elevation guide tube. This assembly consists of an elevating spindle, screw, hand crank, and housing (elevation guide tube). The housing has a latch to secure the collar and shock absorbers to the housing for carrying. The hand crank is turned clockwise to depress, and counterclockwise to elevate.

RIGHT LEG ASSEMBLY

3-13. The right leg assembly has no moving parts. It consists of a foot, tubular steel leg, and hinge attached to the elevating mechanism housing.

LEFT LEG ASSEMBLY

3-14. The left leg assembly consists of a foot, tubular steel leg, hinge attached to the elevating mechanism housing, locking nut, and fine cross-leveling sleeve.

Locking Sleeve

3-15. The locking sleeve is near the spiked foot. It is used to lock the elevation housing in place.

Fine Cross-leveling Nut

3-16. The fine cross-leveling nut above the locking sleeve is used for fine leveling.

SPREAD CABLE

3-17. The spread cable is a plastic-coated steel cable attached to the bipod legs, which controls the spread of the two tubular steel legs. A snap hook is fixed to the cable to secure the bipod legs when they are collapsed for carrying.

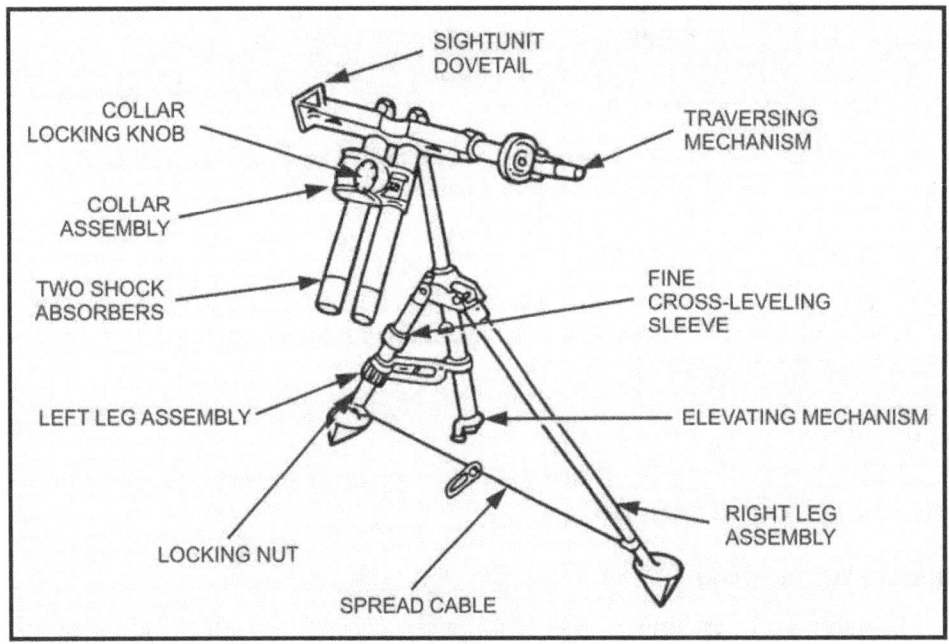

Figure 3-3. Bipod assembly, M170.

BASEPLATE, M7

3-18. The baseplate, M7, (Figure 3-4) is a one-piece, circular, aluminum-forged base. It has a ball socket with a rotating locking cap and a stationary retaining ring held in place by four screws and lock washers. The locking cap rotates 6400 mils, giving the mortar full-circle firing capability. The underside of the baseplate has four spades to stabilize the mortar during firing.

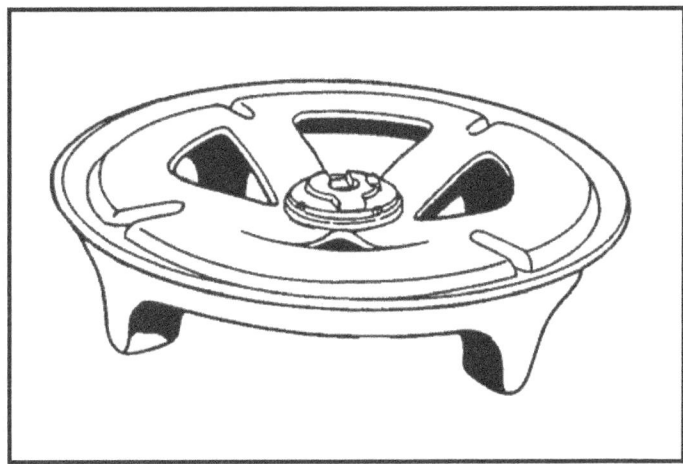

Figure 3-4. Baseplate, M7.

BASEPLATE, M8

3-19. The baseplate, M8, (Figure 3-5) is a one-piece, rectangular, aluminum-forged base. It should be used when the mortar is fired in the handheld mode. The baseplate allows the mortar to be fired 0800 mils left and 0800 mils right of the center of sector for a total sector coverage of 1600 mils. It has a socket in which the cannon can be locked to the baseplate by securing the locking arm. The underside of the baseplate has four spades to strengthen and stabilize the mortar during firing. Two spring-loaded plungers lock the baseplate to the cannon in its carry position.

Figure 3-5. Baseplate, M8.

Chapter 3

SECTION III. OPERATION

Safe operation of the 60-mm mortar requires that training include drill practice on tasks for safe manipulation and effective employment. Crew training achieves the speed, precision, and teamwork needed to deliver responsive and effective fire on target.

PREMOUNT CHECKS

3-20. Before the mortar is mounted, the squad must perform premount checks. Each squad member should be capable of performing all the premount checks.

GUNNER

3-21. The gunner performs premount checks on the mount so that—
- The spread cable is fixed to both legs and taut.
- The clearance on the left leg above the adjusting nut is two fingers in width.
- The locking sleeve is neither too loose nor too tight.
- The traversing bearing is centered.

3-22. When all pieces of equipment are checked, the gunner notifies the section leader by announcing, "All correct."

SQUAD LEADER

3-23. The squad leader performs the premount checks on the cannon so that—
- The cannon is clean both inside and outside.
- The firing pin is visible.
- The spherical projection is clean, and the firing pin is firmly seated.
- The selector switch is on drop-fire mode.

AMMUNITION BEARER

3-24. The ammunition bearer is responsible for the premount checks on the baseplate ensuring that—
- The rotatable socket cap moves freely and has a light coat of oil.
- The ribs and braces are checked for breaks and dents, and the inner ring is secured to the outer ring.

MOUNTING OF THE MORTAR

3-25. The squad mounts the mortar.
(1) The squad leader picks up and places the sight case and two aiming posts at the exact position where the mortar is to be mounted.
(2) The ammunition bearer places the outer edge of the baseplate against the baseplate stake. He aligns the left edge of the cutout portion of the baseplate with the right edge of the baseplate stake. He then rotates the socket cap so that the open end points in the direction of fire.
(3) The gunner picks up the bipod with his left hand on the traversing hand crank and his right hand on the dovetail slot. He moves forward of the baseplate about 12 to 15 inches and faces the baseplate on line with the left edge (gunner's viewpoint) of the baseplate. Dropping down on one knee in front of the bipod, the gunner supports the bipod with his left hand on the gear case. He then detaches the hook and unwraps the cable assembly. The gunner places his left hand on the midsection of the traversing slide and his right hand on the mechanical leg, and he extends the bipod legs the length of the cable assembly. He then aligns the center of the bipod assembly with the center of the baseplate. He ensures that the elevation guide cannon is vertical and the locking nut is hand-tight. The gunner moves to

the mechanical leg side and supports the bipod with his left hand on the shock absorber. He unscrews the collar locking knob to open the collar.

(4) The ammunition bearer picks up the cannon and inserts the spherical projection of the base plug into the socket. He rotates the cannon 90 degrees to lock it to the baseplate. If performed properly, the carrying handle is on the upper side of the cannon, facing skyward.

(5) The gunner pushes down on the shock absorber and raises the collar assembly. The ammunition bearer lowers the cannon and places the lower saddle on the lower part of the collar. The gunner closes the upper part of the collar over the cannon. He replaces the locking knob to its original position and makes it hand-tight. The ammunition bearer cranks the elevation hand crank up 15 to 17 turns.

(6) The gunner takes the sight out of the case and sets a deflection of 3200 mils and an elevation of 1100 mils. He mounts the sight to the mortar by pushing the lock latch on the sight inward. He slides the dovetail on the sight into the dovetail slot on the bipod until firmly seated. The gunner releases the latch. He should tap up on the bottom of the sight to ensure proper seating. He then levels the mortar first for elevation 1100 mils, and then cross-levels. The gunner announces, "(gun number) up," to his squad leader.

SAFETY CHECKS BEFORE FIRING

3-26. The entire squad performs safety checks.

GUNNER

3-27. The gunner ensures that—
- The cannon is locked to the baseplate and the open end of the socket cap points in the direction of fire. The bipod should be connected to either the upper or lower saddle of the cannon.
- The cannon is locked on the collar by the locking knob.
- The locking nut is wrist tight.
- The cable is taut.
- The selector switch on the cannon is on drop-fire.
- There is mask and overhead clearance.

Mask and Overhead Clearance

3-28. Since the mortar is normally mounted in defilade, there could be a mask such as a hill, trees, buildings, or a rise in the ground. Roofs or overhanging tree branches can cause overhead interference. The gunner must be sure the round does not strike any obstruction.

3-29. When selecting the exact mortar position, the squad leader checks quickly for mask and overhead clearance. After the mortar is mounted, the gunner checks it thoroughly. He determines mask and overhead clearance by sighting along the top of the cannon with his eye placed near the base plug. If the line of sight clears the mask, it is safe to fire. If not, he may still fire at the desired range by selecting a charge zone having a higher elevation for that particular range. When firing under the control of an FDC, the gunner reports to the FDC that mask clearance cannot be obtained at a certain elevation.

3-30. Firing is slowed if mask clearance is checked before each firing. Therefore, if the mask is not regular throughout the sector of fire, the minimum mask clearance is determined to eliminate the need for checking on each mission. To do this, the gunner depresses the cannon until the top of the mask is sighted. He then levels the elevation bubble and reads the setting on the elevation scale and elevation micrometer. That setting is the minimum mask clearance. The gunner notifies the squad leader of the minimum mask clearance elevation. Any target that requires that elevation or lower cannot be engaged from that position.

Chapter 3

3-31. Placing the mortar in position at night does not relieve the gunner of the responsibility for checking for mask and overhead clearance.

CREWMAN

3-32. One crewman ensures that the bore is clean; he swabs the bore dry.

3-33. The second crewman ensures that each round is clean, safety pin is present, and ignition cartridge is in proper condition.

SMALL DEFLECTION AND ELEVATION CHANGES

3-34. With the mortar mounted and the sight installed, the gunner makes small deflection and elevation changes.

(1) The gunner lays the sight on the two aiming posts (placed out 50 and 100 meters from the mortar) on a referred deflection of 2800 mils and an elevation of 1100 mils. The mortar is within two turns of center of traverse. The vertical cross line of the sight is on the left edge of the aiming post.

(2) The gunner is given a deflection change in a fire command between 20 and 60 mils. The elevation change announced must be less than 90 mils and more than 35 mils.

(3) As soon as the sight data are announced, the gunner places it on the sight, lays the mortar for elevation, and then traverses onto the aiming post by turning the traversing handwheel and the adjusting nut in the same direction. A one-quarter turn on the adjusting nut equals one turn of the traversing handwheel. When the gunner is satisfied with his sight picture he announces, "Up."

NOTE: The gunner repeats all elements given in the fire command.

(4) After the gunner has announced "Up," the mortar should be checked by the squad leader to determine if the exercise was performed correctly.

LARGE DEFLECTION AND ELEVATION CHANGES

3-35. With the mortar mounted and the sight installed, the squad makes large deflection and elevation changes.

(1) The gunner lays the sight on the two aiming posts (placed out 50 and 100 meters from the mortar) on a referred deflection of 2800 mils and an elevation of 1100 mils.

(2) The gunner is given a deflection and elevation change in a fire command causing the gunner to shift the mortar between 200 and 300 mils and an elevation change between 100 and 200 mils.

(3) As soon as the sight data are announced, the gunner places it on the sight, elevates the mortar until the elevation bubble floats freely, and then centers the traversing bearing. If the elevation is between 1100 to 1511 mils, the cannon is mounted in the lower saddle. If the elevation is between 0800 to 1100 mils, the high saddle is used. If the saddle is changed, the squad leader helps the gunner.

(4) The squad leader moves into position to the front of the bipod, kneels on either knee, and grasps the bipod legs (palms out), lifting until the feet clear the ground enough to permit lateral movement. The gunner moves the mortar as the squad leader steadies it, attempting to horizontally maintain the traversing mechanism. To make the shift, the gunner places the fingers of his right hand in the muzzle (Figure 3-6) and his left hand on the left leg, and moves the mortar until the vertical line of the sight is aligned approximately on the aiming post. When the approximate alignment is completed, the gunner signals the squad leader to lower the bipod by pushing down on the mortar.

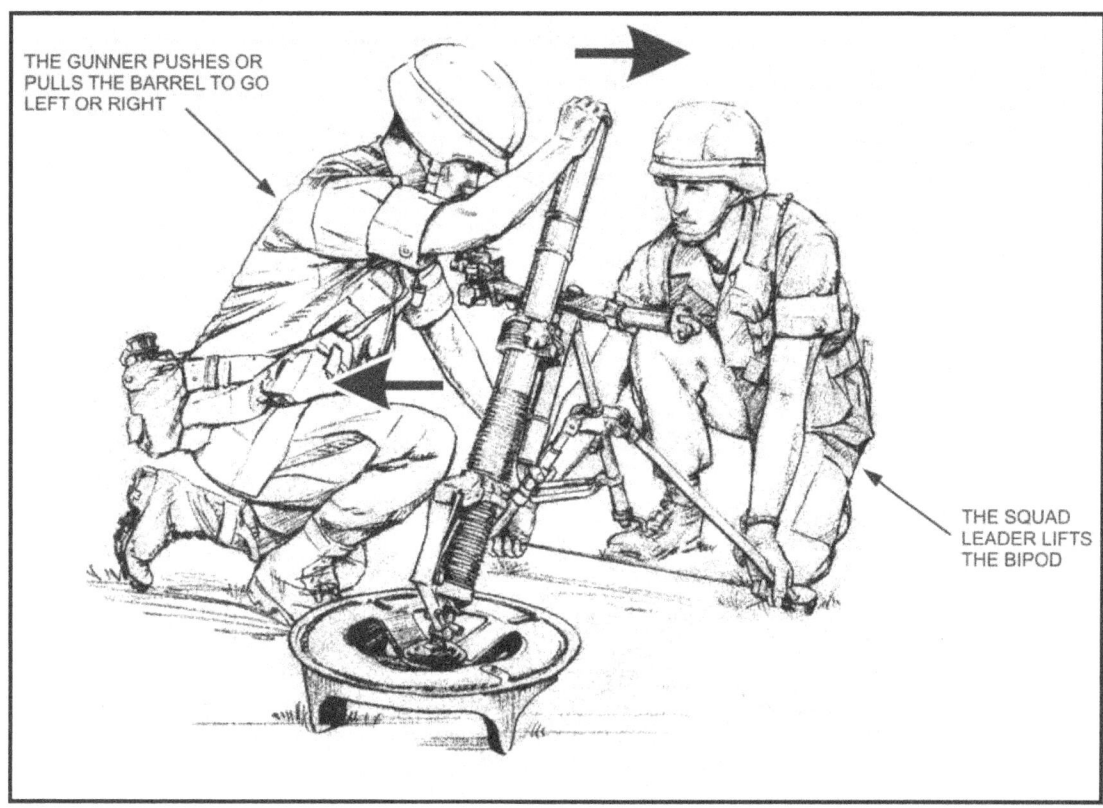

Figure 3-6. Large deflection and elevation changes.

(5) The gunner levels the mortar for elevation, then cross-levels. He continues to traverse and cross-level until the correct sight picture is obtained. The mortar should be within two turns of center of traverse when the exercise is completed.

REFERRING OF THE SIGHT AND REALIGNMENT OF AIMING POSTS

3-36. Referring the sight and realigning aiming posts ensure that all mortars are set on the same data. The section leader, acting as the FDC, has one deflection instead of two.

   (1) The sheaf is paralleled, and each mortar is laid on the correct data.

   (2) The section leader, acting as the FDC, prepares an administrative announcement using the format for a fire command and the hit data of the basepiece as follows:
- "Section."
- "Do not fire."
- "Refer deflection one eight zero zero (1800)."
- "Realign aiming posts."

   (3) The gunners refer their sights to the announced deflection. Each gunner checks his sight picture. If he has an aligned sight picture, no further action is required.

   (4) In laying the mortar for direction, the two aiming posts do not always appear as one when viewed through the sight. This separation is caused by either a large deflection shift of the cannon or by a rearward displacement of the baseplate assembly caused by the shock of firing. When the aiming posts appear separated, the gunner cannot correctly use either one of them as his aiming point. To lay the mortar correctly, he takes a compensated sight picture (Figure 3-7). He traverses the mortar until the sight picture appears with the left edge of the far aiming post, which is placed exactly midway between the left edge of the near aiming post and the vertical line of the sight. This corrects for the displacement.

Chapter 3

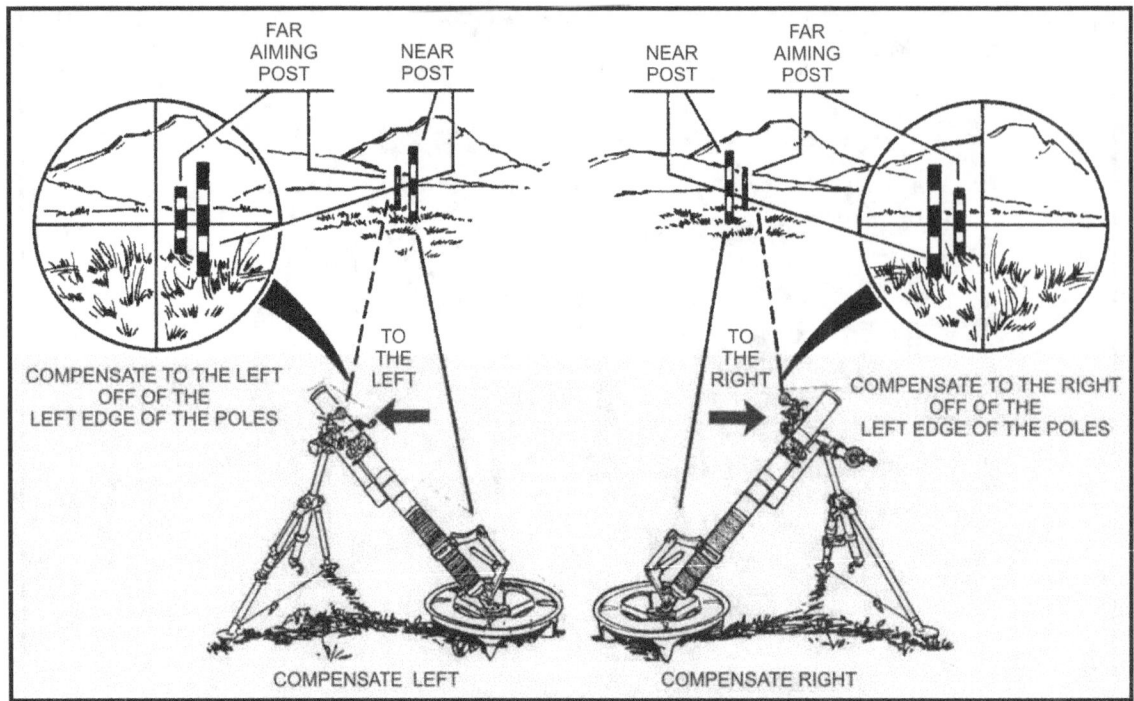

Figure 3-7. Compensated sight picture.

(5) At the first lull in firing, the gunner must determine whether the displacement is caused by traversing the mortar or by displacement of the baseplate assembly. To do this, he places the referred initial deflection on the sight and lays on the aiming posts. If both aiming posts appear as one, the separation is caused by traversing. In this case, he continues to lay the mortar as described and does not realign the aiming posts. When the posts still appear separated, the separation is caused by displacement of the baseplate assembly. He notifies his squad leader, who in turn requests permission from the section leader to realign the aiming posts. To realign the aiming posts using the sightunit, the gunner—

- Places on the sight the deflection originally used to place out the posts.
- Lays the mortar so that the vertical line of the sight is aligned on the left edge of the far aiming post.
- Without shifting the mortar, refers the sight until the vertical cross line falls on the left edge of the near aiming post. This actually measures the angle between the posts.
- With this last deflection set on the sight, re-lays the mortar until the vertical cross line is aligned on the far aiming post.
- Without shifting the mortar, refers the sight again to the original referred deflection used to place out the aiming posts. The line of sight, through the sight, is now parallel to the original line established by the aiming posts.
- Looking through the sight, directs the ammunition bearer to move the aiming posts so that they are realigned with the sight's vertical line. The posts are now realigned to correct the displacement.

NOTE: This procedure is used only when displacement is so great that it is difficult to obtain a compensated sight picture.

MALFUNCTIONS

3-37. Mortarmen must be aware of the following malfunctions.

60-mm Mortar, M224

MISFIRE

3-38. A misfire is a complete failure to fire. It can be caused by a faulty firing mechanism or faulty element in the propelling charge explosive train. A misfire cannot be immediately distinguished from a delay in functioning of the firing mechanism or from a hangfire; therefore, it must be handled with care. All firing malfunctions should be considered a misfire. Mechanical malfunctions can be caused by a faulty wiring pin or by rounds lodged in the cannon because of burrs, excess paint, oversized rounds, or foreign matter in the cannon. Procedures for removing a misfire are discussed in paragraphs 3-41 to 3-43.

HANGFIRE

3-39. A hangfire is a delay in the functioning of a propelling charge explosive train at the time of firing. In most cases, the delay ranges from a split second to several minutes. Thus, a hangfire cannot be distinguished immediately from a misfire.

COOKOFF

3-40. A cookoff is a functioning of one or more of the explosive components of a round chambered in a hot weapon, initiated by the heat of the weapon.

REMOVAL OF A MISFIRE

3-41. The procedures for removing a misfire are different for handheld and conventional modes. The propelling charge may not function for the following reasons:

- Defective ignition cartridge.
- Defective, damaged, or loose firing pin.
- Fouled firing pin or firing pin obstructed by extraneous material.
- Fouled bore.
- Excess oil or water in the bore.
- Misaligned stabilizing fin.
- Foreign matter or excess paint on round.
- Selector switch on SAFE or TRIGGER.

CONVENTIONAL MODE

3-42. Use the following procedures to remove a misfire in the conventional mode.

NOTES: 1. A faulty firing pin assembly on a 60-mm mortar, M224, requires that the firing pin be tightened upon completion of firing when frequent trigger firing using M720-, M721-, or M722-series ammunition. These type rounds usually fire when trigger fired.

2. The squad leader must supervise the removal of the misfire using a copy of the current misfire procedures.

CAUTION
Trigger fire the M720-series rounds with charges 0 and 1 only. If an emergency requires trigger fire above charge 1, the baseplate must be well-seated and extreme caution taken.

(1) When a misfire occurs, any member of the squad immediately announces, "Misfire."

> **WARNING**
>
> **During peacetime live-fire training, the ammunition bearer moves at least 50 meters to the rear of the mortar.**

(2) The entire squad stays with the mortar. If it is obvious to the squad leader that the round has reached the bottom of the cannon and has failed to ignite, the gunner places the selector switch on trigger fire and squeezes the trigger several times to try to clear the round.

(3) If the round fails on trigger fire, the gunner places the selector switch back on drop fire. Staying clear of the muzzle at all times, the squad leader holds the bipod legs to prevent slippage (Figure 3-8). The gunner strikes the cannon sharply several times with the heel of his boot just above the handle assembly. If the round fails to dislodge after trigger firing and kicking, the gunner places the selector switch on SAFE.

Figure 3-8. Kicking the mortar to clear a misfire.

60-mm Mortar, M224

> **WARNING**
>
> During peacetime live-fire training, the gunner and squad leader join the ammunition bearer and wait one minute (in case of a cookoff). After waiting one minute, the gunner returns to the mortar.

(4) The gunner checks for heat by starting from just below the muzzle and working down to the base with his fingertips. If the cannon is too hot to be handled, he cools it with water (or snow) and checks it one minute later. If no water (or snow) is available, the cannon is air-cooled until it can be easily handled with bare hands.

NOTES: 1. Liquids must never be poured into the cannon.

2. During peacetime live-fire training, the gunner signals the squad to come forward once the cannon is cool.

(5) The gunner locks the data down on the sight, then removes the sight and places it in a safe location. Then, he ensures that the barrel has been lowered to its lowest elevation, leaving 0.25 inches (0.64 cm) of inner elevating sleeve showing.

> **WARNING**
>
> Do not move the bipod legs or change the firing saddle until the cannon is in the horizontal position.

(6) The gunner loosens the collar assembly so that the cannon can be rotated. He rotates the barrel 90 degrees so that the flats on the breech cap ball are aligned with the flats on the socket cap. The squad leader places his left leg in front of the nonmechanical leg of the bipod to keep the bipod steady during the misfire removal. After placing his leg in this position, the ammunition bearer places his left hand near the top of the cannon and his right hand on the underside, just below the muzzle. He prevents any part of his body from passing in front of the muzzle.

> **WARNING**
>
> Once the cannon is horizontal, the rear of the cannon must not be lowered back down until the round is extracted. If the round slips down the cannon before extraction, it could ignite, causing death or personal injury.

(7) The gunner continues to raise the cannon so that the base of the cannon is higher than the top. With the muzzle pointing toward the ground, the gunner shakes it slightly to help dislodge the round. As the round starts to clear the muzzle, the ammunition bearer squeezes with the meaty portion of his thumbs against the body of the round—not the fuze—and removes it. If the round fails to come out once the cannon is lifted, the cannon is lowered back to the horizontal. It is removed from the bipod assembly and placed in a designated dud pit. EOD personnel are notified for removal or disposal.

(8) Once the round has been removed, the ammunition bearer inspects the cartridge. The ammunition bearer replaces any safety pins and inspects the primer of the ignition cartridge.

Chapter 3

If dented, the round should not be fired. If the primer has not been dented, the firing pin on the mortar should be checked for proper seating and tightened down if needed.

(9) While the round is being inspected, the gunner lowers the cannon back into the baseplate and remounts the sight unit to the bipod. The ammunition bearer then swabs the bore, and the gunner re-lays the mortar on the previous firing data.

NOTE: If the baseplate moved during the misfire procedure, the mortar must be reciprocally laid.

(10) If the primer on the round has not been dented, the gunner tries to fire the round again. If the same round misfires, he repeats the misfire procedures. If the primer has been dented, he notifies range control (during peacetime) or contacts the unit headquarters for further guidance (during combat).

HANDHELD MODE

3-43. Use the following procedures to remove a misfire in the handheld mode.
(1) When a misfire occurs, any member of the squad immediately announces, "Misfire."
(2) The entire crew stays with the mortar, and the gunner immediately pulls the trigger twice. If the round still fails to function, he announces, "Misfire." The gunner places the selector switch on SAFE and bounces the mortar from at least 6 inches off the ground to dislodge the round. (Disregard if the crew heard the round strike the bottom of the cannon.) The gunner continues to keep the cannon pointed in a safe direction and elevation.

> **WARNING**
>
> **During peacetime live-fire training, the gunner stays with the mortar and all other crew members move at least 50 meters behind the mortar.**

NOTE: The gunner bounces the mortar only if the round is between the muzzle and firing pin.

(3) The gunner places the selector switch back on trigger fire and squeezes the trigger twice—the mortar should fire. If the round does not fire, he places the selector switch on SAFE and supports the mortar cannon with sand bags, logs, or empty ammunition boxes to keep the cannon upright and stable. He ensures the cannon is up and pointing downrange.

> **WARNING**
>
> **During peacetime live-fire training, the gunner joins the rest of the squad and waits one minute. After one minute, the gunner returns to the mortar.**

(4) The gunner checks for heat by starting from just below the muzzle and working down to the base with his fingertips. If the cannon is too hot to be handled, he cools it with water (or snow) and checks it one minute later. If no water (or snow) is available, the cannon is air-cooled until it can be easily handled with bare hands.

NOTES: 1. Liquids must never be poured into the cannon.

2. During peacetime live-fire training, the gunner signals the squad to come forward once the cannon is cool.

(5) Once the cannon is cool, the ammunition bearer places his left hand (fingers and thumb extended and together) near the top of the cannon and his right hand on the underside just below the muzzle. In one smooth motion, the gunner lifts the base of the mortar with the M8 baseplate to the horizontal position. Once the cannon reaches the horizontal position, the ammunition bearer extends the meaty portion of his thumbs over the end of the muzzle. The gunner continues to raise the base of the cannon past the horizontal. With the muzzle pointing downward, the gunner slightly shakes the cannon to help dislodge the round. As the round starts to clear the muzzle, the squad leader catches the round by squeezing his thumbs against the sides of the body—not the fuze—and removes it. If the fuze has safety pins (other than the M734), the ammunition bearer tries to replace them.

WARNING

Once the cannon is horizontal, the rear of the cannon must not be lowered back down until the round is extracted. If the round slips down the cannon before extraction, it could ignite, causing death or personal injury.

NOTE: Removing the firing pin ensures that the mortar will not fire should the round slip down the cannon during the subsequent drill.

(6) The ammunition bearer inspects the primer of the ignition cartridge. If dented, he does not try to fire the round again. If the pins cannot be replaced, the fuze may be armed. He lays the round in the designated dud pit and notifies explosive ordnance disposal (EOD) personnel.

(7) If the round does not come out after lifting the cannon up and shaking it, the gunner returns the cannon to the horizontal. The squad leader places the cannon in the designated dud pit and notifies EOD personnel.

(8) If the primer on the round has not been dented, the gunner lifts the base as high as possible. He shakes the cannon to dislodge any debris and swabs the bore. He tries to fire the round again. If two misfires occur in a row without the primer being dented, the gunner notifies range control (during peacetime) or contacts the unit headquarters for further guidance (during combat).

DISMOUNTING AND CARRYING OF THE MORTAR

3-44. To dismount and carry the mortar, the squad leader commands OUT OF ACTION.

DISMOUNTING

3-45. Squads dismount using the following procedures.

(1) The ammunition bearer retrieves the aiming posts. The gunner removes the sight, places an elevation of 0800 mils and a deflection of 3200 mils on the M64 sightunit, and places it in the case. Then he lowers the mortar to its minimum elevation and backs off one-quarter turn. He then centers the traversing mechanism and unlocks the collar with the collar locking knob.

(2) The squad leader grasps the base of the cannon and turns it 90 degrees (a one-quarter turn), until the spherical projection is in the unlocked position in the baseplate socket. He then lifts

up on the base end of the cannon and removes it from the collar assembly. The ammunition bearer secures the baseplate.

(3) The gunner relocks the collar with the collar locking knob. He moves to the front of the bipod and faces it, kneels on his right knee with his left hand on the gear case, and loosens the locking nut. He tilts the bipod to his left and closes the bipod legs, placing the cable around the legs and rehooking the cable. He stands up, placing his right hand on the sight slot and his left hand on the traversing handwheel.

(4) On the command MARCH ORDER, squad members take the equipment distributed to them by the squad leader and move.

CARRYING

3-46. The mortar can be carried by one or two men for short distances. When the sight is left mounted on the mortar, care must be taken to prevent damaging it. Adhere to the following procedures to properly carry the mortar.

(1) For a one-man carry, the mortar is in the firing position with the mount attached to the cannon at the lower saddle. The elevating mechanism is fully depressed, and the bipod legs are together. The mount is folded back underneath the cannon until the elevating mechanism latches to the collar assembly. The cable is passed through one of the baseplate openings and wrapped around the cannon. The cable is attached to itself, using its snap hook. The carrying handle is used to carry the complete mortar.

(2) For a two-man carry, the M7 baseplate is one load, and the cannon/mount combination is the second load. The mount is attached to the cannon at the lower saddle, and the elevating mechanism is fully depressed. The bipod legs are together, and the bipod is folded up under the cannon until the elevating mechanism latches to the collar assembly. The cable is wrapped around the legs and cannon and hooked onto itself with its snap hook.

NOTE: The carrying position can be in the upper or lower saddle, depending on the mission or enemy situation.

(3) For the handheld mode, the M8 baseplate is left attached to the cannon. The baseplate is rotated 90 degrees to the right and rotated up until the two spring plungers on the front edge of the baseplate body latch onto the protrusion on the right side of the basecap. Then the auxiliary carrying handle is placed in the carrying position.

SECTION IV. AMMUNITION

This section implements STANAG 2321.

The four types of ammunition for the 60-mm mortar are HE, WP, ILLUM, and training. All 60-mm mortar cartridges, except training cartridges, have three major components—a fuze, a body, and a tail fin with a propulsion system assembly.

CLASSIFICATION AND TYPES OF AMMUNITION

3-47. The ammunition used with the M224 mortar can be classified by its use. See the following tables for details on each cartridge used with the M224.

NOTE: Firing Table 60-P-1 details all ammunition used with the M224 mortar.

HIGH-EXPLOSIVE AMMUNITION

3-48. HE ammunition is used against enemy personnel and light materiel targets. Table 3-2 details the HE ammunition that can be used when firing the M224 mortar.

Table 3-2. High-explosive ammunition for the 60-mm mortar, M224.

CARTRIDGE/TYPE	MAXIMUM RANGE (METERS)	FUZE	CHARACTERISTICS AND LIMITATIONS
M720 HE cartridge	3,489	M734 multioption fuze	This cartridge cannot be fired above charge one in the handheld mode and weighs 3.75 pounds.
M720A1 HE cartridge	3,489	M734A1 multioption fuze	This cartridge cannot be fired above charge one in the handheld mode and weighs 3.58 pounds.
M888 HE cartridge	3,520	M935 PD fuze	This cartridge is identical to the M720, except that it has the M935 PD fuze. It cannot be fired above charge one in the handheld mode and weighs 3.75 pounds.
M768 HE cartridge	3,489	M738 PD/ DLY fuze	This cartridge cannot be fired above charge one in the handheld mode and weighs 3.59 pounds.

ILLUMINATION AMMUNITION

3-49. ILLUM ammunition is used during night missions requiring assistance in observation. Table 3-3 details the ILLUM ammunition that can be used when firing the M224 mortar.

Table 3-3. Illumination ammunition for the 60-mm mortar, M224.

CARTRIDGE/TYPE	MAXIMUM RANGE (METERS)	FUZE	CHARACTERISTICS AND LIMITATIONS
M721 ILLUM cartridge	3,200	M776 MTSQ fuze	This cartridge provides an average of 325,000 candlepower for about 40 seconds. It cannot be fired at charge zero in the M224 and weighs 3.75 pounds.
M767 IR ILLUM cartridge	3,489	M776 MTSQ fuze	This cartridge contains an infrared illuminant mix that provides approximately 75 watts/steradian of infrared (IR) light with less than 350 candlepower of visible light. It weighs 3.76 pounds. It is intended to be used to support troops with night vision devices.
M83A3 ILLUM cartridge	951	M65A1-series time fuze	This cartridge provides an average of 250,000 candlepower for at least 30 seconds. It weighs 3.7 pounds.

Chapter 3

SMOKE, WHITE PHOSPHORUS AMMUNITION

3-50. Smoke, WP ammunition is used as a screening, signaling, or incendiary agent. Table 3-4 details the smoke WP ammunition that can be used when firing the M224 mortar.

Table 3-4. Smoke, white phosphorus ammunition for the 60-mm mortar, M224.

CARTRIDGE/TYPE	MAXIMUM RANGE (METERS)	FUZE	CHARACTERISTICS AND LIMITATIONS
M722 smoke WP cartridge	3,200	M745 PD fuze	This cartridge is used as bulk WP for spotting/marking. It weighs 3.7 pounds.
M722A1 smoke WP cartridge	3,200	M783 PD/ DLY fuze	This cartridge is used as bulk WP for spotting/marking. It weighs 3.79 pounds.
M302A1/A2 smoke WP cartridge	1,629	M527, M935, or M936 PD fuze	This cartridge is restricted to training use and weighs 4.08 pounds.

TRAINING PRACTICE AMMUNITION

3-51. Training practice (TP) ammunition is used for training when service ammunition is not available or there are restrictions on the use of service ammunition for training. Table 3-5 details the TP ammunition that can be used when firing the M224 mortar.

Table 3-5. Training practice ammunition for the 60-mm mortar, M224.

CARTRIDGE/TYPE	MAXIMUM RANGE (METERS)	FUZE	CHARACTERISTICS AND LIMITATIONS
M766 SRTC	525	M779 PD practice fuze	This cartridge weighs 2.9 pounds. The range of this cartridge can be reduced by removing increment plugs from the projectile body. This cartridge produces a flash and a bang on impact. It can also be recovered, refurbished, and reused.
M769 FRTC	3,500	M775 PD practice fuze	This non-dud producing training cartridge is ballistically similar to the M720-series HE cartridge. It weighs 3.82 pounds. The M775 practice fuze simulates the M734-series fuze for realistic ammunition preparation procedures.

FUZES

3-52. A fuze is a device to explode a projectile at the time and under the circumstances required. Mortar fuzes are located at the top of the cartridge and have four possible actions: SQ acts on impact, DLY acts 0.05 seconds after impact, variable time (VT) acts a certain distance above the ground, and MT acts after a preset number of seconds have elapsed after being fired. Many modern fuzes have more than one action. The following describes the type of fuzes available on ammunition fired by the M224.

> **WARNING**
>
> Do not try to disassemble any fuze.

MULTIOPTION FUZE, M734

3-53. This fuze (Figure 3-9) can be set to function as a proximity burst (1 to 4 meters above the surface), near-surface burst (0 to 1 meter above the surface), impact/superquick burst (on the surface), or delay burst (0.05 seconds).

3-54. It is set by hand without using a tool. Once set, an M734's setting can be changed many times before firing without damaging the fuze.

3-55. This fuze has no safety pins or wires.

3-56. If a cartridge set for PROX fails to burst at the proximity distance above the target, it automatically bursts at 0 to 1 meter (0 to 3 feet) above the target. If a cartridge set for NSB fails to burst at the near-surface distance above the target, it automatically bursts on impact. If a cartridge set for impact fails to burst on impact, it automatically bursts 0.05 seconds after impact.

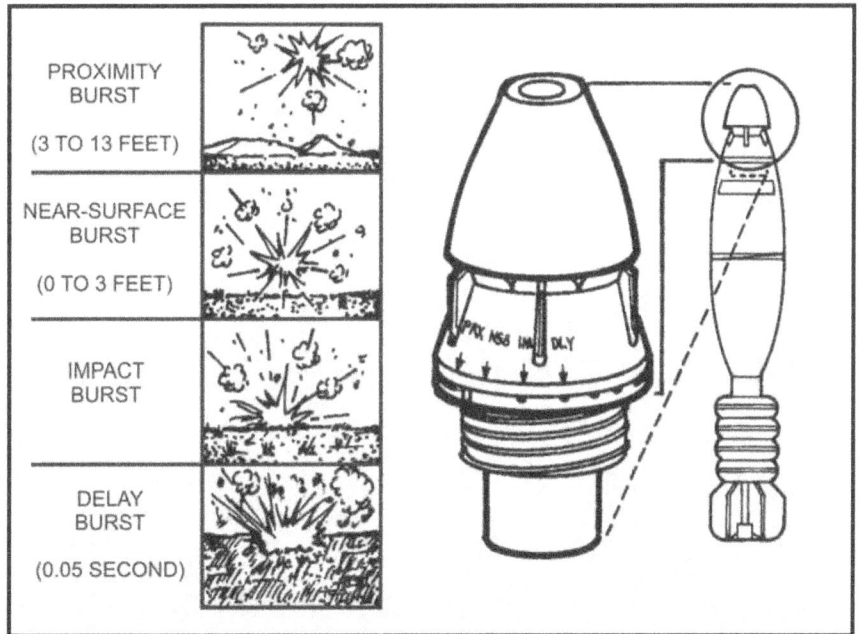

Figure 3-9. Multioption fuze, M734.

Multioption Fuze, M734A1

3-57. The air-powered M734A1 multioption fuze has four selectable functions:
- PRX 120.
- PRX 60/81.
- IMP.
- DLY.

3-58. In HE PROX mode, the height of burst (HOB) remains constant over all types of targets. The IMP mode causes the round to function on contact with the target and is the first backup function for either PROX setting. In the DLY mode, the fuze functions about 30 to 200 milliseconds after target contact. The DLY mode is the backup for the IMP and PROX modes. The IMP and DLY modes have not changed from the current M734 multioption fuze.

3-59. Radio frequency jamming can affect the functioning of PROX fuzes. Radio frequency jamming initiates a gradual desensitizing of the fuze electronics to prevent premature fuze function. Once the fuze is out of jamming range, the fuze electronics recover and function in the PROX mode if the designed HOB has not been passed. To limit the time of fuze radio frequency radiation, the proximity turn-on is controlled by an apex sensor that does not allow initiation of the fuze proximity electronics until after the apex of the ballistic trajectory has been passed.

3-60. In compliance with the safety requirements of military standard 1316C, the M734A1 uses ram air and setback to provide two independent environment sensors.

Point-Detonating Fuze, M935

3-61. The M935 PD fuze can be set to function as SQ or DLY.

3-62. It is preset for the SQ function, and the ammunition handler verifies that the selector slot is aligned with the "SQ" mark on the ogive. The ammunition handler selects DLY by turning the slot clockwise until it is aligned with the "D" marking on the ogive.

3-63. Its standard pull-wire safety is removed immediately before firing.

Point-Detonating Fuzes, M527-Series and M936

3-64. These fuzes act only on impact.

Point-Detonating /Delay Fuze, M783

3-65. This fuze can be set to function as SQ or DLY.

Point-Detonating Fuze, M745

3-66. Although there are markings on the fuze for other settings, it acts only on impact.

Time Fuze, M65-Series

3-67. The M65 is a fixed time fuze and has a safety wire.

Mechanical Time Superquick Fuze, M776

3-68. Using a fuze setter, the ammunition handler sets the time on the fuze, and once fired, the cartridge explodes after that amount of time has elapsed. If the mechanical time fails, the fuze acts on impact. The safety wire is removed just before firing.

Point-Detonating Fuze, M779, and Practice Fuze, M775

3-69. These fuzes are used with M766 and M769 training cartridges only.

> **WARNINGS**
>
> Do not fire cartridges with M525 or M527 PD fuzes if the fuze makes a buzzing sound when the safety pins are removed. Check the fuze for the presence of the bore-riding pin after removing the safety pin and do not fire the cartridge if it is missing. Notify EOD.
>
> Adequate fragmentation cover must be taken when firing cartridges for distances less than 300 meters
>
> Consult the TM on firing temperature limits.

CARTRIDGE PREPARATION

3-70. Cartridges are prepared as close to the actual firing period as possible.

UNPACKING

3-71. Standard A cartridges are individually packed in fiber containers with eight cartridges in a metal container. Use proper tools to open ammunition containers. Unpack the cartridge and remove any packing material. The moisture-resistant seal of the shipping container should not be broken until the ammunition is to be used. When a large number of cartridges (15 or more for each squad) are prepared before a combat mission, the cartridges may be removed from the shipping container and the propellant increments adjusted. The fin assemblies should then be reinserted into the container to protect the propelling charges.

INSPECTING FIN ASSEMBLIES

3-72. Inspect the cartridge fin assembly for any visible damage and retighten loose fin assemblies as required. Cartridges with damaged fins are turned in to the ammunition supply point (ASP).

PREPARING TO FIRE

3-73. When the ammunition bearer receives information in a fire command, he prepares the ammunition for firing. The number of cartridges, type of cartridge, fuze setting, and charge are all included in the fire command. To apply the data, the ammunition bearer selects the proper cartridge, sets the fuze, and adjusts (removes or replaces) the number of propelling charges on the quantity of cartridges called for in the fire command. He also inspects each cartridge for cleanliness and serviceability. Any safety wires are pulled only just before firing.

> **WARNING**
>
> For protection, a cartridge prepared but not fired should be returned to its container, increment end first. The pull wire on the M888 fuze must be replaced before returning it to its container.

Chapter 3

ADJUSTING THE PROPELLING CHARGE

3-74. The ammunition bearer adjusts the number of charges immediately in a fire for effect (FFE) mission. In an adjust-fire mission, he prepares the cartridge and delays adjusting the charges until the FFE is entered because the charge may change during the subsequent adjustments of fire. Cartridges are shipped with a complete propelling charge consisting of the ignition cartridge and four charges.

3-75. The fire command includes the number of charges on the cartridge.

3-76. If the cartridge is fired at less than full charge, the ammunition handler removes the number of charges based on the fire command.

- Charge zero is the ignition charge only
- Charge 1 has the ignition charge and 1 increment
- Charge 2 has the ignition charge and 2 increments
- Charge 3 has the ignition charge and 3 increments
- Charge 4 has the ignition charge and 4 increments

3-77. If the charge is less than charge 4, slide the remaining increments towards the rear until they are positioned against the fins.

3-78. Place the excess increments in an empty ammunition box for protection and close the lid during firing. Dispose of the excess increments in accordance with local range regulations.

WARNINGS

Propelling charges are not interchangeable. Do not substitute one model for another. Do not mix lots.

Never fire cartridges with a greater number of propelling charges than authorized for the ammunition and weapon.

UNFIRED CARTRIDGES

3-79. If a cartridge is not fired, replace the safety wire (if it was removed from the fuze) and reset the fuze. Re-install the propellant increments so that the cartridge has a full charge. Repack the cartridge in its original packaging.

CARE AND HANDLING

3-80. Ammunition is manufactured and packed to withstand all conditions normally encountered in the field. However, moisture and high temperature can damage ammunition. Also, explosive elements in primers and fuzes are sensitive to strong shock and high temperature. Complete cartridges being fired should be handled with care. Adhere to the following guidelines for proper ammunition care and handling.

- Do not throw or drop cartridges.

WARNING

Do not walk on, tumble, drag, throw, roll, or drop ammunition. Ensure that ammunition is kept in original container until ready for use. Do not combine WP and HE in storage. Maintain compatibility and quantity of ammunition, as outlined in TB 43-0250.

60-mm Mortar, M224

- Protect ammunition from mud, dirt, sand, water, snow, and direct sunlight. Cartridges must be free of such foreign matter before firing. Ammunition that is wet or dirty should be wiped off at once.
- Always store ammunition under cover. When this is not possible, raise it at least 6 inches (15 centimeters) off the ground and cover with a double thickness of tarpaulin. Dig trenches around the ammunition pile for drainage. WP cartridges must be stored with the fuze end up.
    - In combat, store ammunition underground such as in bunkers.
    - In the field, use waterproof bags, ponchos, ground cloths, and dunnage to prevent deterioration of ammunition. Ensure that ammunition does not become water-soaked.
    - In arctic weather, store the ammunition in wooden boxes or crates. Place the boxes or crates on pallets and cover them with a double thickness of tarpaulin.
    - Cover fin assemblies and propelling charges with the fiber container or end cap. Stack cartridges on top of empty ammunition boxes and cover them with plastic sheets.
- WP cartridges must be handled carefully.
    - Store WP cartridges at temperatures below 111 degrees Fahrenheit to prevent the WP from melting. When temperatures are above 111 degrees Fahrenheit, store the WP with the fuze-end up. This prevents the WP filler from forming voids away from the cartridge's designed center of mass and causes an erratic flight of the cartridge.
    - Store WP cartridges away from other types of ammunition.
- Do not handle duds. Follow local range procedures to dispose of them.

> **WARNING**
>
> **Duds are cartridges that have been fired but have not exploded. Duds are dangerous and should not be handled by anyone other than a member of the EOD team.**

This page intentionally left blank.

Chapter 4
81-mm Mortar, M252

The 81-mm mortar, M252, delivers timely, accurate fires to meet the requirements of supported troops. This chapter discusses assigned personnel duties, crew drill, mechanical training, and characteristics of the mortar.

SECTION I. SQUAD AND SECTION ORGANIZATION AND DUTIES

Each member of the infantry mortar squad has principle duties and responsibilities. (See FM 7-90 for a discussion of the duties of the platoon headquarters.)

ORGANIZATION

4-1. If the mortar squad and section are to operate quickly and effectively in accomplishing their mission, mortar squad members must be proficient in individually assigned duties. Correctly applying and performing these duties enables the mortar section to perform as an effective fighting team. The platoon leader commands the platoon and supervises the training of the elements. He uses the chain of command to assist him in effecting his command and supervising duties.

DUTIES

4-2. The mortar squad consists of four men (Figure 4-1):
- Squad leader.
- Gunner.
- Assistant gunner.
- Ammunition bearer.

SQUAD LEADER

4-3. The squad leader stands behind the mortar where he can command and control his squad. He supervises the emplacement, laying, and firing of the mortar, and all other squad activities.

GUNNER

4-4. The gunner stands to the left side of the mortar where he can manipulate the sight, elevating handwheel, and traversing handwheel. He places firing data on the sight and lays the mortar for deflection and elevation. He makes large deflection shifts by shifting the bipod assembly and keeps the bubbles level during firing.

ASSISTANT GUNNER

4-5. The assistant gunner stands to the right of the mortar, facing the cannon and ready to load. In addition to loading, he swabs the bore after every 10 rounds and after each fire mission. He may assist the gunner in shifting the mortar when the gunner is making large deflection changes.

AMMUNITION BEARER

4-6. The ammunition bearer stands to the right rear of the mortar. He maintains the ammunition for firing, prepares the ammunition and passes it to the assistant gunner, and provides local security for the mortar position. He also acts as the squad driver.

Chapter 4

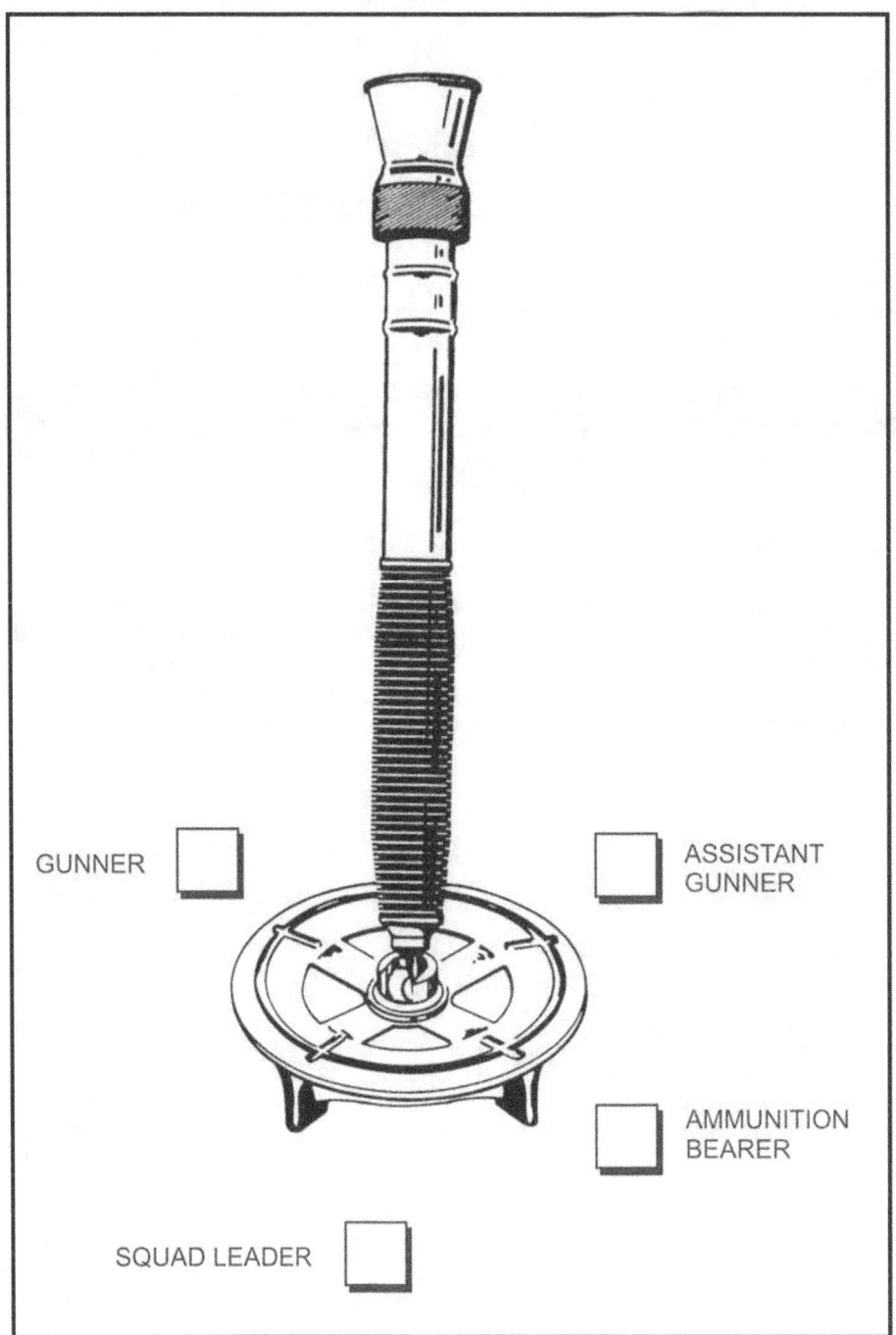

Figure 4-1. Position of squad members.

81-mm Mortar, M252

SECTION II. COMPONENTS

The 81-mm mortar, M252, is a smooth-bore, muzzle-loaded, high angle-of-fire weapon. The components of the mortar consist of a cannon, mount, and baseplate. This section discusses the characteristics and nomenclature of each component (Figure 4-2).

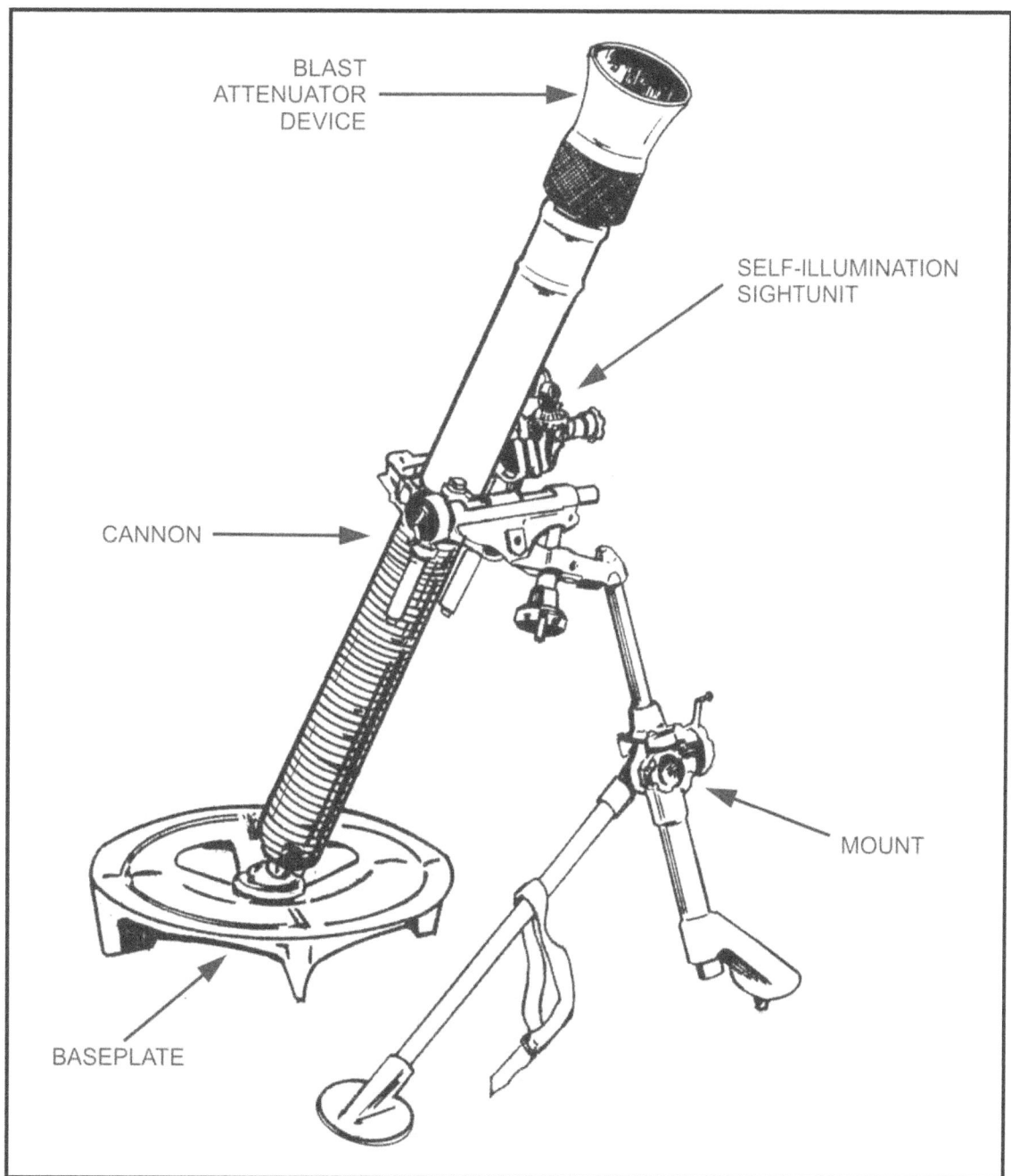

Figure 4-2. 81-mm mortar, M252.

Chapter 4

TABULATED DATA

4-7. The tabulated data for the 81-mm mortar, M252, are shown in Table 4-1.

Table 4-1. Tabulated data for the 81-mm mortar, M252.

Weight (pounds) System Cannon (with blast attenuator device) Mount M3A1 Baseplate M64A1 Sightunit	93 / 35 / 27 / 29 / 2.25			

Ammunition	HE	Smoke	Illum	Practice
Ready to fire	9.4	10.6	9.1	9.4
In single container	12.0	13.8	12.4	12.5
In three-round pack	57.0	63.0	60.0	60.0

Elevation	
Elevation (approximate)	800 to 1515 mils
For each turn of elevation drum (approximate)	10 mils
Traverse	
Right or left from center (approximate)	100 mils (10 turns)
For each turn of traversing handwheel (approximate)	7 mils
Range	
Minimum to maximum	83 to 5,608 meters
Rate of fire (with 800 series ammunition)	
Sustained	15 rounds per minute indefinitely
Maximum	30 rounds per minute for 2 minutes
HE fuze options	M821 w/multioption fuze M734 M889 w/PD fuze M935
Sight	M64A1, lightweight, self-Illuminating

CANNON ASSEMBLY, M253

4-8. The cannon assembly consists of the cannon that is sealed at the lower end with a removable breech plug, which houses a removable firing pin (Figure 4-3). At the muzzle end is a cone-shaped BAD that is fitted to reduce noise. The BAD is removed only by qualified maintenance personnel.

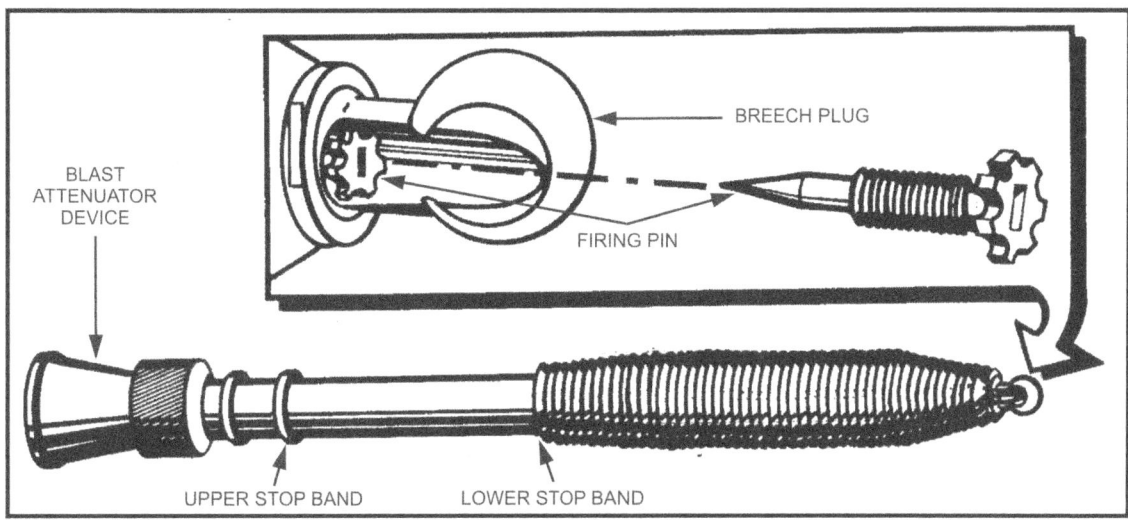

Figure 4-3. Cannon assembly, M253.

MOUNT, M177

4-9. The mount consists of elevating and traversing mechanisms and a bipod (Figure 4-4).

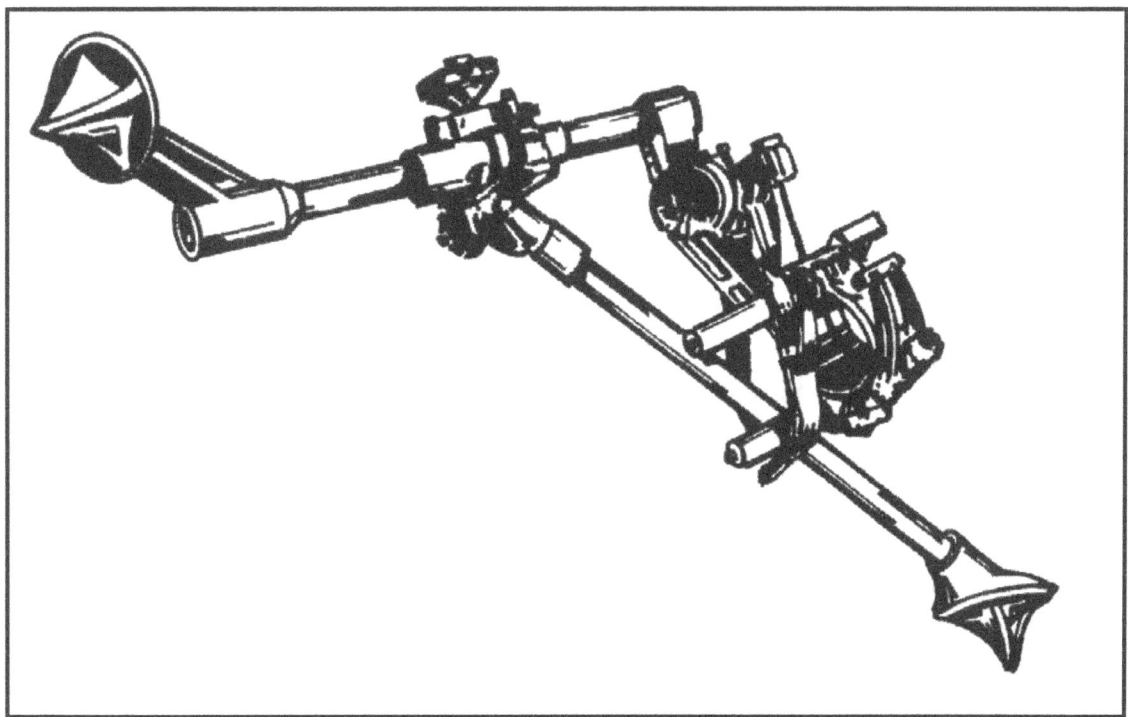

Figure 4-4. Mount, M177, in folded position.

Chapter 4

BIPOD

4-10. The bipod provides front support for the cannon and carries the gears necessary to lay the mortar. The cannon clamp, which consists of an upper and lower clamp, is situated at the top. The upper clamp is fitted with a locking arrangement that consists of a curved handle and a spring-loaded locking rod that is ball-shaped at its lower end. The lower clamp is shaped and bored on each side to house the buffer cylinders. On the right side, the clamp is recessed to receive the ball end of the locking rod. A safety latch located at the side of the recess is used to secure the ball.

TRAVERSING MECHANISM

4-11. The sight bracket is attached to the buffer carrier, which is fitted to the traversing screw assembly. Attached to the right of the screw is the traversing handwheel. The traversing screw assembly is fitted to the clamp assembly, which is pivoted in the center on an arm attached to the elevating leg. Attached to the arm is the cross-leveling mechanism, which is attached to the clamp assembly at its upper end.

ELEVATING MECHANISM

4-12. The elevating shaft is contained in the elevating leg; to the left of the elevating leg is the elevating handwheel. A plain leg is fitted to a stud on the elevating leg and is secured by a leg-locking handwheel. A spring-loaded locating catch is behind the elevating gear housing, which locates the plain leg in its supporting position for level ground. A securing strap is attached to the plain leg for securing the bipod in the folded position. Both legs are fitted with a disk-shaped foot with a spike beneath to prevent the mount from slipping.

BASEPLATE, M3A1

4-13. The baseplate (Figure 4-5) is constructed of one piece and supports and aligns the mortar for firing. During firing, the breech plug on the cannon is sealed and locked to the rotatable socket in the baseplate.

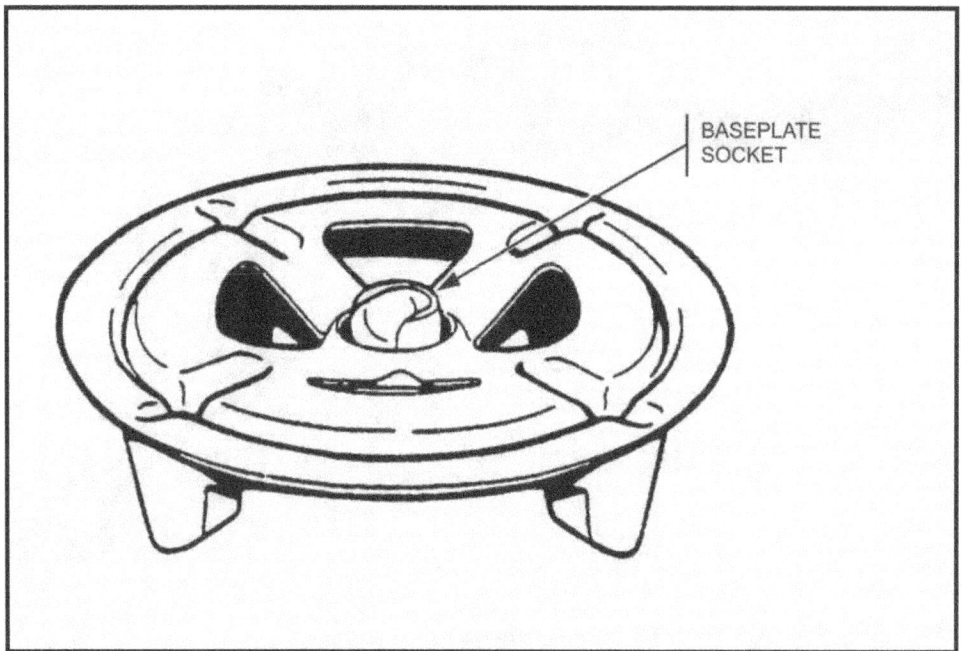

Figure 4-5. Baseplate, M3A1.

81-mm Mortar, M252

SECTION III. OPERATION

This section contains information on how to prepare the 81-mm mortar, M252, for firing; how to conduct safety checks; and what actions to apply to remove the cartridge from the cannon if a misfire should occur during firing.

PREMOUNT CHECKS

4-14. Before the mortar is mounted, the squad must perform premount checks. Each squad member should be able to perform all of the following premount checks.

GUNNER

4-15. The gunner checks the baseplate and ensures that—
- The rotating socket is free to move in a complete circle.
- The ribs and braces have no breaks, cracks, or dents.
- The retaining ring is correctly located, securing the rotating socket to the baseplate.

ASSISTANT GUNNER

4-16. The assistant gunner checks the bipod and ensures that—
- The cannon clamps are clean and dry.
- The cannon carrier is centered.
- The securing strap is correctly located, securing the cannon clamps and buffers to the plain leg.
- The leg-locking handwheel is hand-tight.
- Four inches of elevation shaft are exposed, and the shaft is not bent.

AMMUNITION BEARER

4-17. The ammunition bearer checks the cannon and ensures that—
- The cannon is clean and free from grease and oil, both inside and out.
- The breech plug is screwed tightly to the cannon.
- The firing pin is secured correctly.
- The BAD is secured correctly.

SQUAD LEADER

4-18. The squad leader supervises the squad and is responsible for laying out the equipment as shown in Figure 4-6.

Chapter 4

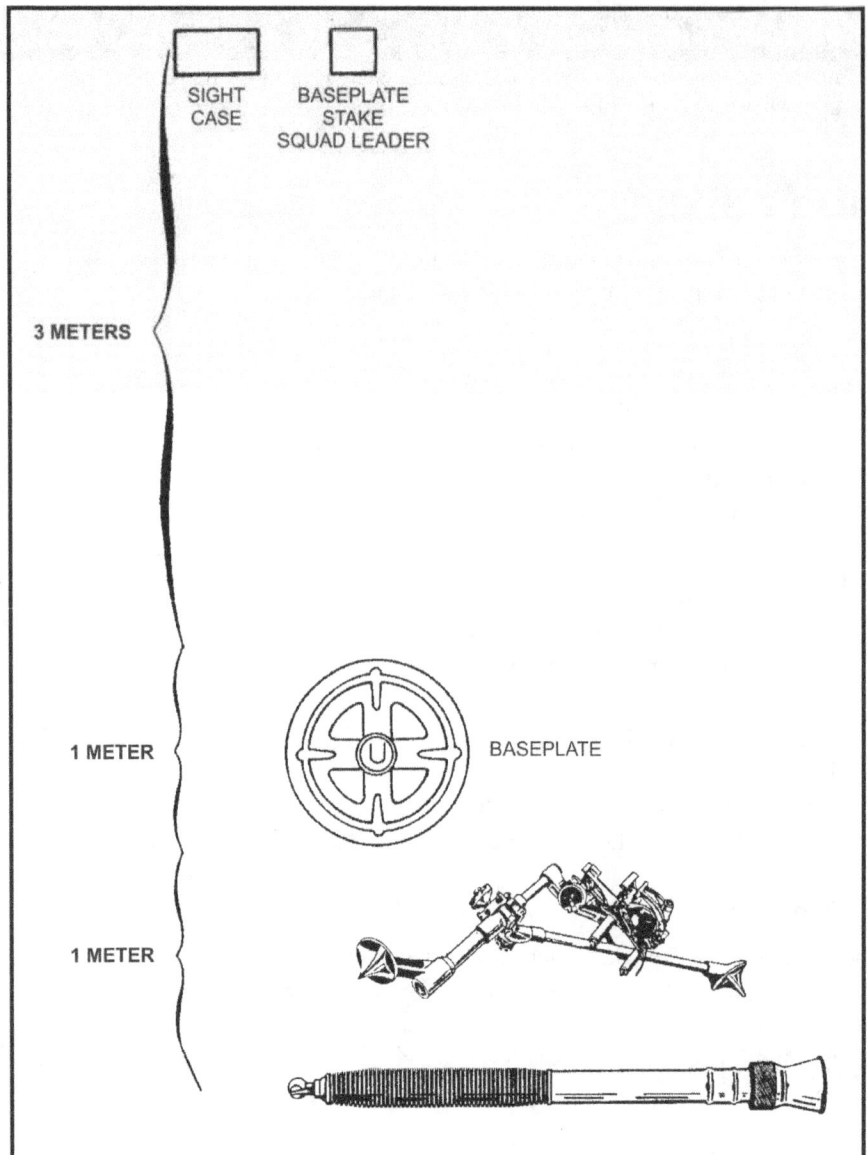

Figure 4-6. Layout of equipment.

MOUNTING OF THE MORTAR

4-19. The squad adheres to the following procedures to mount the mortar.
   (1) The squad leader picks up the sight case and the two aiming posts and moves to the exact position where the mortar is to be mounted. He places the sight case and aiming posts to the left front of the mortar position. The squad leader points to the exact spot where the mortar is to be mounted. He indicates the initial direction of fire by pointing in that direction and commands ACTION.
   (2) The gunner places the outer edge of the baseplate against the baseplate stake so that the left edge of the cutaway portion of the baseplate is aligned with the right edge of the stake (Figure 4-7). He rotates the socket so that the open end is pointing in the direction of fire. During training, the gunner may use the driving stake from the aiming post case.

NOTE: The squad leader indicates the direction of fire when mounting.

81-mm Mortar, M252

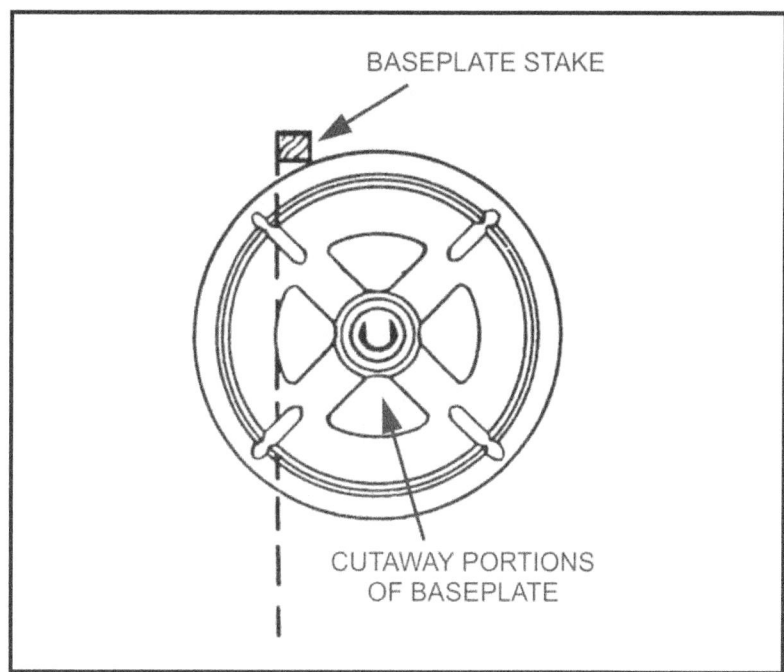

Figure 4-7. Baseplate placed against baseplate stake.

(3) When the baseplate is in position, the ammunition bearer lowers the breech plug into the rotating socket and rotates the cannon a quarter of a turn to lock it. He ensures that the firing pin recess is facing upward. He stands to the rear of the baseplate and supports the cannon until the bipod is fitted.

(4) The assistant gunner lifts the bipod and stands it on its elevating leg so that the elevating handwheel is to the rear and the plain leg is to the front. He releases the securing strap, loosens the leg-locking handwheel, and lowers the plain leg until the locating catch engages in the recess. The leg-locking handwheel must then be tightened by hand, ensuring the teeth on either side are correctly meshed.

(5) The assistant gunner exposes 8 inches (200 millimeters) of elevation shaft, leaving the elevation handwheel unfolded. He opens the cross-level handwheel, traversing handwheel, and cannon clamp.

(6) The assistant gunner carries the bipod to the front of the cannon and places the bipod feet on the ground 12 to 15 inches in front of the baseplate and astride the line of fire. He positions the lower cannon clamp against the lower stop band on the cannon and secures the upper cannon clamp. He must ensure that the ball-shaped end of the locking rod is secured in its recess by the locking latch.

(7) The gunner removes the sight from its case, mounts it on the mortar, and sets a deflection of 3200 mils and an elevation of 1100 mils. He levels all bubbles.

Chapter 4

SAFETY CHECKS BEFORE FIRING

4-20. The following safety checks must be enforced before firing the mortar.

GUNNER

4-21. The gunner ensures that—
- The cannon is locked to the baseplate, and the open end of the socket points in the direction of fire.
- The firing pin recess faces upwards.
- The bipod locking latch is locked, securing the cannon clamps.

NOTE: See TM 9-1015-249-10 for more on loading and firing.

- The leg-locking handwheel is tight.
- Mask clearance and overhead clearance are sufficient.

Mask and Overhead Clearance

4-22. Since the mortar is normally mounted in defilade, there could be a mask such as a hill, tree, building, or rise in the ground. Overhead interference can be branches of trees or roofs of buildings. In any case, the gunner must ensure that the cartridge does not strike an obstacle.

4-23. In selecting the exact mortar position, the leader looks quickly for mask clearance and overhead interference. After the mortar is mounted, the gunner makes a thorough check.

4-24. The gunner determines mask and overhead clearance by sighting along the cannon with his eye near the breech plug. If the line of sight clears the mask, it is safe to fire. If not, he may still fire at the desired range by selecting a charge zone having a higher elevation. When firing under the control of an FDC, he reports to the FDC that mask clearance cannot be obtained at a certain elevation.

4-25. Firing is slowed if mask clearance must be checked before each firing, but this can be eliminated if minimum mask clearance is determined. This is accomplished by depressing the cannon until the highest section of the mask is clear. The gunner levels the elevation bubble by turning the elevation micrometer knob and reading the setting on the elevation scale and elevation micrometer—this setting is the minimum mask clearance. The squad leader notifies the FDC of the minimum mask clearance elevation. Any target that requires that elevation, or a lower one, cannot be engaged from that position.

4-26. If the mask is not regular throughout the sector of fire, the gunner determines the minimum mask clearance as previously described. Placing the mortar in position at night does not relieve the gunner of the responsibility of checking for mask clearance and overhead interference.

ASSISTANT GUNNER

4-27. The assistant gunner cleans the bore and swabs it dry.

AMMUNITION BEARER

4-28. The ammunition bearer ensures that each cartridge is clean, the safety pin is present, and the ignition cartridge is in good condition.

SMALL DEFLECTION AND ELEVATION CHANGES

4-29. With the mortar mounted and the sight installed, the gunner makes small deflection and elevation changes.
    (1) The gunner lays the sight on the two aiming posts (placed out 50 and 100 meters from the mortar) on a referred deflection of 2800 mils and an elevation of 1100 mils. The mortar is

within two turns of center of traverse. The vertical cross line of the sight is on the left edge of the aiming point.

(2) The gunner is given a deflection change in a fire command between 20 and 60 mils inclusive. The elevation change announced must be less than 90 mils and more than 35 mils.

(3) As soon as the sight data are announced, the gunner places it on the sight, lays the mortar for elevation, and traverses onto the aiming post by turning the traversing handwheel and adjusting nut in the same direction. A one-quarter turn on the adjusting nut equals one turn of the traversing handwheel. When the gunner is satisfied with his sight picture, he announces, "Up."

NOTE: All elements given in the fire command are repeated by the squad.

(4) After the gunner has announced, "Up," the squad leader checks the mortar to determine if the crew drill was performed correctly.

LARGE DEFLECTION AND ELEVATION CHANGES

4-30. With the mortar mounted and the sight installed, the squad makes large deflection and elevation changes.

(1) The gunner lays the sight on the two aiming posts (placed out 50 and 100 meters from the mortar) on a referred deflection of 2800 mils and an elevation of 1100 mils.

(2) The gunner is given a deflection and elevation change in a fire command causing the gunner to shift the mortar between 200 and 300 mils for deflection and between 100 and 200 mils for elevation.

(3) As soon as the sight data are announced, the gunner places it on the sight. The gunner exposes 8 inches (200 millimeters) of elevation shaft and centers the buffer carrier. This ensures a maximum traversing and elevating capability after making the movement.

(4) The assistant gunner moves into position to the front of the bipod on his right knee and grasps the bipod legs (palms up), lifting until they clear the ground enough to permit lateral movement. The gunner moves the mortar as the assistant gunner steadies it. The assistant gunner tries to maintain the traversing mechanism on a horizontal plane. To make the shift, the gunner places the fingers of his right hand in the muzzle and his left hand on the left leg. He moves the mortar until the vertical line of the sight is aligned approximately on the aiming post. When the approximate alignment is completed, the gunner signals the assistant gunner to lower the bipod by pushing down on the mortar.

(5) The gunner rough-levels the cross-level bubble by making the bubble float from side to side. Then, he checks the sight picture. If he is not within 20 mils of a proper sight picture, the gunner and assistant gunner must make another large shift before continuing.

(6) The gunner centers the elevation bubble. He lays for deflection, taking the proper sight picture. The mortar should be within two turns of center of traverse when the task is complete.

(7) The open end of the socket must continue to point in the direction of fire. Normally, it can be moved by hand, although this may be difficult to do if the mortar is moved through a large arc. If required, the gunner/assistant gunner lowers the cannon so that the breech plug engages with the open end of the socket, and he uses the cannon as a lever to move the socket.

(8) The cannon clamps can be moved along the cannon to counter large changes in elevation, which may preclude moving the bipod. It is especially useful if the baseplate sinks deep into the ground during prolonged firing. Upon completion of any bipod movement on the cannon, the gunner ensures that the firing pin recess is facing upward.

(9) On uneven ground without a level surface for the bipod, the gunner can adjust the plain leg. While the assistant gunner supports the cannon, the gunner slackens the leg-locking handwheel, releases the locating catch, and positions the plain leg. The leg-locking handwheel must then be tightened, ensuring the teeth are correctly meshed.

REFERRING OF THE SIGHT AND REALIGNMENT OF AIMING POSTS USING THE M64 SIGHT

4-31. Referring and realigning aiming posts ensure that all mortars are set on the same data. The section leader, acting as FDC, has one deflection instead of two or more.

(1) The mortar is mounted and the sight is installed. The sight is laid on two aiming posts (placed out 50 and 100 meters from the mortar) on a referred deflection of 2800 mils and an elevation of 1100 mils. The mortar is within two turns of center of traverse. The gunner is given an administrative command to lay the mortar on a deflection of 2860 or 2740 mils. The mortar is then re-laid on the aiming posts using the traversing crank.

(2) The gunner is given a deflection change between 5 and 25 mils, either increasing or decreasing from the last stated deflection, and the command to refer and realign aiming posts.

EXAMPLE

An example of referring and realigning aiming posts begins with the command REFER DEFLECTION TWO EIGHT SEVEN FIVE (2875), REALIGN AIMING POSTS.

Upon receiving the command REFER, REALIGN AIMING POSTS, the mortar squad performs two simultaneous actions. The gunner places the announced deflection on the sight (without disturbing the lay of the weapon) and looks through the sightunit. At the same time, the ammunition bearer realigns the aiming posts. He knocks down the near aiming post and proceeds to the far aiming post. Following the arm-and-hand signals of the gunner (who is looking through the sightunit), he moves the far aiming post so that the gunner obtains an aligned sight picture. Then, he performs the same procedure to align the near aiming post.

MALFUNCTIONS

4-32. See paragraphs 3-37 to 3-40 of Chapter 3 for a detailed discussion of malfunctions.

REMOVAL OF A MISFIRE

4-33. Use the following procedures to remove a misfire.

NOTE: The squad leader immediately alerts the FDC of the misfire. The squad leader must supervise the removal of the misfire using a printed copy of the current misfire procedures.

(1) When a misfire occurs, any member of the squad immediately announces, "Misfire."

WARNING

During peacetime live-fire training, all personnel, except the gunner, move 50 meters or farther to the rear of the mortar.

(2) Ensuring that he does not stand directly behind the cannon, the gunner kicks the cannon several times with his heel in an attempt to dislodge the round (Figure 4-8). If the round fires, the mortar is re-laid on the aiming point and firing continues. If the round does not fire, the gunner tests the cannon for heat. If the cannon is cool enough to handle, the crew removes the round as described below.

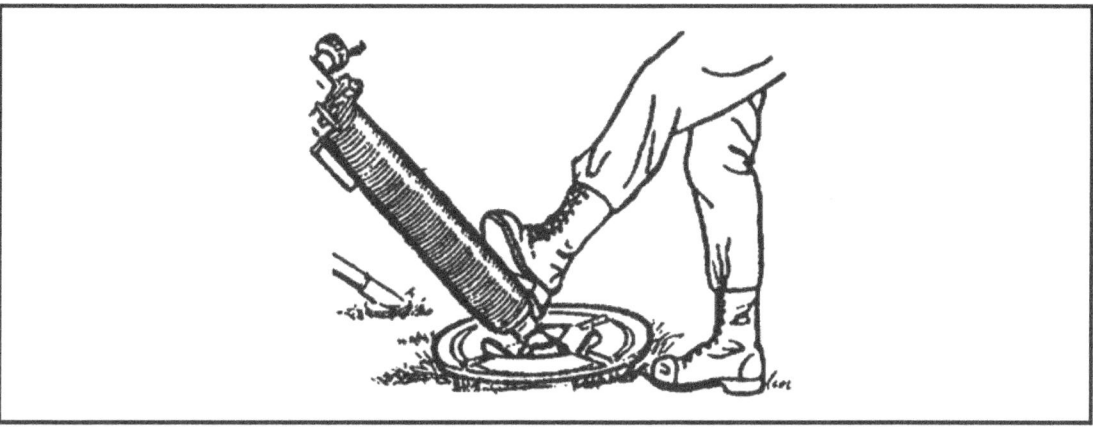

Figure 4-8. Kicking the mortar to dislodge the round.

WARNING

During peacetime live-fire training, the gunner joins the squad members behind the mortar and waits one minute (in case of a cookoff). After waiting one minute, the gunner returns to the mortar.

(3) The gunner checks for heat by starting from just below the muzzle and working down to the base with his fingertips. If the cannon is too hot to be handled, he cools it with water (or snow) and checks it one minute later. If no water (or snow) is available, the cannon is air-cooled until it can be easily handled with bare hands.

NOTES: 1. Liquids must never be poured into the cannon.

2. During peacetime live-fire training, the gunner signals the squad to come forward once the cannon is cool.

(4) When the mortar is cool enough to handle, the gunner removes the firing pin by turning the firing pin wrench counterclockwise (Figure 4-9). If necessary to provide easier access to the firing pin, the gunner depresses the cannon until the firing pin can be completely removed from the breech cap.

NOTE: Removing the firing pin ensures that the mortar will not fire should the round slip down the cannon during the subsequent drill.

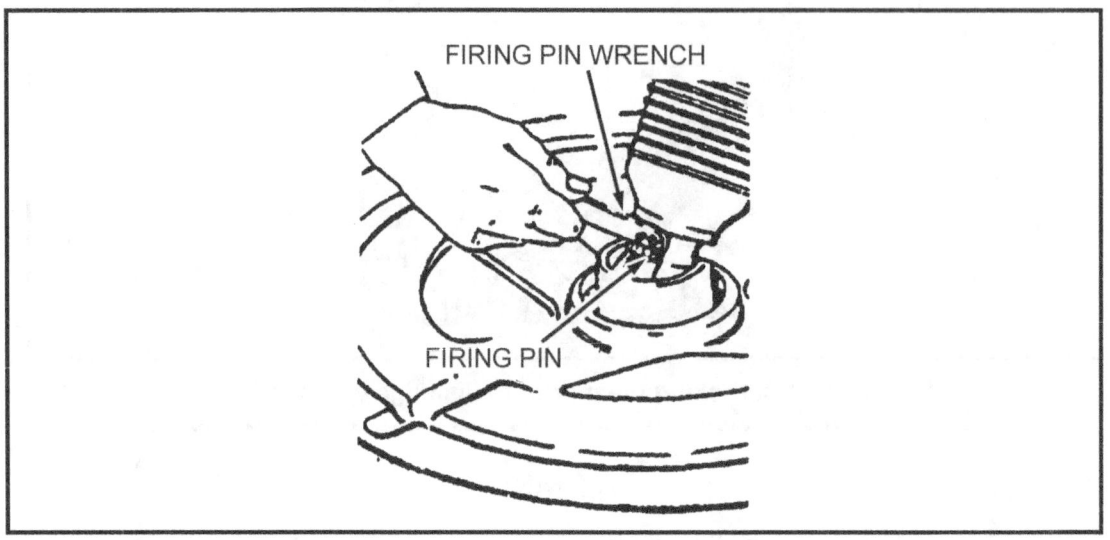

Figure 4-9. Removing the firing pin.

(5) After removing the firing pin, the gunner shouts, "Firing pin is removed," and hands the firing pin to the squad leader. The gunner locks the data down on the sight, then removes the sight and places it in a safe location. Then, the gunner unlocks and loosens the cannon clamp just enough to rotate the cannon to unlock the breech plug from the rotating socket of the baseplate. He then relocks the cannon clamp.

(6) The gunner grasps both ends of the traverse screw assembly and supports the mortar during the subsequent drill. The assistant gunner places his right hand palm up (1 inch from the muzzle end) under the BAD and his left hand palm down (1 inch from the muzzle end) on top (Figure 4-10). He places his thumbs alongside the forefingers, being careful to keep both hands away from the muzzle. The ammunition bearer puts both hands on the cooling fins under the cannon and slowly lifts the cannon until it is horizontal. He must not stand directly behind the mortar.

WARNING

Once the cannon is horizontal, the rear of the cannon must not be lowered back down until the round is extracted. If the round slips down the cannon before extraction, it could ignite, causing death or personal injury.

81-mm Mortar, M252

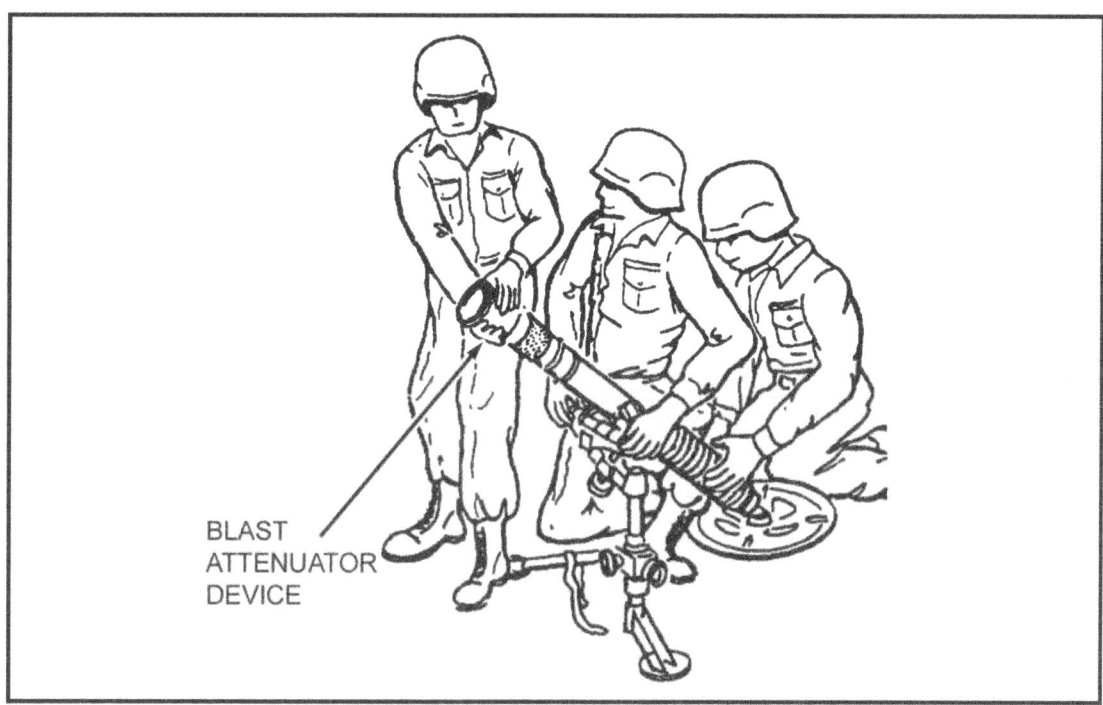

Figure 4-10. Raising the cannon to a horizontal position.

(7) When the cannon reaches the horizontal position, the assistant gunner moves both thumbs over the muzzle. When the fuze of the misfired round reaches the BAD, the assistant gunner stops the round with his thumbs (without touching the fuze) and carefully removes it from the cannon (Figure 4-11). The ammunition bearer shakes the cannon to dislodge any remnants of the last round fired and lowers the cannon into the rotating socket of the baseplate.

Figure 4-11. Removing the round from the cannon.

7 December 2007   FM 3-22.90   4-15

Chapter 4

(8) The assistant gunner passes the round to the ammunition bearer, who inspects it for the cause of the misfire. If the primer of the ignition cartridge is dented, the ammunition bearer attempts to replace the safety wire and places the round in a marked, safe location for disposal by ordnance personnel. If the primer is not dented, the round may be used again. Before attempting to fire, the firing pin must be replaced and the bore swabbed. The sight unit must then be replaced. The gunner will ensure that the correct firing data is indexed on the sight and that the mortar is relayed on the aiming posts.

4-34. If the above procedure fails to remove the misfire, the cannon must be kept horizontal, removed from the bipod, and laid horizontally on the ground in the dud pit until it can be turned over to ordnance personnel.

DISMOUNTING OF THE MORTAR

4-35. To dismount the mortar, the squad leader commands, OUT OF ACTION. At this command, the squad proceeds as follows:

(1) The gunner removes the sight and places it in the case with 3200 deflection and 0800 elevation indexed.

(2) The ammunition bearer holds the cannon until the assistant gunner removes the mount. Then, the ammunition bearer rotates the cannon a quarter of a turn to unlock it from the socket and places it in an area designated by the squad leader, and retrieves the aiming posts.

(3) The assistant gunner disengages the cannon clamps and moves the bipod from the immediate area of the mortar position. The cannon clamps are then closed.

(4) With the clamps facing away from him, the assistant gunner traverses the buffer carrier to the traversing handwheel and folds the handle. He exposes 1 inch (25 millimeters) of the cross-level shaft and folds the handle. He exposes 4 inches (100 millimeters) of the elevation shaft and folds the handle. Finally, the assistant gunner loosens the leg-locking handwheel, presses the spring-loaded locating catch, and raises the plain leg behind the buffer cylinders until it touches the traversing handwheel.

(5) He tightens the leg-locking handwheel (ensuring the teeth are correctly meshed) and fastens the securing strap over the arm and around the buffers.

(6) The gunner returns the baseplate to the area designated by the squad leader.

(7) The squad leader picks up the aiming posts and sight. At the command, MARCH ORDER, the squad places the mortar, equipment, and ammunition in the squad vehicle and trailer.

SECTION IV. AMMUNITION

This section implements STANAG 2321.

This section describes the ammunition that can be fired from the 81-mm mortar, M252. The four types of ammunition for the 81-mm mortar are HE, WP or RP smoke, ILLUM, and training.

CLASSIFICATION AND TYPES OF AMMUNITION

4-36. Ammunition for the M252 is typed according to its use.

IDENTIFICATION

4-37. All mortar cartridges are painted to prevent rust and to identify their type.

Markings on Container

4-38. The markings on an ammunition container indicate its contents. Additional information is included on an ammunition data card inside each container.

Color Code

4-39. All rounds are colored according to their type.

Markings on Rounds

4-40. Each round is stenciled with the ammunition lot number, type of round, type of filler, and caliber.

AMMUNITION LOT NUMBER

4-41. Each ammunition lot is assigned an ammunition lot number, which is marked on each cartridge and packing container. It is used for records such as reports on condition, malfunctions, and accidents.

FUNCTIONING

4-42. Fins around the tail of each cartridge stabilize it in flight and cause it to strike fuze-end first. The propelling charge consists of an ignition cartridge and removable propellant increments. The ignition cartridge (with primer) is fitted into the base of the fin shaft. The removable increments are fitted onto or around the shaft, depending on their type. When the cartridge is dropped down the cannon (fin-end first), the ignition cartridge strikes the firing pin and detonates, causing a flash that passes through the radial holes in the shaft. This ignites the propellant increments, which produce rapidly expanding gases that force the cartridge from the barrel. The obturator band ensures equal muzzle velocities in hot or cold cannons by keeping gases inside the cannon until the cartridge has fired. When fired, the cartridge carries the ignition cartridge with it, leaving the mortar ready for the next cartridge.

AUTHORIZED CARTRIDGES

4-43. The following tables outline the authorized cartridges for the 81-mm mortar, M252.

Chapter 4

High-Explosive Ammunition

4-44. HE ammunition is used against enemy personnel and light materiel targets for both fragmentation and blast effects. Table 4-2 details the HE ammunition that can be used when firing the M252 mortar.

Table 4-2. High-explosive ammunition for the 81-mm mortar, M252.

CARTRIDGE/TYPE	MAXIMUM RANGE (METERS)	FUZE	CHARACTERISTICS AND LIMITATIONS
M821, M821A1, and M821A2 HE cartridges	5,608 (M821) 5,790 (M821A1 and M821A2)	M734 multioption (M821/M821A1) fuze M734A1 multioption (M821A2) fuze	These cartridges are constructed from ductile cast iron and contain about 1.5 pounds of cyclonite (RDX)/trinitrotoluene (TNT) composition explosive. The aluminum tail assembly has six integral fins equally spaced around the rear, which stabilize the round in flight. Each cartridge weighs 9.03 pounds and is provided with the M223 propelling charge, consisting of four horseshoe-shaped increments fitted around the tail.
M889 and M889A1 HE cartridges	5,540 (M889) 5,850 (M889A1)	M935 PD fuze	These cartridges are identical to the M821, except that they have M935 PD fuzes. They both weigh 9.03 pounds.
M374A2 and M374A3 HE cartridge	4,595 (M374A2) 4,789 (M374A3)	M567 PD fuze M524-series PD fuze M526 PD fuze M532 PROX fuze	These cartridges are constructed from malleable cast iron and contain about 2.10 pounds of composition B explosive. The M374A2 weighs 9.12 pounds, and the M374A3 weighs 9.05 pounds.

Illumination Ammunition

4-45. ILLUM ammunition is used during night missions requiring assistance in observation. Table 4-3 details the ILLUM ammunition that can be used when firing the M252 mortar.

Table 4-3. Illumination ammunition for the 81-mm mortar, M252.

CARTRIDGE/TYPE	MAXIMUM RANGE (METERS)	FUZE	CHARACTERISTICS AND LIMITATIONS
M853A1 ILLUM cartridge	4,900	M772 MTSQ fuze	This cartridge has a cylindrical body that contains an illuminating candle and parachute assembly. It provides 600,000 candlepower illumination for 50 to 60 seconds. It uses the M219 propelling charge and weighs 9.02 pounds.
M816 IR ILLUM cartridge	4,925	M772 MTSQ fuze	This cartridge is identical to the M853 cartridge, except that it contains an infrared illuminant mix for use with standard night vision devices. It weighs 9.25 pounds. It is designed to provide infrared illumination which cannot be detected by the human eye.
M301-series ILLUM cartridge	2,150	M84-series time fuze	This cartridge has a cylindrical body that contains an illuminating candle and parachute assembly. It weighs 9.1 pounds. The round has a burst height of 600 meters and illuminates an area of about 1,200 meters for about 60 seconds. It uses the M185 propelling charge, which consists of eight increments fitted into the spaces between the fins and held by a propellant holder. The increments are individually wrapped with waterproof bags that are not to be removed. The cartridge must not be fired below charge 3.

Chapter 4

Smoke, White Phosphorus and Red Phosphorus Ammunition

4-46. Smoke, WP and RP ammunition is used as a screening, signaling, or incendiary agent. Table 4-4 details the smoke WP and RP ammunition that can be used when firing the M252 mortar.

Table 4-4. Smoke, white phosphorus ammunition for the 81-mm mortar, M252.

CARTRIDGE/TYPE	MAXIMUM RANGE (METERS)	FUZE	CHARACTERISTICS AND LIMITATIONS
M819 smoke RP cartridge	4,950	M772 PD MTSQ fuze	This cartridge has a cylindrical body and contains red phosphorus that produces smoke on contact with air. The cartridge uses the M218 propelling charge, which consists of four horseshoe-shaped increments fitted around the tail. It weighs 10.37 pounds.
M375A2 smoke WP cartridge	4,595	M524-series PD fuze	The white phosphorus inside this cartridge produces white smoke on contact with air. The cartridge uses the M205 propelling charge and weighs 9.12 pounds. This round must be stored in a vertical position.

Training Practice Ammunition

4-47. *TP* ammunition is used for training when service ammunition is not available or there are restrictions on the use of service ammunition for training. Table 4-5 details the TP ammunition that can be used when firing the M252 mortar.

Table 4-5. Training practice ammunition for the 81-mm mortar, M252.

CARTRIDGE/TYPE	MAXIMUM RANGE (METERS)	FUZE	CHARACTERISTICS AND LIMITATIONS
M880 SRTC	458	M775 PD practice fuze	This short-range practice cartridge can be used with the M252 or the 120-mm mortar (when used with an M303 or M313 81-mm mortar insert). It weighs 6.8 pounds. It is used on a 1 to 10 scale range. The cartridge's range can be reduced by removing plastic plugs from the projectile body. It can be refurbished and reused.
M879 FRTC	5,600	M751 PD practice fuze	This cartridge is matched ballistically to M821A1/889A1 HE cartridges. It weighs 8.98 pounds. The M751 practice fuze operates like the M734 multioption fuze. On impact, it produces a flash, a bang, and smoke.

FUZES

4-48. The types of fuzes described in this paragraph are PD, PROX, MT, multioption, and dummy.

POINT-DETONATING FUZES

4-49. All PD fuzes are SQ and, therefore, detonate on impact.

Fuze, M935

4-50. The M935 fuze has two function settings: IMP and a 0.05-second DLY. It is set using the bladed end of the M18 fuze wrench. The M935 is fitted with a standard pull wire and safety pin that are removed immediately before firing.

Fuze, M524-Series

4-51. The M524-series fuze has two function settings: SQ and DLY. When set at DLY, the fuze train causes a 0.05-second delay before functioning. When set at SQ, the fuze functions on point impact or graze contact. The fuze contains a delayed arming feature that ensures it remains unarmed and detonator safe for a minimum of 1.25 seconds and a maximum of 2.50 seconds of flight. To prepare for firing, the slot is aligned in the striker with SQ or DLY using the M18 fuze wrench. The safety pull wire is removed just before inserting the cartridge into the mortar.

NOTE: If, upon removal of the safety wire, a buzzing sound in the fuze is heard, the round should not be used. The round is still safe to handle and transport if the safety wire is reinserted.

> **WARNING**
>
> If the plunger safety pin (upper pin) cannot be reinserted, the fuze may be armed. An armed fuze must not be fired since it will be premature. It should be handled with extreme care and EOD personnel notified immediately. If it is necessary to handle a round with a suspected armed fuze, personnel must hold the round vertically with the fuze striker assembly up.

Fuze, M526-Series

4-52. The M526-series fuze has an SQ function only. It is fitted with a safety wire and pin that are removed immediately before firing.

Fuze, M567

4-53. The M567 is a selective SQ or 0.05-second delay impact fuze. It comes preset to function as SQ, and the selector slot should align with the "SQ" mark on the ogive. To set for delay, the selector slot should be rotated clockwise until it is aligned with the "D" mark on the ogive. An M18 fuze wrench is used to change settings. The fuze has a safety wire that must be removed before firing.

Chapter 4

PROXIMITY FUZES

4-54. A PROX fuze is an electronic device that detonates a projectile by means of radio waves sent out from a small radio set in the nose of the projectile. The M532 fuze is a radio Doppler fuze that has a PROX or SQ function. An internal clock mechanism provides nine seconds of safe air travel (610 to 2,340 meters along trajectory for charge 0 through 9, respectively). Once set to act as an IMP fuze, the mechanism cannot be reset for PROX. The fuze arms and functions normally when fired at any angle of elevation between 0800 and 1406 mils at charges 1 through 9. The fuze is not intended to function at charge 0. However, at temperatures above 32 degrees Fahrenheit and at angles greater than 1068 mils, the flight time is sufficient to permit arming. To convert the fuze from PROX to SQ, the top of the fuze must be rotated 120 degrees (one-third turn) in either direction. This action breaks an internal sheet pin and internal wire, thereby disabling the proximity function.

Disposal Precautions

4-55. PROX-fuzed short cartridges that are duds contain a complete explosive train and impact element. They should not be approached for 5 minutes or disturbed for at least 30 minutes after misfiring. After 30 minutes, the dud is still dangerous but can be approached and removed carefully or destroyed in place by qualified disposal personnel. If the situation allows for a longer waiting period, the dud can be considered safe for handling after 40 hours.

Burst Height

4-56. The principal factors affecting height of burst are the angle of approach to the target and the reflectivity of the target terrain. The air burst over average types of soil ranges from 1 to 6 meters, depending on the angle of approach. High angles of approach (near vertical) give the lowest burst heights. Light tree foliage and light vegetation affect the height of burst only slightly, but dense tree foliage and dense vegetation increase the height of burst. Target terrain such as ice and dry sand gives the lowest burst heights, whereas water and wet ground give the highest burst heights.

Crest Clearance

4-57. Close approach to crests, trees, towers, large buildings, parked aircraft, mechanized equipment, and similar irregularities causes functioning at heights greater than average level. When targets are beyond such irregularities, a clearance of at least 30 meters should be allowed to ensure maximum effect over the target area.

Climatic Effects

4-58. The fuzes may be used for day or night operations. They function normally in light rain; however, heavy rain, sleet, or snow can cause an increase in the number of early bursts. At extreme temperatures (below -40 degrees Fahrenheit and above 125 degrees Fahrenheit), it is not unusual to experience an increase in malfunctions proportionate to the severity of conditions.

Care, Handling, and Preservation

4-59. PROX fuzes withstand normal handling without danger of detonation or damage when in their original packing containers or when assembled to projectiles in their packing containers.

WARNING

The explosive elements in primers and fuzes are sensitive to shock and high temperatures. Boxes containing ammunition should not be dropped, thrown, tumbled, or dragged.

Installation

4-60. The fuze should already be fitted to the cartridge. If not, the cartridge is placed on its side, and the closing plug is removed using an M18 fuze wrench. (The handle of the wrench is turned counterclockwise.) The fuze threads and fuze well threads are inspected for damage. The fuze is screwed into the cartridge body, and it is seated and secured using an M18 wrench. No visible gap should be between the fuze and cartridge body.

> **WARNING**
>
> Do not use the fuze if the thread(s) is damaged. Do not use the cartridge if the fuze well is damaged or if the explosive is visible on the thread.

MECHANICAL TIME FUZES

4-61. These fuzes use a clockwork mechanism to delay functioning for a specific time.

Fuze, M772A1

4-62. The M772A1 fuze is an MTSQ. It can be set from 3 to 55 seconds at half-second intervals. The settings are obtained from the range tables and are applied using a wrench (number 9239539) or a 1 3/4-turn open-end wrench. The safety wire must be removed before firing.

Fuze, M84

4-63. The M84 fuze is a single-purpose, powder-train, MT fuze used with the 81-mm M301A1 and M301A2 illuminating cartridges. It has a time setting of up to 25 seconds. The fuze consists of a brass head, body assembly, and expelling charge. The fuze body is graduated from 0 to 25 seconds in 1-second intervals; 5-second intervals are indicated by bosses. The 0-second boss is wider and differs in shape from the other body bosses; the safe setting position is indicated by the letter "S" on the fuze body. The adjustment ring has six raised ribs for use in conjunction with fuze setter, M25, and a setting indicator rib (marked SET) about half the height and width of the other six ribs. Safety before firing is provided by a safety wire, which must be removed just before firing.

Fuze, M84A1

4-64. The M84A1 fuze is a single-purpose, tungsten-ring, MT fuze used with the 81-mm M301A3 ILLUM cartridge. It has a time setting of up to 50 seconds. All other features are the same as the M84 fuze.

MULTIOPTION FUZES

4-65. The M734 and M734A1 are the only multioption fuzes used with the 81-mm mortar.

Fuze, M734

4-66. This fuze has four function settings:
- PRX causes the cartridge to explode between 3 and 13 feet above the ground.
- NSB causes the cartridge to explode up to 3 feet above the ground.
- IMP causes the cartridge to explode on contact.
- DLY incorporates a 0.05-second delay in the fuze train before exploding the cartridge.

Chapter 4

4-67. No tools are needed to set the fuze, and the setting can be changed several times without damaging the fuze. It has no safety pins or wires, which reduces preparation time. If the fuze does not function as set, it automatically functions at the next lower setting.

Fuze, M734A1

4-68. The air-powered M734A1 multioption fuze has four selectable functions:
- PRX 120.
- PRX 60/81.
- IMP.
- DLY.

4-69. In HE PROX mode, the HOB remains constant over all types of targets. The impact mode causes the round to function on contact with the target and is the first backup function for either PROX setting. In the DLY mode, the fuze functions about 30 to 200 milliseconds after target contact. The DLY mode is the backup for the IMP and PRX modes. The IMP and DLY modes have not changed from the current M734 multioption fuze.

4-70. Radio frequency jamming can affect the functioning of PROX fuzes. Radio frequency jamming initiates a gradual desensitizing of the fuze electronics to prevent premature fuze function. Once the fuze is out of jamming range, the fuze electronics recover and function in the PROX mode if the designed HOB has not been passed. To limit the time of fuze radio frequency radiation, the proximity turn-on is controlled by an apex sensor that does not allow initiation of the fuze proximity electronics until after the apex of the ballistic trajectory has been passed.

4-71. In compliance with the safety requirements of military standard 1316C, the M734A1 uses ram air and setback to provide two independent environment sensors.

DUMMY FUZES

4-72. The M751 and the M775 are used with the 81-mm mortar. There are two types of M751 fuzes:
- Type 1 resembles the M734 fuze.
- Type 2 resembles the M935 fuze.

4-73. The M751 is fitted with a smoke charge that operates on impact. The safety/packing clip should be removed when the cartridge is unpacked.

FUZE WRENCH

4-74. The fuze wrench, M18, assembles the fuze to the cartridge, and the bladed tip on the end sets PD-type fuzes. The fuze setting wrench (NSN: 5120 00 203 4801) sets M772 MT and M768 time fuzes. It engages the 1 3/4-inch flats on the setting ring or the fuze head. The fuze setter, M25, sets M84-series time fuzes. Notches in the setter engage ribs in the setting ring of the fuze.

CARTRIDGE PREPARATION

4-75. The propellant train (except the training cartridge) consists of an ignition cartridge and propellant charges. The ignition cartridge has a percussion primer and is assembled to the end of the fin assembly. The propelling charge is contained in four horseshoe-shaped, felt-fiber containers or nine wax-tested, cotton cloth, bag increments. The propelling charges are assembled around the fin assembly shaft.

4-76. Cartridges are shipped with a complete propelling charge, an ignition cartridge, and a primer. Firing tables are used to determine the correct charge for firing. Remaining increments are repositioned toward the rear of the tail fin assembly when firing the cartridge with horseshoe-shaped increments at less than full charge.

> NOTE: Charge 0—Ignition cartridge only.
>
> Charge 1—Ignition cartridge and one increment.
>
> Charge 2—Ignition cartridge and two increments.
>
> Charge 9—Ignition cartridge and nine increments.

4-77. Increments removed from cartridges before firing should be placed in a metal or wooden container located at least 25 meters away from the firing vehicle or position. Excess increments should not accumulate near the mortar positions but are removed to a designated place of burning and destroyed. Units should follow specific range SOPs to dispose of unused increments. The following is one way to dispose of them.

(1) Select a place at least 100 meters from the mortar position, parked vehicles, and ammunition points.

(2) Clear all dead grass or brush within 30 meters around the burning place. Do not burn increments in piles—spread them in a train 1 to 2 inches deep, 4 to 6 inches wide, and as long as necessary.

(3) From this train, extend a starting train that will burn against the wind of single increments laid end to end. End this starting train with not less than 1 meter of inert material (dry grass, leaves, or newspapers).

(4) Ignite the inert material.

(5) Do not leave unused increments unburned in combat operational areas. The enemy will use them.

CARE AND HANDLING

4-78. Ammunition is made and packed to withstand all conditions ordinarily encountered in the field. However, since explosives are affected by moisture and high temperature, they must be protected.

4-79. Before-firing checks include the following:

- Ammunition should be free of moisture, rust, and dirt.
- The fin and fuze assembly must be checked for tightness and damage.
- Charges must be kept dry.
- Extra increments are removed if the cartridge is to be fired with less than full charge.
- With the exception of a few unused increments (within the same ammunition lot number) as replacements for defective increments, excess powder should be removed from the mortar position.
- The primer cartridge is checked for damage or dampness.
- When opening an ammunition box, the ammunition bearer ensures the box is horizontal to the ground, not nose- or fin-end up. After the bands are broken and the box opened, the rounds should be removed by allowing them to roll out along the lid of the box (Figure 4-12). After the rounds have been removed, they should always be handled with two hands to prevent accidental dropping. Dropping may cause the propellant charges to ignite, causing bodily injuries.

WARNING

Incidents occurring from mishandling 300-series ammunition have resulted in minor burns to the hands and legs.

Chapter 4

Figure 4-12. Correct way to open an ammunition box.

NOTE: The floating firing pin within the primer has approximately 1/16 of an inch to move around. This may cause the firing pin to ignite the charges if the cartridge is dropped on the fin end (Figure 4-13).

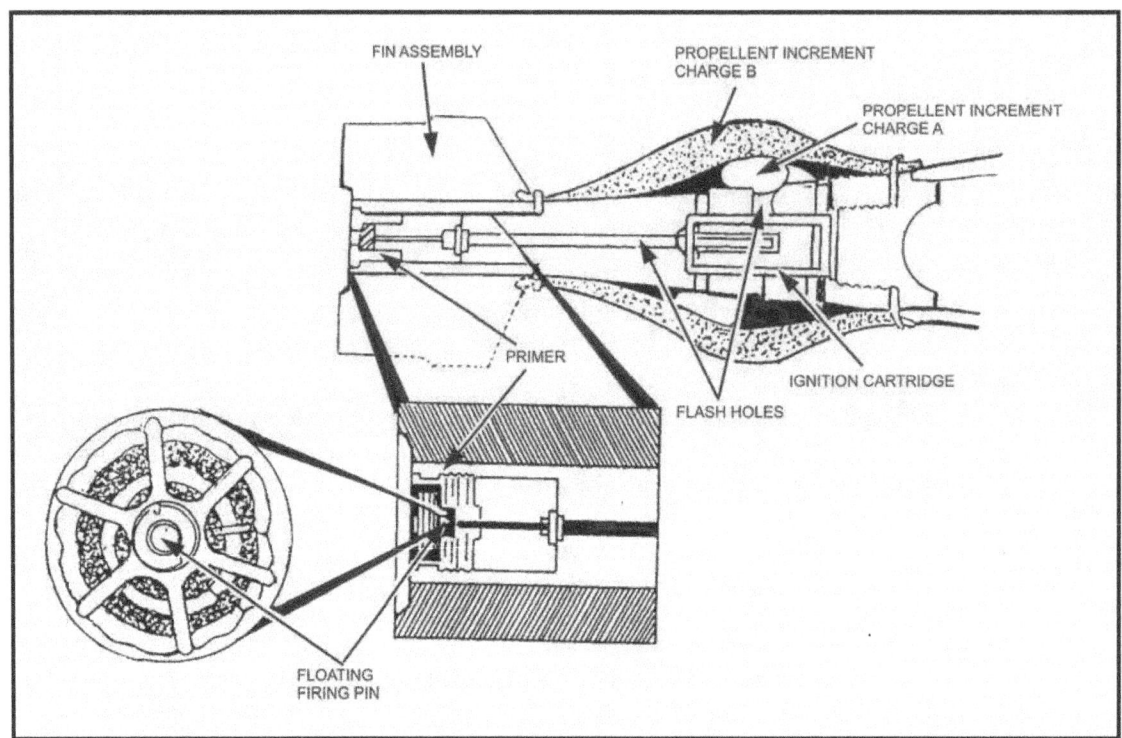

Figure 4-13. Floating firing pin.

- To help minimize the occurrence of short rounds and duds, unpackaged ammunition that has been dropped should not be fired. It should be destroyed in accordance with standard procedures.

4-80. Complete cartridges are always handled with care. The explosive elements in primers and fuzes are sensitive to shock and high temperature.

81-mm Mortar, M252

4-81. The moisture-resistant seal of the container is broken when the ammunition is to be used. When a large number of cartridges are needed for a mission, they may be removed from the containers and prepared. Propelling charges are covered or protected from dampness or heat.

4-82. The ammunition is protected from mud, sand, dirt, and water. If it gets wet or dirty, it must be wiped off at once. The powder increments should not be exposed to direct sunlight. Keeping ammunition at the same temperature results in a more uniform firing.

4-83. The pull wire and safety wire are removed from the fuze just before firing. When cartridges have been prepared for firing, but are not used, all powder increments and safety wires are replaced. The cartridges are returned to their original containers. These cartridges are used first in subsequent firing so that once-opened stocks can be kept to a minimum.

WARNING

Do not handle duds; the fuzes could be armed. Duds are extremely dangerous. Do not move or turn them. To dispose of duds, immediately call the nearest EOD unit.

4-84. Ammunition should be stored under cover. If it is necessary to leave the ammunition uncovered, it should be raised on dunnage at least 6 inches above the ground. The pile is covered with a double thickness of tarpaulin. Trenches are dug to prevent water from flowing under the pile. Phosphorus liquefies at 111.4 degrees Fahrenheit; therefore, WP cartridges are stored with the fuze end up to protect against uneven rehardening of the filler. Otherwise, an air cavity can form on one side of a cartridge when the filler rehardens and cause instability in flight. RP cartridges (M252 only) are stored the same as HE cartridges.

WARNING

When firing HE ammunition less than 400 meters, personnel must have adequate cover for protection from fragments.

This page intentionally left blank.

Chapter 5
120-mm Mortars, M120 and M121

The 120-mm mortar provides close-in and continuous indirect fire support to maneuver forces. The 120-mm mortar also provides increased range and lethality over 81-mm and 60-mm mortars, but the mortar and ammunition are heavier. There are three versions of the 120-mm mortar: the towed M120, the vehicle-mounted M121, and the Stryker-mounted RMS6-L recoilless mortar. This chapter discusses the organization, capabilities, and operations of the M120 and M121, 120-mm mortars. TM 9-2320-311-10-12 covers the RMS6-L in detail.

SECTION I. SQUAD ORGANIZATION AND DUTIES

A mortar squad maintains and fires a single 120-mm mortar, and each member has principal duties and responsibilities.

ORGANIZATION

5-1. For the mortar section to operate effectively, each squad member must be proficient in his individual duties. By performing those duties as a team member, he enables the mortar squad and section to perform as a fighting team. The squad, section, and platoon leaders trains and leads their units.

DUTIES

5-2. The mortar squad for the M120 and M121 consists of four men (Figure 5-1):
- Squad leader.
- Gunner.
- Assistant gunner.
- Ammunition bearer.

5-3. The following sections address their principal duties.

SQUAD LEADER

5-4. The *squad leader* stands behind the mortar where he can command and control his squad. He supervises the emplacement, laying, and firing of the mortar, and all other squad activities.

GUNNER

5-5. The *gunner* stands to the left side of the mortar where he can manipulate the sight, elevating handwheel, and traversing handwheel. He places firing data on the sight and lays the mortar for deflection and elevation. He makes large deflection shifts by shifting the bipod assembly and keeps the bubbles level during firing.

ASSISTANT GUNNER

5-6. The *assistant gunner* stands to the right of the mortar, facing the cannon and ready to load. In addition to loading, he swabs the bore after every 10 rounds or after each fire mission. He assists the gunner in shifting the mortar when the gunner is making large deflection changes.

Chapter 5

AMMUNITION BEARER

5-7. The *ammunition bearer* stands to the right rear of the mortar. He maintains and prepares the ammunition for firing, and passes prepared ammunition to the assistant gunner. He also acts as the squad driver.

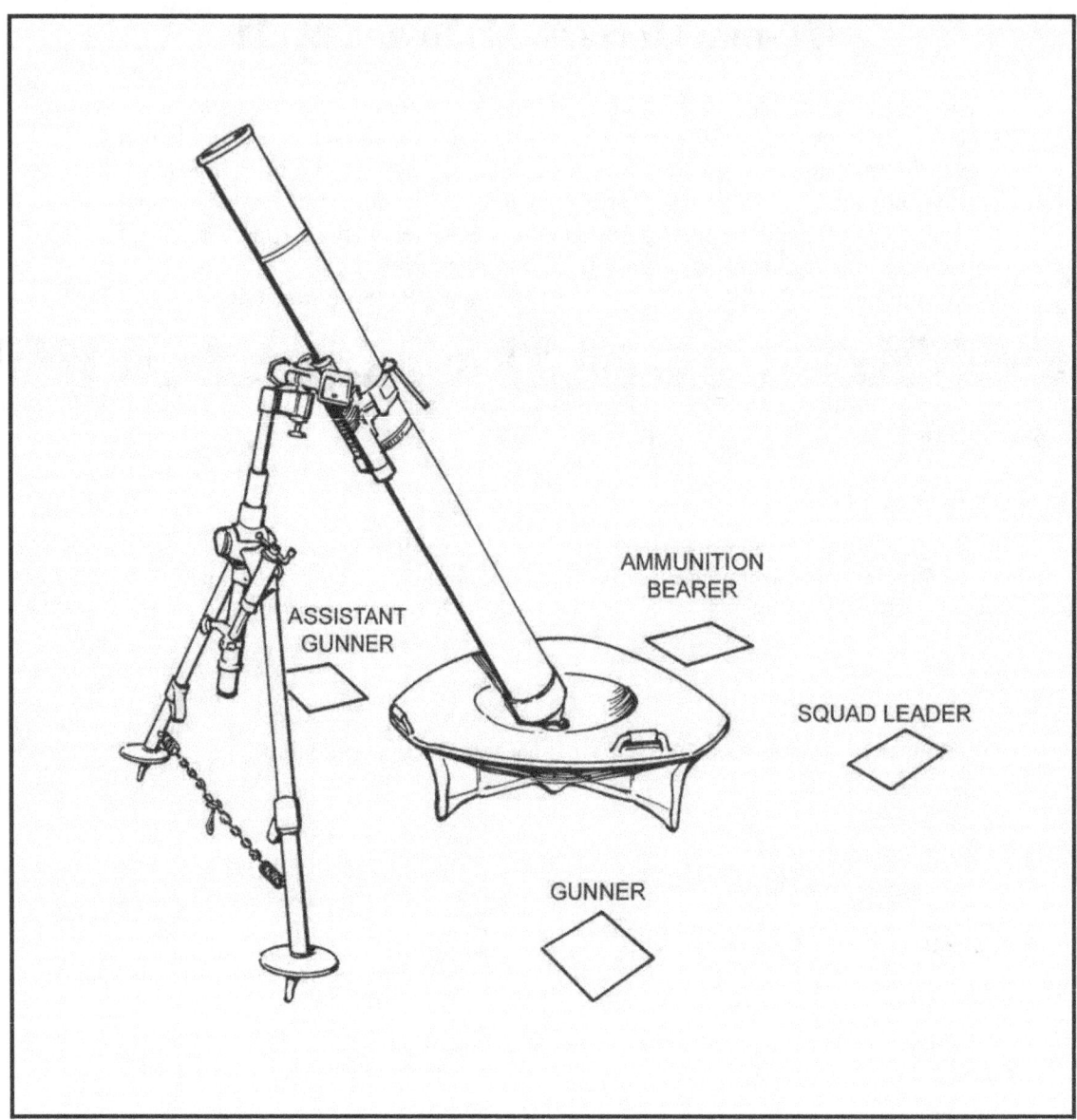

Figure 5-1. Position of squad members.

SECTION II. COMPONENTS

This section contains the technical data and description of each component of the 120-mm mortar (Figure 5-2). The mortar is a smooth-bore, muzzle-loaded, crew-served, high angle-of-fire weapon. It consists of a cannon assembly, bipod assembly, and baseplate. The 120-mm mortar is designed to be employed in all phases and types of land warfare, and in all weather conditions. (See TM 9-1015-250-10 for detailed information.)

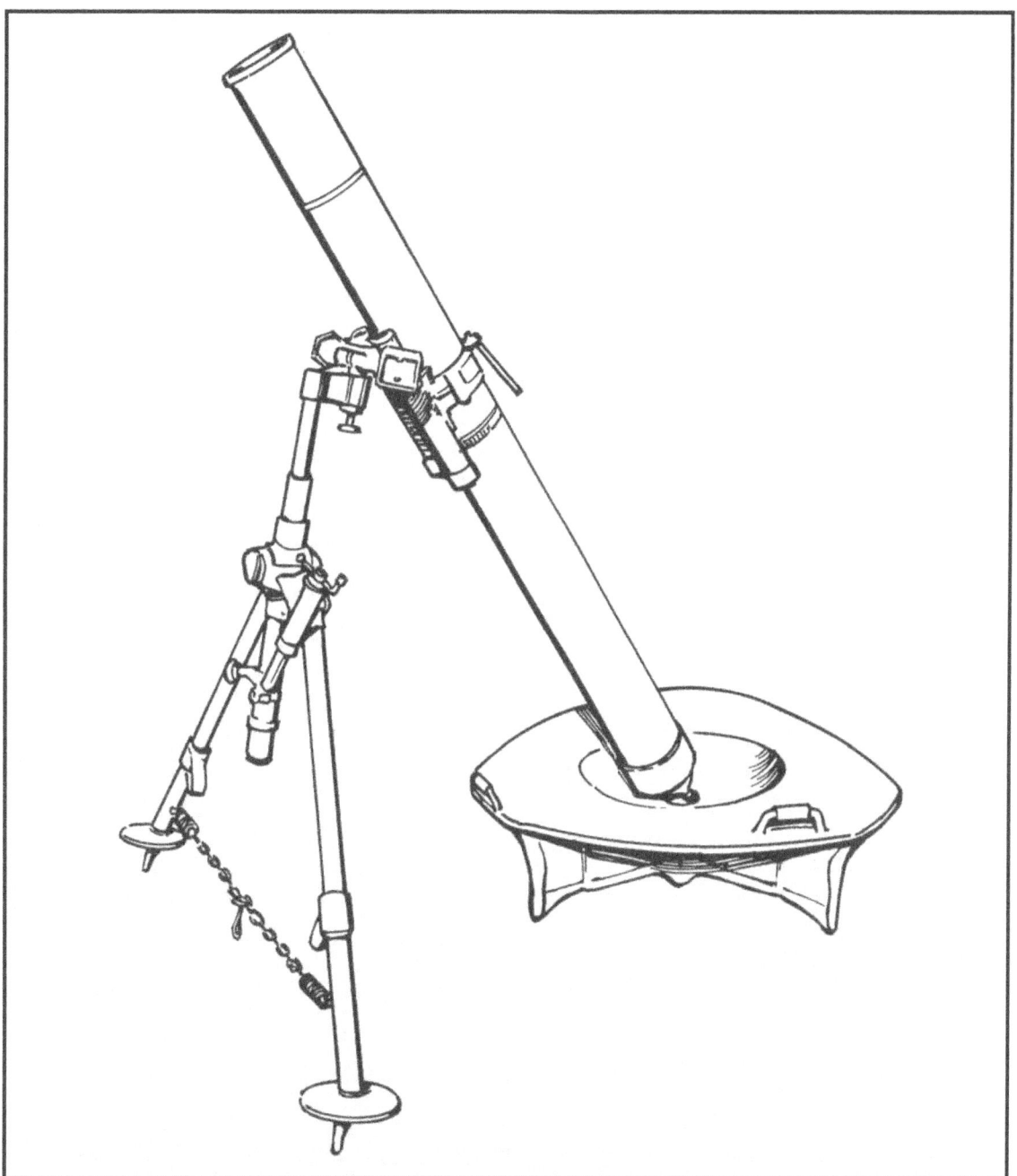

Figure 5-2. 120-mm mortar.

Chapter 5

TABULATED DATA FOR THE 120-mm MORTAR

5-8. The tabulated data for the 120-mm mortar are as shown in Table 5-1.

Table 5-1. Tabulated data for the 120-mm mortar.

Weights (pounds)	
M298 cannon	110.0
Bipod	
M190 (ground-mounted)	70.0
M191 (ground-/carrier-mounted)	68.0
M9 baseplate	136.0
M67 sightunit	2.9
M1100 trailer	399.0
Elevation (approximate mils)	
Ground-mounted	0710 to 1510
Carrier-mounted	0750 to 1510
For each turn of elevation crank	5
Traverse (approximate mils)	
Right or left from center using traversing wheel	136
With extension	316
One turn of traversing wheel	5
Range (meters)	
Maximum	7,200
Minimum	200
Rate of Fire (rounds per minute)	
Maximum	16 (first minute)
Sustained	4 (indefinitely)
Bursting radius (meters)	70

CANNON ASSEMBLY, M298

5-9. The cannon assembly consists of two parts: the tube and the breech cap (Figure 5-3). The bottom end of the tube is threaded to form a seat, which functions as a gas seal and centers the breech cap.

5-10. The breech cap screws into the base end of the tube with the front end of the breech cap mating to the seat on the tube, forming a gas-tight metal seal. The external rear portion of the breech cap is tapered and has a ball-shaped end. This end is cross-bored to help the user insert or remove the breech cap and lock it into firing position.

5-11. The breech cap houses the firing pin, which can be removed during misfire procedures as a safety measure. The squad leader must physically possess the firing pin before the mortar is considered safe.

NOTES: 1. The cannon and breech cap are serial-numbered identically. They should not be interchanged.

2. Currently, two styles of breech cap are fielded with M120 and M121 mortar systems. The old style of breech cap uses a safety mechanism to place the weapon in SAFE or FIRE mode. This breech cap must be replaced with the new style of breech cap, which uses a removable firing pin. The following paragraph covers procedures for operating the old style of breech cap.

5-12. The old style of breech cap utilizes a safety mechanism that has two positions: "F" for FIRE (firing pin protrudes) and "S" for SAFE (firing pin is withdrawn).

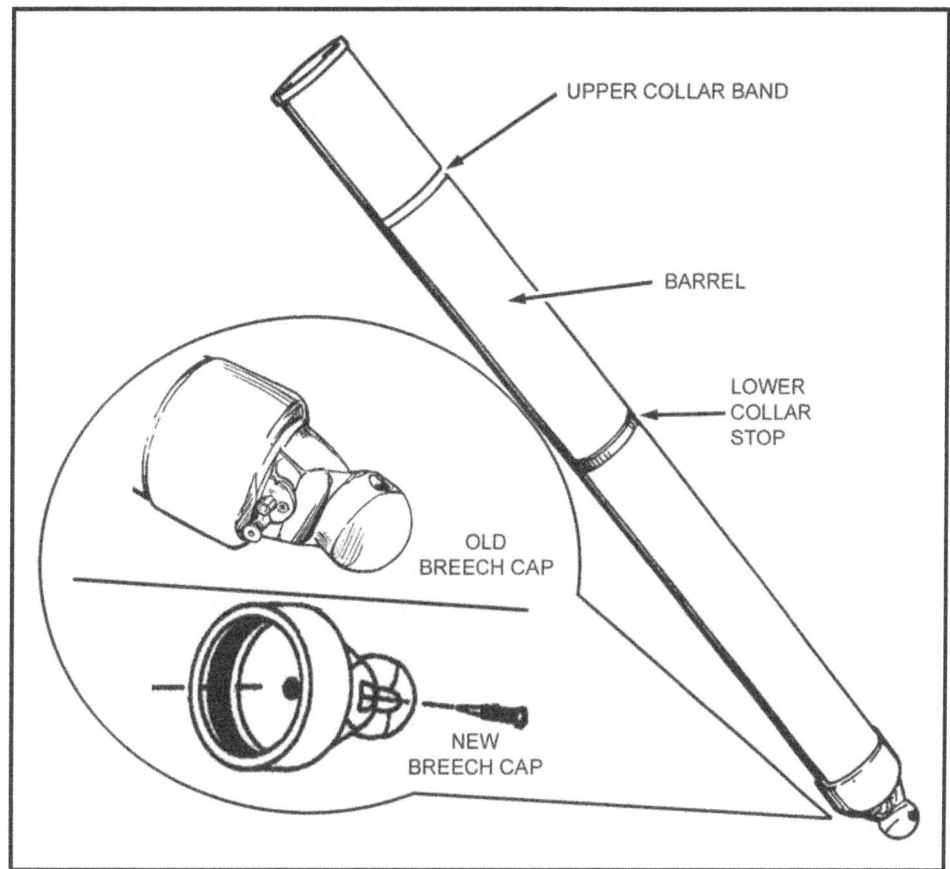

Figure 5-3. Cannon, M298, with old and new styles of breech cap.

BIPOD ASSEMBLY, M191 (CARRIER-/GROUND-MOUNTED)

5-13. The bipod assembly, M191, for the carrier-mounted M121 mortar (Figure 5-4) consists of the following main parts:
- Bipod leg extensions.
- Cross-leveling mechanism.
- Traversing gear assembly.
- Traversing extension assembly.
- Elevating mechanism.
- Buffer housing assembly.
- Buffer mechanism.
- Clamp handle assembly.
- Cross-leveling locking knob.
- Chain assembly.

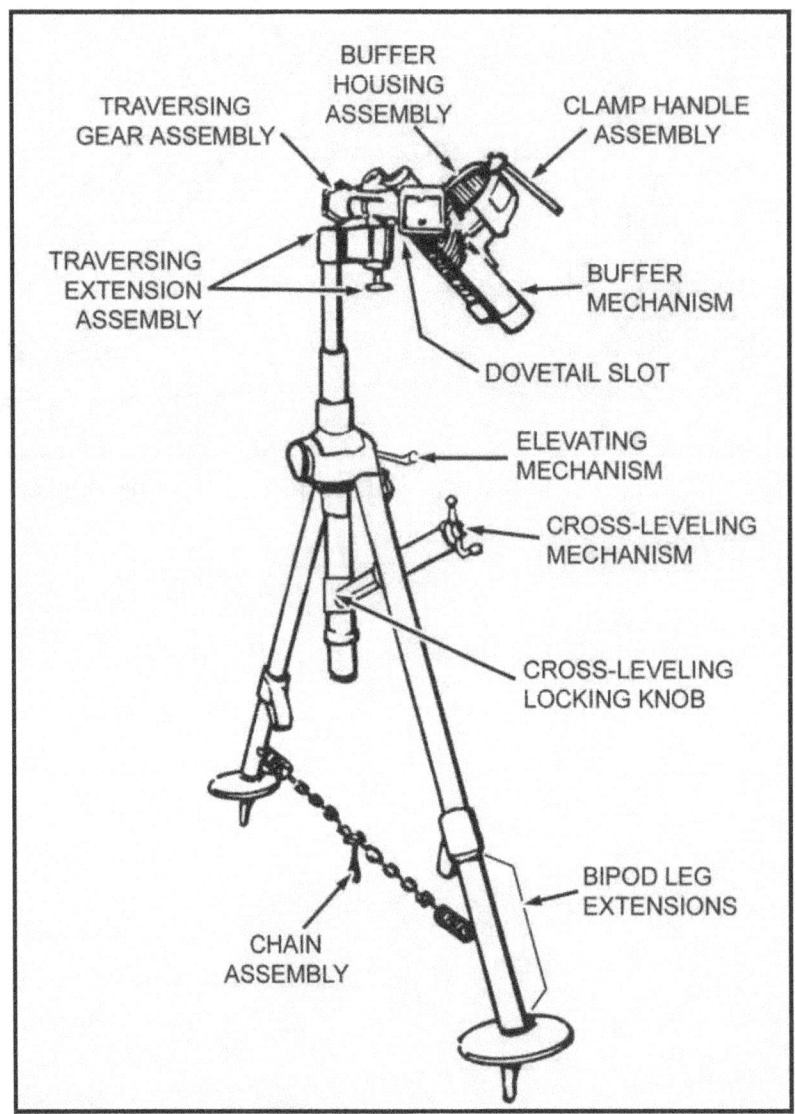

Figure 5-4. Bipod assembly, M191 (carrier-/ground-mounted).

BIPOD ASSEMBLY, M190 (GROUND-MOUNTED)

5-14. The bipod assembly, M190, for the ground-mounted M120 mortar (Figure 5-5) consists of the following main parts:
- Clamp handle assembly.
- Buffer housing assembly.
- Dovetail slot.
- Elevating mechanism.
- Cross-leveling locking knob.
- Cross-leveling mechanism.
- Chain assembly.
- Traversing extension assembly.
    - Traversing gear assembly.
    - Bipod legs.

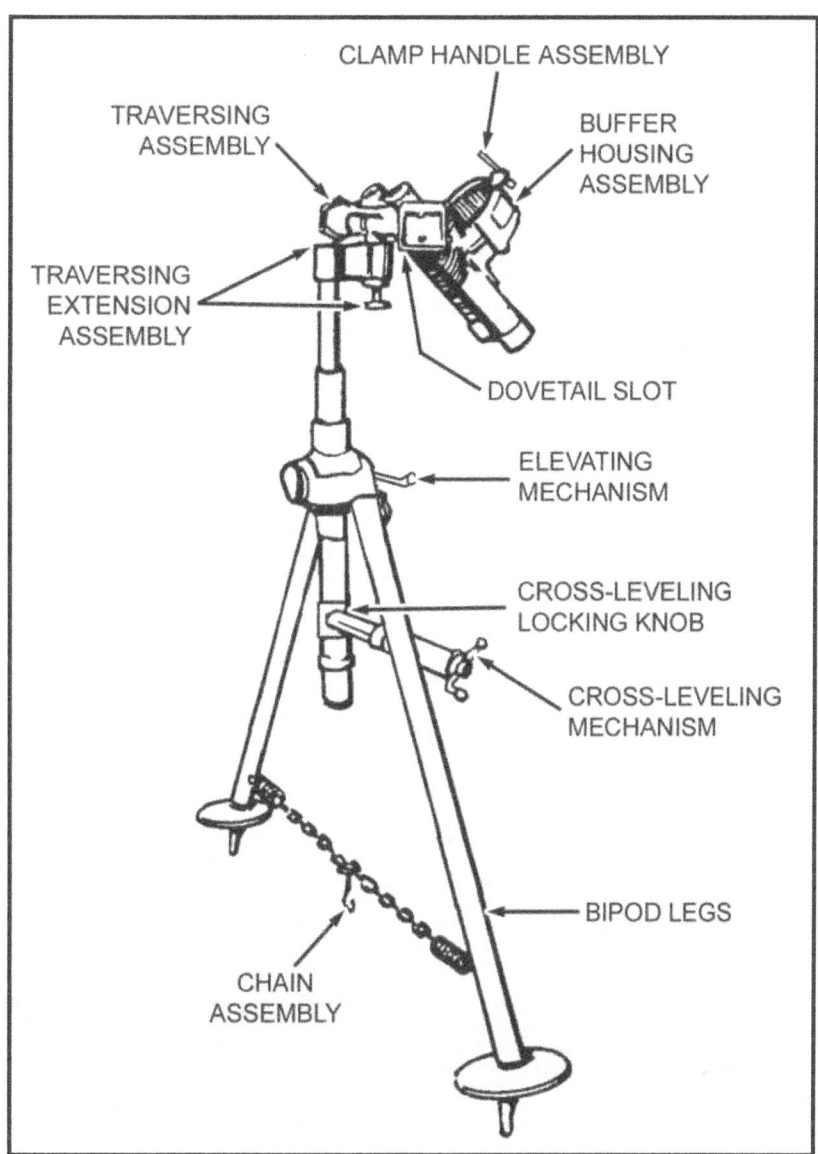

Figure 5-5. Bipod assembly, M190 (ground-mounted).

Chapter 5

BASEPLATE, M9

5-15. The baseplate (Figure 5-6) is shaped like a rounded triangle. It has a socket that enables a full 360-degree traverse without moving the baseplate. It also has legs (spades) under the baseplate, two carrying handles, and one locking handle.

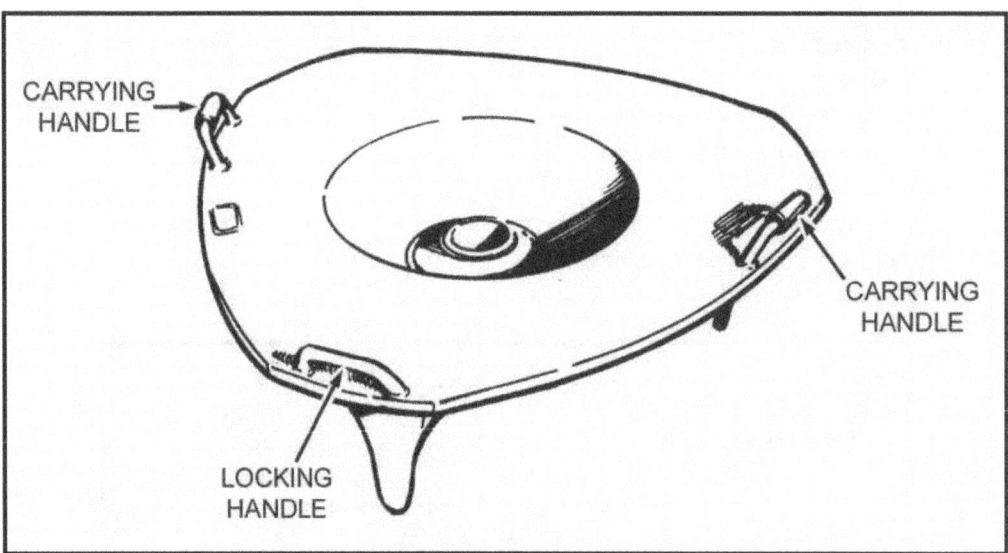

Figure 5-6. Baseplate, M9.

SECTION III. OPERATIONS

This section explains how to place the mortar into action by ground-mounting the weapon system from the trailer position; how to conduct safety checks; and how to perform misfire procedures.

PREMOUNT CHECKS

5-16. Before the mortar is mounted, the squad performs premount checks.

MORTAR CANNON

5-17. Check—
- The BAD for cracks, rust, and missing parts.
- The cannon for cracks, rust, and missing, dented, or damaged parts.
- The breech assembly to ensure that it is tight.
- The white line on the breech cap to ensure that it aligns with the white line on the cannon when fully assembled.
- Around the firing pin or breech cap for evidence of gas leakage.
- The serial numbers on the barrel and breech cap to ensure that they match.
- The breech assembly for bulges, dents, or visible cracks.

5-18. If the mortar cannon is equipped with the old style of breech cap, check—
- The safety mechanism to ensure that it aligns with the white line on the cannon when fully assembled.
- The safety mechanism to ensure that it locks (under spring tension) into the FIRE and SAFE detents.
- Around the safety mechanism for evidence of gas leakage.

> **WARNING**
>
> For M298 mortar cannons equipped with the old style of breech cap:
>
> Incorrect assembly of the safety mechanism can cause the firing pin to be exposed when in the SAFE position ("S" showing) and retracted when in the FIRE position ("F" showing). This will cause a misfire and may cause the user to unknowingly expose the firing pin during misfire procedures.
>
> Perform the following steps to check the safety mechanism for proper operation:
>
> 1. Unscrew the breech cap from the barrel.
>
> 2. With the safety mechanism in the FIRE position ("F" showing, the firing pin should be extended), check the firing pin protrusion height with the firing pin gauge. Do not use the cannon if the firing pin is defective.
>
> 3. Check the retraction of the firing pin with the safety mechanism in the SAFE position ("S" showing, the firing pin should be recessed).
>
> 4. If the safety mechanism functions properly, keep the weapon on SAFE and reassemble the cannon. Ensure that the white stripe on the cannon lines up with the safety mechanism.
>
> 5. If the safety mechanism does not function properly, disassemble it. If serviceable, reassemble the breech cap and lubricate, as required.
>
> For more detailed information, refer to TM 9-1015 250-10.)

BIPOD ASSEMBLY

5-19. Check—
- The bipod assembly for cracks, broken welds, or loose, missing, or damaged parts.
- To ensure the buffer housing assembly operates properly and securely holds the cannon.
- To ensure the traversing gear assembly, elevating mechanism, and cross-leveling mechanism operate smoothly and without binding through their entire range of travel.
- The buffer mechanism by pulling down on both housing tubes at the same time and ensuring that they return to their original position when released.

MORTAR BASEPLATE

5-20. Check—
- The socket for broken edges, cracks, and corrosion.
- The baseplate for cracks and broken welds.

Chapter 5

SIGHTUNIT, M67

5-21. Check—
- Sources lit by tritium for proper illumination.
- The eyeshield for damage.
- Lenses for scratches, smears, moisture, cracks, and other obstructions.
- The reticle for clarity.
- Level vials to ensure that they are not cracked, broken, or loose in their mountings.
- Level vial covers to ensure that they are present.
- Elevation and deflection knobs to ensure that they move freely over their entire range of movement.
- All scales and index lines to ensure that they are clear and distinct.
- Elevation and deflection knobs for movement after locking knobs are tightened. They should move no more than ± one mil.
- The coarse deflection scale and azimuth control dial to ensure that they rotate freely when depressed and return to their original position under spring tension when released.
- The mortar sight latch to ensure that it secures the sight to the mortar, is not loose, and has no cracks.
- Mounting surfaces to ensure that they are free of burrs and nicks.
- Radiation warning data plates to ensure that they are present and are not damaged.

MORTAR TRAILER, M1100

5-22. Check for—
- Broken welds and damaged, deformed, or missing parts.
- Proper operation of the towing eye, trailer frame, tires, and all mounting assemblies.

PLACING A GROUND-MOUNTED 120-mm MORTAR INTO ACTION

5-23. The mortar must be ground-mounted within 1 minute and 15 seconds, and the following conditions met:
- The sight is set with a deflection of 3200 mils and an elevation of 1100 mils.
- All bubbles are centered within the outer red lines.
- The traversing extension is locked in the center position.
- The bearing is the center of traverse.
- The cannon is locked into the baseplate with the white line up (on top).
- The bipod cannon clamp is positioned and locked.
- The bipod locking knob is hand-tight.

NOTE: Left and right are in relation to the mortar's direction of fire.

5-24. The squad uses the following procedures to place a ground-mounted 120-mm mortar into action.
 (1) The driver/ammunition bearer exits the mortar carrier and moves to the driver's side of the mortar trailer hitch. At the same time, the assistant gunner secures the aiming posts, exits the vehicle, and moves to the passenger's side of the mortar trailer hitch.
 (2) Together, they unhook the trailer from the vehicle and position the trailer at the firing position with the baseplate toward the direction of fire. The ammunition bearer then removes the muzzle plug.
 (3) Once in position, the assistant gunner and ammunition bearer raise the trailer until the baseplate rests on the ground. Then, the assistant gunner releases the trailer, while the

ammunition bearer continues to hold the trailer in place. The assistant gunner moves to the right side of the mortar to assist in mounting the mortar.

(4) The gunner exits the vehicle with the sight and places it on the left side of the mortar. He then moves to the left side of the mortar and removes the lock release lever pin. He releases the mortar baseplate, and unhooks the bipod chain from the eye on the bipod leg and drops it. He then loosens the cross-leveling locking knob.

> **WARNING**
>
> **Stay clear of the mortar baseplate to avoid injury from sudden release.**

(5) The assistant gunner releases the lock release lever and the clamping catch, and swings the trailer bridge assembly out of the way. He raises the bipod legs, rotates them 180 degrees, and spreads them until they are fully extended and the spread cable is taut.

(6) With the gunner standing on the left and the assistant gunner on the right, they grasp the traversing mechanism and bipod legs. With the assistant gunner holding the bipod legs just above the spread cable, they pull and guide the cannon away from the trailer. The gunner and assistant gunner guide the bipod legs to a point about 2 feet in front of the baseplate and lower them to the ground.

(7) Once the bipod is placed on the ground, the assistant gunner tightens the cross-leveling locking knob. The ammunition bearer moves the trailer away from the mortar position, to a point selected by the squad leader.

(8) The gunner unlocks the clamp handle assembly. Now standing to the rear and straddling the cannon, he grasps under the recoil buffer assembly and pulls down, sliding the cannon clamp down the cannon until it rests against the lower collar stop.

(9) The gunner makes sure that the white lines on the cannon and the buffer housing assembly are aligned. He retightens the clamp handle assembly until it clicks. He then ensures that the firing pin is properly seated. If the mortar cannon is equipped with the old style of breech cap, the gunner places the safety mechanism on FIRE ("F" showing). The assistant gunner checks for slack in the spread cable and ensures that the traversing mechanism is within four turns of the center of traverse. The assistant gunner also ensures that the traversing extension is locked in the center position and that the cross-leveling locking knob is hand-tight.

NOTE: Part of the white line on the buffer housing assembly must overlap the white line on the cannon.

(10) The gunner removes the sight from the sight box and uses it to index a deflection of 3200 mils and an elevation of 1100 mils. He places the sight in the dovetail slot of the bipod. The gunner and assistant gunner then level the mortar for elevation and deflection. When the gunner is satisfied with the lay of the mortar, he announces, "Gun up." The mortar is now mounted and ready to be laid.

PERFORMING SAFETY CHECKS ON A GROUND-MOUNTED 120-mm MORTAR

5-25. Specific safety checks must be performed before firing mortars. Most can be made visually. The gunner is responsible for physically performing the checks under the squad leader's supervision.

5-26. To perform safety checks, the gunner—
(1) Checks for mask and overhead clearance.
- To determine mask clearance, the gunner lowers the cannon to 0800 mils elevation. He places his head near the base of the cannon and sights along the top of the cannon for obstructions through the full range of traverse.
- To determine overhead clearance, the gunner raises the cannon to 1511 mils elevation. He places his head near the base of the cannon and sights along the top of the cannon for obstructions through the full range of traverse.

NOTE: If at any point in the full range of traverse, both at minimum or maximum elevation, an obstruction is found, the gunner raises or lowers the cannon until the round will clear the obstruction when fired. He turns the sight elevation micrometer knob until the elevation bubble is level. He reads the elevation at this point and reports the deflection and elevation to the squad leader who, in turn, reports this information to the FDC.

(2) Ensures the cannon is locked to the baseplate.
- He locks the cannon onto the socket of the baseplate with the white line on the cannon facing up.
- He aligns the white line on the cannon with the white line on the clamp handle assembly.
(3) Checks the buffer housing assembly to ensure that it is locked. He checks this by loosening the clamp handle assembly about 1/4 of a turn and retightening it until a metallic click is heard.
(4) Checks the cross-leveling locking knob to ensure it is hand-tight.
(5) Checks the spread cable to ensure that it is taut.
(6) Checks the firing pin to ensure that it is properly installed in the breech cap. If the mortar cannon is equipped with the old style of breech cap, the gunner checks the safety mechanism to ensure that it is in the FIRE position ("F" showing).

PERFORMING SMALL DEFLECTION AND ELEVATION CHANGES ON A GROUND-MOUNTED 120-mm MORTAR

5-27. The gunner receives deflection and elevation changes from the FDC in the form of a fire command. If a deflection change is required, it precedes the elevation change.

NOTE: Small deflection and elevation changes are greater than 20 mils but less than 60 mils for deflection and greater than 30 mils but less than 90 mils for elevation.

(1) The gunner sets the sight for deflection and elevation.
- He places the deflection on the sight by turning the deflection micrometer knob until the correct 100-mil deflection mark is indexed on the coarse deflection scale. He continues to turn the deflection micrometer knob until the remainder of the deflection is indexed on the deflection micrometer scale.
- The gunner places the elevation on the sight by turning the elevation micrometer knob until the correct 100-mil elevation mark is indexed on the coarse elevation scale. He continues to turn the elevation micrometer knob until the remainder of the elevation is indexed on the micrometer scale.
(2) The gunner lays the mortar for deflection.
- After the deflection and elevation are indexed on the sight, the gunner floats the elevation bubble.

- He turns the elevating hand crank to elevate or depress the mortar until the bubble in the elevation level vial starts to move. This initially rough lays the mortar for elevation.
- The gunner looks through the sight and traverses to realign on the aiming post. He traverses half the distance to the aiming post, then cross-levels.
- Once the vertical cross line is near the aiming posts (about 20 mils), the gunner checks the elevation vial. If required, he re-lays for elevation by elevating or depressing the elevating mechanism. He makes final adjustments using the traversing handwheel and cross-levels by traversing half the distance and cross-leveling.
- When the vertical cross line is within 2 mils of the aiming posts, all bubbles are leveled, and the sight is set on a given deflection, the mortar is laid.

NOTE: If the given deflection exceeds left or right traverse, the gunner may choose to use the traversing extension assembly. This gives him an added number of mils in additional traverse to avoid moving the bipod. To use the traversing extension, the gunner pulls down on the traversing extension locking knob and shifts the cannon left or right. He re-locks the traversing extension locking knob by ensuring it is securely seated.

PERFORMING LARGE DEFLECTION AND ELEVATION CHANGES ON A GROUND-MOUNTED 120-mm MORTAR

5-28. The gunner receives deflection and elevation changes from the FDC in the form of a fire command. If a deflection change is required, it will always precede the elevation change. The gunner lays the mortar for large deflection and elevation changes.

NOTE: Large deflection and elevation changes are greater than 200 mils but less than 300 mils for deflection and greater than 100 mils but less than 200 mils for elevation.

(1) The gunner receives a deflection and elevation change in the form of an initial fire command.

NOTE: All elements of the fire command are repeated by the gun squad.

(2) As soon as the gunner receives the data, he places it on the sight and elevates or depresses the mortar to float the elevation bubble.

(3) The assistant gunner positions himself in front of the bipod. He squats slightly with his legs spread shoulder-width apart and supports his elbows on his knees. He grabs the bipod legs and lifts them until the bipod clears the ground.

(4) The gunner moves the mortar by placing his right hand over the clamp handle assembly and his left hand on the bipod leg. He pushes or pulls the bipod in the direction desired until the vertical cross line is within 20 mils of the aiming posts. Once completed, the gunner directs the assistant gunner to lower the bipod. He then floats the deflection bubble and looks into the sight to see if he is within 20 mils of his aiming posts.

(5) The gunner and assistant gunner level the mortar for elevation. If after leveling the mortar for elevation, the vertical cross line of the M67 sight is within 20 mils of the aiming post, he would then center the deflection bubble and take up the proper sight picture by traversing half the distance to the aiming posts and cross-leveling.

(6) The assistant gunner observes the gunner traversing to ensure that he stays within four turns of center-of-traverse. Should the gunner traverse away from center-of-traverse, the assistant gunner advises and instructs the gunner to center back up. The gunner center traverses, and with the help of the assistant gunner, he shifts the bipod again and repeats steps (3) through (5).

Chapter 5

> NOTE: After leveling the mortar, if the vertical cross line of the M67 sight is not within 20 mils of the aiming post, then steps (3) through (5) are repeated.

(7) The gunner makes minor adjustments as necessary and does a final check of the bubbles and center-of-traverse, and announces, "Up."

MALFUNCTIONS ON A GROUND-MOUNTED 120-mm MORTAR DURING PEACETIME

5-29. See paragraphs 3-37 to 3-40 of Chapter 3 for a detailed discussion of malfunctions.

REFERRING OF THE SIGHT AND REALIGNMENT OF AIMING POSTS DURING PEACETIME

5-30. Referring and realigning aiming posts ensures that all mortars are set on the same data. The FDC has one deflection instead of two or more. During peacetime operations, the squad uses the following procedures to refer the sight and realign the aiming posts.

(1) The mortar is mounted and the sight is installed. The sight is laid on two aiming posts (placed out 50 and 100 meters from the mortar) on a referred deflection of 2800 mils and an elevation of 1100 mils. The mortar is within two turns of center of traverse. The gunner is given an administrative command to lay the mortar on a deflection of 2860 or 2740 mils. The mortar is then re-laid on the aiming posts using the traversing crank.

(2) The gunner is given a deflection change between 5 and 25 mils, either increasing or decreasing from the last stated deflection, and the command to refer and realign aiming posts.

REMOVAL OF A MISFIRE ON A GROUND-MOUNTED 120-mm MORTAR

5-31. The following procedures are used when a misfire occurs while using the ground-mounted mortar.

> NOTE: The squad leader immediately alerts the FDC of the misfire. The squad leader must supervise the removal of the misfire using a printed copy of the current misfire procedures.

REMOVING A MISFIRED CARTRIDGE USING A CARTRIDGE EXTRACTOR

5-32. The following steps are taken from the time a misfire is identified to the time the misfired cartridge is extracted using a cartridge extractor.

(1) When a misfire occurs, all crewmembers shout, "MISFIRE!"

> **WARNING**
>
> During peacetime live-fire training, all personnel, except the gunner, move 100 meters or farther to the rear of the mortar.

(2) The gunner ensures that the cross-leveling locking knob is as tight as possible, stands to the left rear of the mortar, and kicks the barrel with the heel of his boot in an attempt to dislodge the round. If the round fires, the mortar is swabbed and re-laid on the aiming point, and the firing mission continues.

> **WARNING**
>
> During peacetime live-fire training, if the round does not fire after the gunner kicks the barrel, the gunner joins the crew and waits one minute to avoid personal injury due to cookoff. After a minute, the gunner returns to the mortar.

(3) The gunner checks the cannon for heat by touching it with his fingertips, starting just below the muzzle and working down to the base. If the cannon is too hot to be handled, he cools it with water or snow and checks it one minute later. If no water or snow is available, the cannon is air-cooled until it can be easily handled with bare hands.

NOTES: 1. Liquids must never be poured into the cannon.

2. During peacetime live-fire training, the gunner signals the squad to come forward once the cannon is cool.

(4) When the mortar is cool enough to handle, the gunner removes the firing pin by turning the firing pin wrench counterclockwise. If necessary to provide easier access to the firing pin, the gunner depresses the cannon until the firing pin can be completely removed from the breech cap.

NOTE: Removing the firing pin ensures that the mortar will not fire should the round slip down the cannon during the subsequent drill.

(5) After removing the firing pin, the gunner shouts, "Firing pin is removed," and hands the firing pin to the squad leader. If the mortar cannon is equipped with the old style of breech cap, the gunner places the safety mechanism on SAFE ("S" showing) and shouts, "The safety mechanism is in the SAFE position." The squad leader confirms that the gunner has performed the correct actions, and misfire procedures continue. The gunner then depresses the barrel to its lowest elevation and backs off a half turn. The squad leader confirms that the firing pin has been removed.

> **WARNING**
>
> When depressing the elevation using the elevating handwheel, ensure that no metal-to-metal contact is made.

> **WARNINGS**
>
> 1. Keep your head and body away from the front of the mortar when removing a misfire.
>
> 2. Do not stand directly behind the mortar when removing a misfire.
>
> 3. Do not open the buffer housing assembly when removing a misfire, except when using the barrel tip method.

Chapter 5

(6) When the M67 sight is installed, the gunner locks the deflection/elevation on the sight unit using the locking knobs and places the sightunit in the carrying case. The squad leader confirms that the gunner's actions were performed correctly.

(7) The assistant gunner inspects the catches on the cartridge extractor to ensure that they are in the latest configuration (as indicated by a 1/8 inch [0.32 cm] hole in the face of the catch) and to be sure that the catches are free of burrs, wear, rust, or corrosion that would impair proper function. Then, the assistant gunner tests each catch to ensure free operation and that each catch will snap into its original position. If the cartridge extractor fails to meet inspection standards, the cartridge extractor is not mission capable and must not be used. The squad leader confirms that the assistant gunner's actions were performed correctly.

NOTE: If the cartridge extractor is not mission capable and there are no other cartridge extractors available, or if there are no functional artillery cleaning staffs available, use the barrel tip method to remove the misfired cartridge (see paragraphs 5-33 to 5-35).

(8) The assistant gunner ensures that the artillery cleaning staff assembly section sleeves are fully extended and tightly locked so that the staff assembly will not extend or retract. If the section sleeves can't be tightly locked, the cartridge extractor assembly is not mission capable and must not be used. The squad leader confirms that the assistant gunner's actions were performed correctly.

(9) The assistant gunner attaches the cartridge extractor securely to the extended artillery cleaning staff assembly. The squad leader confirms that the assistant gunner's actions were performed correctly.

WARNINGS

1. **To avoid serious injury, do not stand in front of or behind the barrel.**

2. **When removing the cartridge, do not stand directly in front of the barrel.**

---

(10) The assistant gunner rotates the artillery cleaning staff assembly until the cartridge extractor is secure against it (Figure 5-7). Then, he inserts the cartridge extractor into the barrel slowly (hand to hand) until it rests on the cartridge and will descend no further.

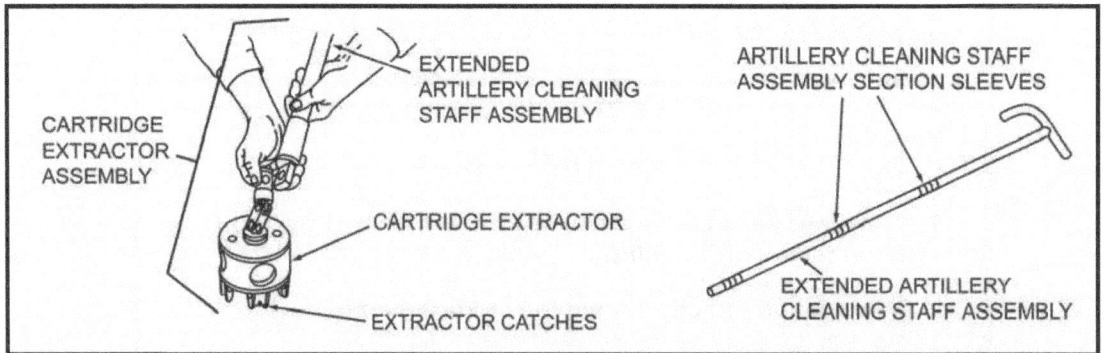

Figure 5-7. Rotating the artillery cleaning staff.

(11) The assistant gunner slowly rotates the cartridge extractor until it grasps the cartridge. As the cartridge extractor grasps the cartridge, the assistant gunner feels the spring-loaded catches click into the holes on the cartridge body (Figure 5-8). The assistant gunner continues to rotate the cartridge extractor an additional quarter of a turn.

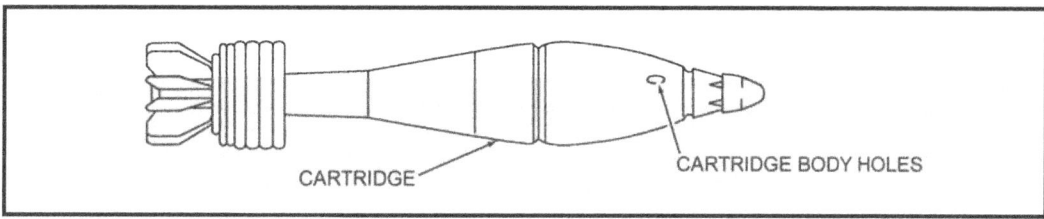

Figure 5-8. Holes in the cartridge body.

NOTE: The assistant gunner should feel much more resistance to rotating the extractor. If not, the cartridge extractor has not successfully connected to the cartridge, and the assistant gunner must continue to slowly rotate the cartridge extractor until it grasps the cartridge. If the assistant gunner still feels no extra resistance, he removes the extractor assembly from the barrel and uses the barrel tip method to remove the misfired cartridge (see paragraphs 5-33 to 5-35).

(12) When the assistant gunner feels resistance, he slightly relaxes his grip on the artillery cleaning staff assembly for a moment. Then, without rotating, he slowly pulls the staff assembly slightly to ensure that the cartridge is grasped. If the assistant gunner feels resistance, he, with the gunner's hands at the muzzle, withdraws the cartridge from the barrel with the cartridge extractor, pulling the artillery staff with both hands (Figure 5-9). When the assistant gunner pulls the staff as high as he is able, he slides his lower hand down the staff and then slides his upper hand to his lower hand (hand to hand), maintaining positive control at all times. He repeats this action until the cartridge is exposed.

NOTE: If the assistant gunner can grasp the cartridge but can't withdraw it, he uses the barrel tip method (see paragraphs 5-33 to 5-35). If the assistant gunner feels no resistance, he removes the cartridge extractor assembly and uses the barrel tip method (see paragraphs 5-33 to 5-35).

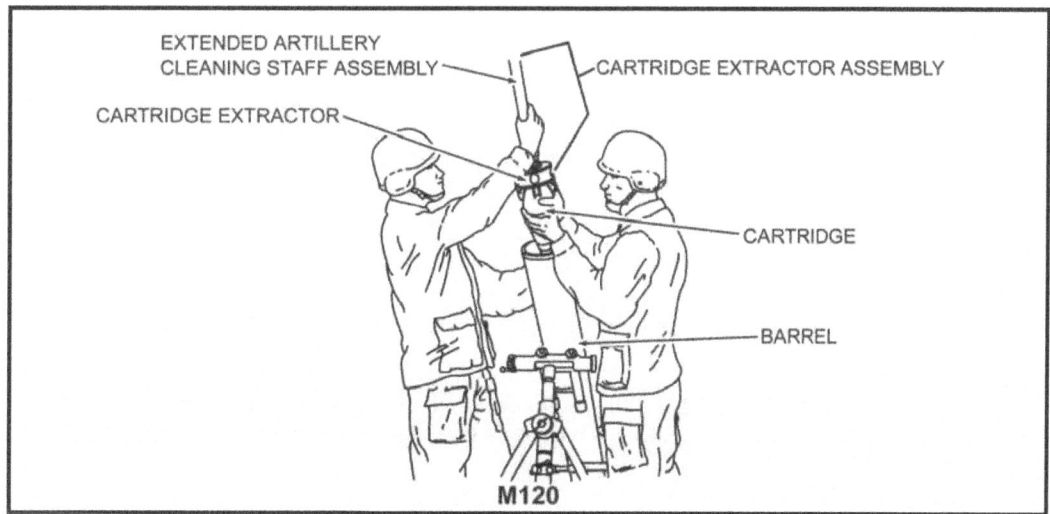

Figure 5-9. Withdrawing the cartridge from the barrel, M120.

Chapter 5

> **WARNING**
>
> 1. When removing the cartridge, do not touch the primer.
>
> 2. Care must be taken to ensure that the extractor catches are not depressed while removing the round from the barrel.

NOTE: Due to the weight of the round, the gunner may assist during the entire extraction process.

(13) The gunner grasps the body of the cartridge as it comes out of the barrel. The gunner and assistant gunner secure the artillery staff, extractor, and misfired cartridge. The ammunition bearer assists in releasing the cartridge from the extractor by pressing all four extractor catches at the same time (Figure 5-10).

NOTE: Do not place the cartridge on the ground when removing the cartridge extractor.

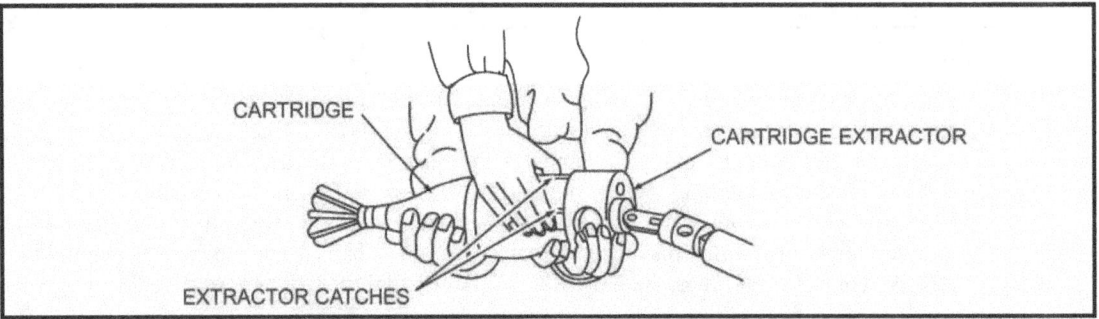

Figure 5-10. Pressing the extractor catches.

(14) The gunner inspects the cartridge to see if the primer has been dented and gives the cartridge to the ammunition bearer. The ammunition bearer attempts to replace the safety wire (if applicable), places the cartridge in the dud pit, and tags the cartridge. The safety officer notifies EOD.

NOTE: If the primer has been dented, perform steps (4) through (8) of paragraph 5-35 before continuing with step (17). If the primer hasn't been dented, proceed to the next step. If two consecutive misfires occur and the primer has not been dented, perform steps (4) through (8) of paragraph 5-35.

(15) The assistant gunner swabs the barrel; the gunner inserts the firing pin into the breech cap and rotates the gunner's display into position; and the mission is continued.

REMOVING A MISFIRED CARTRIDGE USING THE BARREL TIP METHOD

5-33. The barrel tip method is used as circumstances dictate. Follow the steps outlined in paragraph 5-34 only if the extractor assembly is unserviceable or if extractor methods have failed. The steps outlined in paragraph 5-35 apply when the round is grasped by the cartridge extractor, but can't be removed.

5-34. Follow these steps to remove a misfired cartridge if the extractor assembly is unserviceable or if extractor methods have failed.

(1) The gunner and assistant gunner carefully lower the bipod assembly into the lowest position in low range.

(2) The assistant gunner supports the bipod assembly, with his left hand grasping the left side of the traversing mechanism and his right hand grasping the right side of the traversing mechanism.

(3) The gunner cradles the barrel with his right arm near the muzzle.

(4) The ammunition bearer cradles the barrel with his right arm above the buffer housing assembly and unlocks the clamp handle assembly with his left hand, releasing the buffer housing assembly.

> **WARNING**
>
> Once the cannon is horizontal, the rear of the cannon must not be lowered back down until the round is extracted. If the round slips down the cannon before extraction, it could ignite, causing death or personal injury.

(5) The gunner and ammunition bearer lift (approximately 60 degrees) and then rotate the barrel so that the white line is in the down position and the breech cap can be removed from the socket. Then, they carefully remove the barrel from the breech cap socket and raise it to the horizontal position.

(6) Keeping the barrel horizontal and pointed in the direction of fire, the gunner, ammunition bearer, and crewmembers hold the barrel.

(7) The assistant gunner places the meaty portions of his thumbs over the edges of the muzzle, grasping the barrel with his fingers.

> **WARNING**
>
> 1. When removing the cartridge, do not touch the primer, and do not stand directly in front of the barrel.
>
> 2. Care must be taken to ensure that the extractor catches are not depressed while removing the round from the barrel.

(8) At the assistant gunner's command, crewmembers lift the cannon's breech cap assembly, causing the cartridge to slide down to the assistant gunner's hands.

(9) The assistant gunner removes the cartridge, inspects the cartridge to see if the primer has been dented, attempts to replace the safety wire (if applicable), places the cartridge in the dud pit, and tags the cartridge. The safety officer notifies EOD.

NOTE: If the primer has been dented, perform steps (4) through (8) of 5-35 before continuing with step (10). If the primer has not been dented, proceed to the next step.

(10) The assistant gunner swabs the bore; the gunner inserts the firing pin into the breech cap; and the barrel is returned to action.

5-35. Follow these steps if the round is grasped by the cartridge extractor, but cannot be removed.

(1) The assistant gunner loosens the sleeve on the artillery cleaning staff assembly, depresses the ball-bearing spring lock, slowly lowers the upper staff section into the lower staff section, and tightens the sleeve.

(2) Perform steps (1) through (8) of paragraph 5-34.

(3) Keeping the barrel horizontal and pointed in the direction of fire, the gunner, assistant gunner, and crewmembers carry the barrel (with the extractor still attached to the round) to the dud pit. The safety officer notifies EOD.

NOTE: The following steps are part of the breech cap removal. These steps are used before firing, during firing, if two or more misfires occur, and after firing is completed. If the primer has been dented and time permits, perform steps (4) through (8).

(4) The assistant gunner and ammunition bearer place the barrel on two empty ammunition boxes and stabilize the barrel. The gunner inserts the breech cap removal tool into the cross bore of the breech cap and taps the end of the tool with a hammer to turn it clockwise (Figure 5-11). Then, he unscrews and removes the breech cap assembly from the barrel, and wipes away any debris from the inner part of the breech cap.

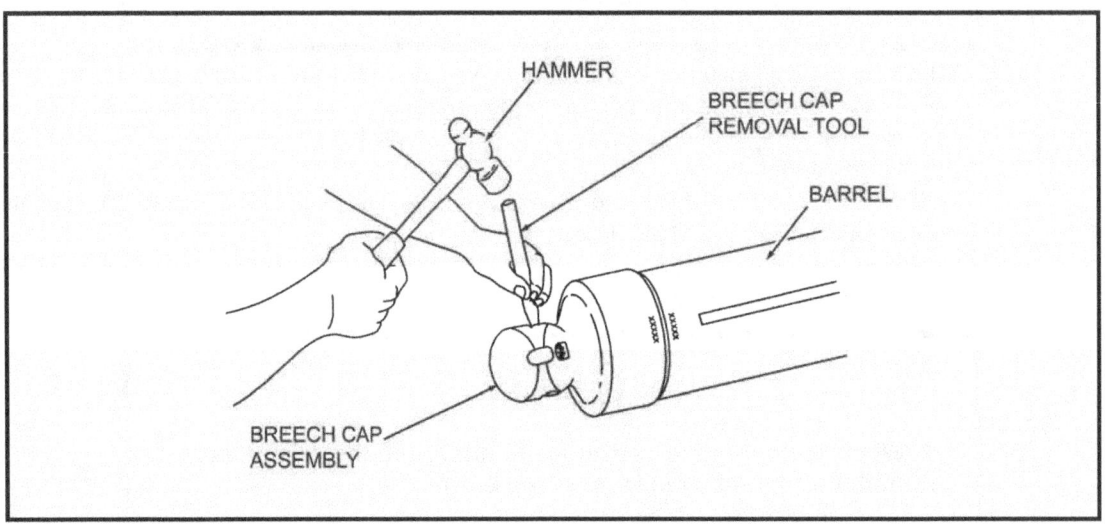

Figure 5-11. Removing the breech cap assembly from the barrel.

(5) The gunner inspects the firing pin housing on the breech cap assembly and removes any dirt or debris. The gunner replaces the firing pin by hand, ensuring that the threads are aligned; turns the firing pin clockwise until it is hand tight; and then tightens it further using the firing pin wrench.

(6) Using the firing pin gauge, the gunner measures the protrusion of the firing pin. The firing pin must protrude above the MIN and below the MAX cutouts. The gunner replaces the firing pin if it is defective.

(7) Using a wiping rag, the gunner cleans the inner threads of the breech cap assembly and outer threads of the barrel. The gunner applies a light coat of grease to the threads of the breech cap assembly and installs the breech cap on the barrel. He tightens the breech cap assembly using the breech cap removal tool found in the breech cap bore and backs off a quarter of a turn using a jerking motion to tighten.

(8) The barrel is returned to action.

NOTE: The white stripe on the barrel must align with the firing pin.

(9) Both the squad leader and section leader confirm that the crew performed the actions correctly.

(10) The assistant gunner swabs the barrel; the gunner rotates the gunner's display into position; and the mission is continued.

LOADING AND FIRING OF THE GROUND-MOUNTED 120-mm MORTAR

5-36. This paragraph explains the procedures for loading and firing a ground-mounted 120-mm mortar.

(1) The FDC issues a fire command to the squad leader.

(2) The squad leader records and issues the fire command to the squad.

(3) The squad repeats the fire command.

(4) The ammunition bearer prepares the round in accordance with the fire command.

(5) The squad leader inspects the round before it is passed to the assistant gunner. The ammunition bearer holds the round with both hands (palms up) near each end of the round body (not on the fuze or the charges).

(6) The assistant gunner checks the round for correct charges, fuze tightness, and fuze setting.

(7) When both the gun and the round(s) have been determined safe and ready to fire, the squad leader gives the following command to the FDC: NUMBER (NUMBER OF MORTAR) GUN, UP.

(8) The ammunition bearer holds the round with the fuze pointed to his left. By pivoting his body to the left, the assistant gunner accepts the round from the ammunition bearer with his right hand under the round and his left hand on top of the round.

(9) Once the assistant gunner has the round, he keeps two hands on it until it is fired.

> **CAUTION**
> The assistant gunner is the only member of the mortar squad who loads and fires the round.

(10) The squad leader commands, HANG IT, FIRE in accordance with the method of fire given by the FDC.

(11) The assistant gunner holds the round in front of the muzzle at about the same angle as the cannon. At the command, HANG IT, the assistant gunner guides the round into the cannon (tail end first) to a point beyond the narrow portion of the body (about three-quarters of the round) being careful not to hit the primer or charges or disturb the lay of the mortar.

(12) Once the round is inserted into the cannon the proper distance, the assistant gunner shouts, "Number (number of mortar) gun, hanging."

(13) At the command, FIRE, the assistant gunner releases the round by pulling both hands down and away from the outside of the cannon. The assistant gunner ensures that he does not take his hands across the muzzle of the cannon as he drops the round.

(14) Once the round is released, the gunner and assistant gunner take a full step toward the rear of the weapon while pivoting their bodies so that they are both facing away from the blast.

(15) The assistant gunner pivots to his left and down toward the ammunition bearer, ready to accept the next round to be fired (unless major movements of the bipod require him to lift and clear the bipod off the ground).

(16) Subsequent rounds are fired based on the FDC fire commands.

(17) The assistant gunner ensures the round has fired safely before he attempts to load the next round.

(18) The assistant gunner does not shove or push the round down the cannon. The round slides down the cannon under its own weight, strikes the firing pin, ignites, and fires.

(19) The assistant gunner and gunner, as well as the remainder of the mortar crew, keep their upper body below the muzzle until the round fires to avoid muzzle blast.

(20) During an FFE, the gunner tries to level all bubbles between each round ensuring his upper body is away from the mortar and below the muzzle when the assistant gunner announces, "Hanging," for each round fired.

(21) The assistant gunner informs the squad leader when all rounds for the fire mission are expended, and the squad leader informs the FDC when all of the rounds are completed. For example, "NUMBER TWO GUN, ALL ROUNDS COMPLETE."

TAKING THE 120-mm MORTAR OUT OF ACTION

5-37. To take the 120-mm mortar out of action, the squad leader commands, OUT OF ACTION. Then, each member of the squad does the following:

(1) The ammunition bearer retrieves the aiming posts, places them in their case, and puts the case on the right side of the mortar.

(2) The gunner places a deflection of 3800 mils and an elevation of 0800 mils on the sight.

(3) The gunner removes the sight from the dovetail slot and places it in the sight case. He places the sight mount cover back on and secures it with the snap button.

NOTE: Left and right are in relation to the mortar's direction of fire.

(4) The gunner loosens the clamp handle assembly.

(5) The gunner slides the buffer housing assembly up the cannon until the white line on the clamp handle assembly is aligned with the white line on the cannon near the muzzle. The gunner tightens the buffer housing assembly until he hears a metallic click. He continues turning the handle until it is parallel to the cannon.

(6) The gunner centers the traversing extension and the traversing mechanism.

(7) The gunner lowers the elevation so that about four fingers (three to four inches) of the elevating mechanism remains exposed.

(8) The assistant gunner loosens the cross-leveling locking knob, while the gunner steadies the cannon.

(9) The ammunition bearer positions himself behind the trailer. He positions the trailer behind the baseplate with the towing eye almost straight up. The assistant gunner helps after he retrieves the aiming posts.

(10) The ammunition bearer holds the trailer in place with the cannon cradle touching the top of the breech cap. He blocks the wheels so that the trailer cannot roll.

(11) The assistant gunner releases the clamping catch on the trailer bridge assembly and swings the bridge assembly out of the way.

(12) The gunner and the assistant gunner grasp the traversing mechanism and bipod legs, and then swing the cannon and bipod over the baseplate and onto the trailer. The assistant gunner loosens the cross-leveling locking knob and assists in raising the bipod legs. Then, he joins the legs and rotates them 180 degrees.

(13) With the trailer still in the upright position, the assistant gunner makes sure that the upper collar stop is placed just forward of the upper cannon bracket on the trailer.

(14) The assistant gunner holds and lifts the baseplate onto the trailer as the ammunition bearer returns the trailer to the towing position, lifting the entire mortar onto the trailer.

(15) The gunner attaches the hook on the spread cable to the eye on the bipod leg. He adjusts the elevation so that the bipod legs fit under the trailer bridge assembly. Then, he retightens the locking knob.

(16) The assistant gunner closes the trailer bridge assembly. He locks the baseplate with the lock release lever and secures it with the pin. He then secures the trailer bridge assembly with the clamping catch.

(17) The ammunition bearer emplaces the muzzle cap.

(18) The ammunition bearer moves to the driver's side of the towing eye. The assistant gunner moves to the passenger's side of the towing eye. Together, they move the trailer to the vehicle and hook the trailer to the towing pintle of the vehicle. They hook up the trailer light cable and safety chains.

(19) The gunner secures the sight case.

(20) The assistant gunner secures the aiming post case.

(21) The squad leader ensures that all equipment is accounted for and properly secured.

(22) The squad leader announces, "Number two gun, up."

SECTION IV. MORTAR CARRIER, M1064A3

This section is a guide for training mortar units equipped with the M1064A3-series mortar carrier for mounting the M121 mortar. The procedures and techniques used for a mounted mortar are different from the ground-mounted mortar.

DESCRIPTION

5-38. The M1064A3 carrier (Figure 5-12 and Figure 5-13) is an M113A3 armored personnel carrier modified to carry the 120-mm mortar, M121, on a specially designed mount. It is fully tracked, highly mobile, and armor protected. It can be transported by air and is able to propel itself across water obstacles.

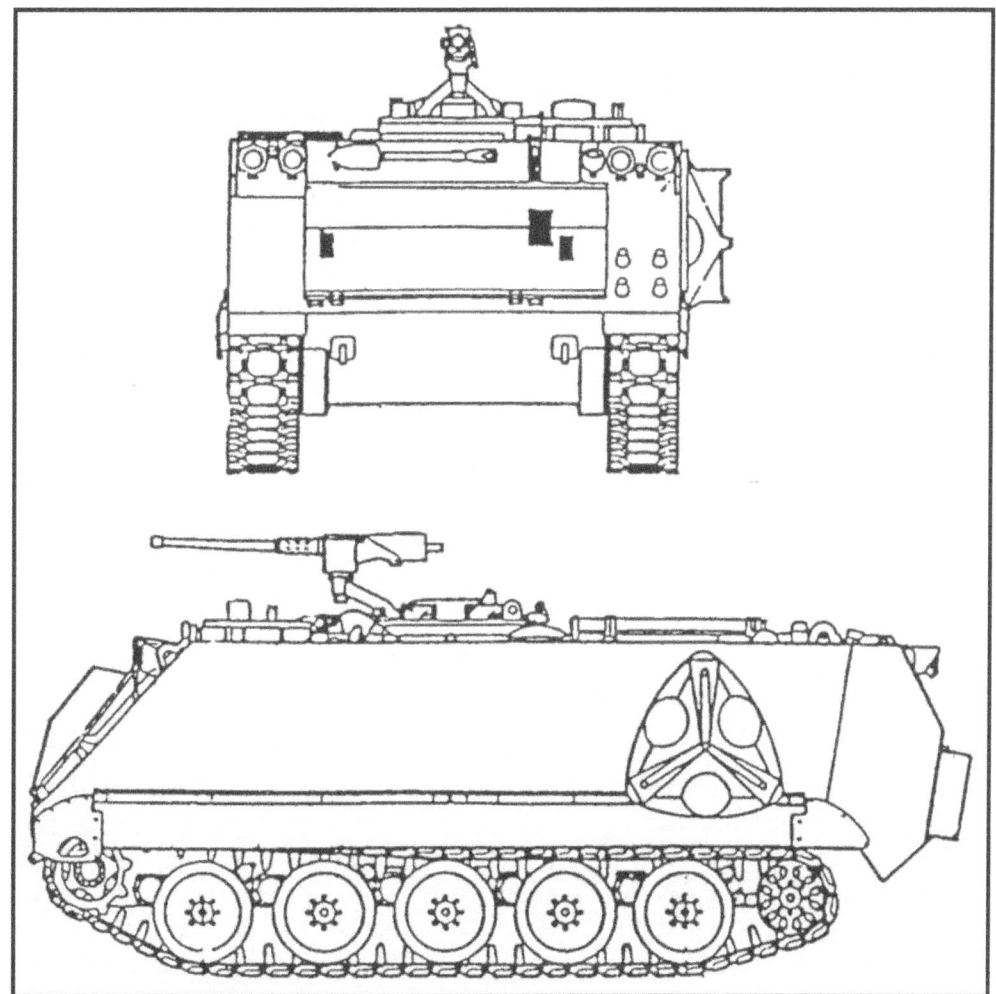

Figure 5-12. Mortar carrier, M1064A3, front and side view.

Chapter 5

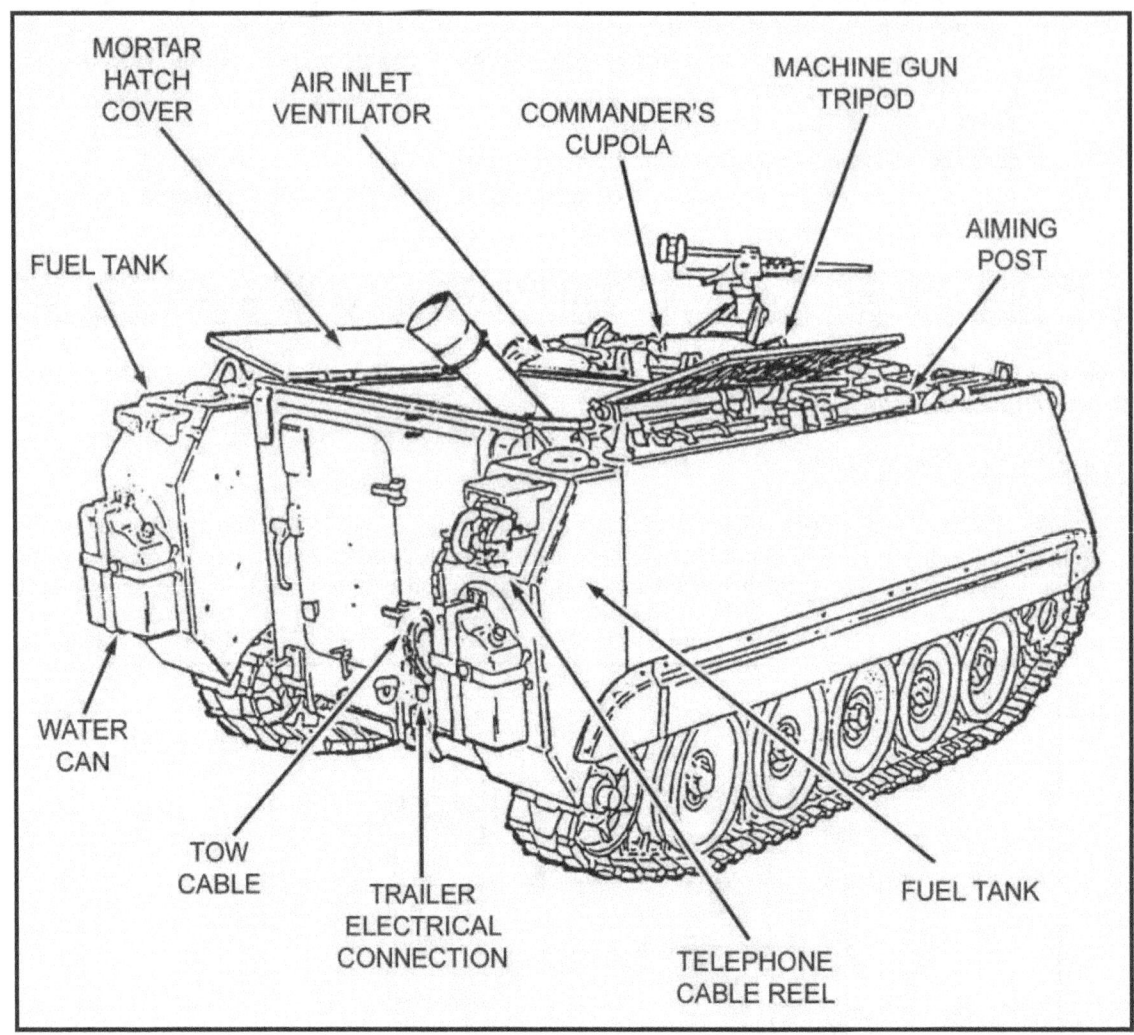

Figure 5-13. Mortar carrier, M1064A3, rear view.

TABULATED DATA FOR THE M1064A3 CARRIER

5-39. Table 5-2 shows the tabulated data for the M1064A3 carrier and the mortar capabilities when mounted on a turntable.

Table 5-2. Tabulated data for the mortar carrier, M1064A3.

GENERAL
Weight:
    Combat loaded   28,240 pounds
    Air transportable   23,360 pounds
Ground pressure, combat loaded   8.97 psi
Personnel capacity   4
Fuel tank capacity   95 gallons

PERFORMANCE
Speed:
    On land   40 mph
    In water, with track   3.6 mph
Cruising range   300 miles
Turning radius   Pivot to infinite
Slope   60 percent
Side slope   40 percent
Trench crossing   66 inches
Vertical wall climbing   24 inches
Gross horsepower-to-weight ratio   19.6 hp/ton

ARMAMENT
.50 caliber machine gun   2,000 rounds
120-mm mortar   69 ready rounds (45 horizontal, 24 vertical). Type of rounds carried depends on the tactical mission.

MORTAR CAPABILITIES MOUNTED ON TURNTABLE

TRAVERSING LIMITS
Right of center with traverse extension   858 mils
Left of center with traverse extension   808 mils

TOTAL TRAVERSE
Capability from extreme left to extreme right
    Without traversing extension   1486 mils
    With traversing extension   1666 mils

ELEVATION LIMITS (LEVEL)
Maximum   1510 mils
Minimum   0750 mils

MORTAR AND VEHICULAR MOUNT

5-40. The mortar is mounted on its vehicular mount, and a clamping support is provided to hold the mortar during travel. A bipod support assembly provides attachment mounting for the M191 bipod and secures it in a locking position during travel. The breech socket provides a base in which the weapon rests. The mortar is provided with a sight extension arm assembly, which is received by the socket of the coupling and sight mount assembly. The gunner uses the sight extension arm to sight on his aiming point above the hull of the vehicle. The extension must be removed before moving the mortar to prevent wear on the sight mount's coupling gears.

MAINTENANCE

5-41. Care and cleaning of the mortar, instruments, and equipment are the duty and responsibility of the mortar squad. Care and cleaning of the carrier-mounted mortar are the same as for the ground-mounted mortar. All maintenance records and lubricating procedures for the 120-mm mortar and the mount, M191, can be found in TM 9-1015-250-10. (For maintenance procedures for the M1064A3 carrier, see TM 9-2350-277-10.)

SECTION V. OPERATION OF A CARRIER-MOUNTED 120-mm MORTAR

The mortar section is the fire unit for the mortar platoon. When a position is occupied, mortars are emplaced 75 meters apart making the section front (distance between flank mortars) about 150 meters. The mortars are numbered 1, 2, and 3 (from right to left) when facing the direction of fire. The squad of the mortar carrier consists of four members—squad leader, gunner, assistant gunner, and ammunition bearer/driver.

PREMOUNT CHECKS

5-42. The premount checks for the carrier-mounted mortar are the same as in paragraphs 5-16 to 5-22.

PLACING A CARRIER-MOUNTED 120-mm MORTAR INTO ACTION

5-43. To place the mortar into action, the crew performs the following actions upon the squad leader's command, ACTION.

> NOTE: If the weapon system cannot be leveled at any elevation given, the gunner is authorized to go to high range (low range) and move the buffer housing assembly along the cannon between the circular white line and the lower collar stop without moving past the vertical position until the gun system can be leveled.

(1) The gunner pulls the chain on the center cargo hatch releasing and folding it over onto the right hatch ensuring that the hatch locks into place. He then pulls the chain on the right cargo hatch releasing and folding it over and secures it.

(2) Once the gunner has the center and right cargo hatch secured, the assistant gunner pulls the chain on the left cargo hatch releasing and folding the hatch over ensuring that it is locked into place.

(3) The assistant gunner unlocks the clamping support assembly.

(4) The gunner loosens the clamp handle assembly and grasps the buffer housing assembly and pulls it down until it is flush with the lower collar stop.

(5) The gunner ensures that the white line on the cannon aligns with the white line on the buffer housing assembly. He then tightens the clamp handle assembly until he hears a metallic click. If the mortar cannon is equipped with the old style of breech cap, the gunner places the weapon on FIRE ("F" showing).

(6) The assistant gunner then passes the sight extension to the gunner and removes the muzzle cover.

(7) The gunner places the sight extension into the dovetail slot on the bipod and secures it.

(8) The gunner mounts the sight onto the sight extension, places a deflection of 3200 mils and an elevation of 1100 mils onto the sight, and levels the deflection bubble with the cross-leveling handwheel on the bipod.

(9) The assistant gunner places the BAD onto the cannon and secures it and levels the elevation bubble.

(10) The gunner then ensures that the traversing mechanism is within four turns of center, and the traversing extension is centered on the bipod.

(11) When the gunner is satisfied that the mortar is mounted correctly, he announces, "Up" so that the squad leader can inspect the weapon system.

(12) The carrier-mounted 120-mm mortar is now placed into action and ready to be reciprocally laid.

MOUNTING OF THE MORTAR FROM A CARRIER- TO A GROUND-MOUNTED POSITION

5-44. The procedures for placing the mortar into action by mounting it from a carrier- to a ground-mounted position are described herein.

NOTE: Left and right are in relation to the mortar's direction of fire.

(1) The driver lowers the ramp and dismounts from the mortar carrier.

(2) The gunner pulls down on the chain for the center hatch and folds it over, then secures it to the right cargo hatch. He pulls down on the chain for the right cargo hatch and folds both hatches over, then secures them in place. The assistant gunner pulls down on the chain for the left cargo hatch and folds it over, then secures it in place.

(3) The squad leader dismounts the carrier and shows the crew where he wants the mortar to be mounted and indicates the direction of fire.

(4) The driver/ammunition bearer and assistant gunner release the baseplate by removing the safety pin and pushing the handle up. Together they tilt the baseplate out and lift it from the lower brackets.

(5) The driver/ammunition bearer and assistant gunner place the baseplate at the firing position while the gunner retrieves the sight and aiming posts and places them on the left side of the baseplate.

(6) The assistant gunner holds the bipod while the gunner unlocks the clamp handle assembly. The assistant gunner lowers the bipod until it rests on the ramp. At the same time, the driver/ammunition bearer assembles the aiming posts.

(7) The assistant gunner releases the safety pins and rotates the handles until the arrows are facing each other and then pulls the handles out, releasing the bipod.

(8) The assistant gunner installs the bipod leg extensions onto the bipod and secures them with the safety pins.

(9) The assistant gunner and driver/ammunition bearer secure the bipod and carry it to the firing position. They place it about 2 feet in front of the baseplate.

CAUTION
Damage may occur to the turntable socket if the following procedures are not followed.

(10) The driver/ammunition bearer removes the muzzle plug while the gunner releases the clamping support assembly. Together they raise the cannon to a 60-degree angle and rotate it until the white line is facing the turntable. They then lift straight up removing the cannon from the socket.

(11) The gunner and driver/ammunition bearer carry the cannon to the baseplate. With the white line on the cannon facing the ground, they tilt the cannon to a 60-degree angle. They carefully insert the cannon into the baseplate socket. They rotate the cannon until the white line is facing skyward and then lower it onto the buffer housing assembly ensuring the white line on the barrel aligns with the white line on the buffer housing assembly. The gunner then

Chapter 5

slides the buffer housing assembly down the barrel until it is flush with the lower collar stop on the cannon.

(12) The gunner tightens the clamp handle assembly until a metallic click is heard.

(13) If the mortar cannon is equipped with the old style of breech cap, the gunner places the safety mechanism on FIRE ("F" showing).

(14) The gunner places the sight on the weapon and indexes a deflection and elevation.

PERFORMING SAFETY CHECKS ON A CARRIER-MOUNTED 120-mm MORTAR

5-45. Specific safety checks must be performed before firing mortars. Most can be made visually. The gunner is responsible for physically performing the checks under the squad leader's supervision.

NOTE: The buffer housing assembly may need to be adjusted to complete this task.

5-46. To perform safety checks on a carrier-mounted 120-mm mortar, the gunner ensures that—
- There is no mask and overhead clearance.
- The breech assembly is locked in the turntable socket and the white line on the cannon bisects the white line on the buffer housing assembly.
- The M191 bipod assembly is locked to the turntable mount and the arrows on the mount handles are pointed down (vertical).
- The safety pins are installed.
- The buffer housing assembly is secured to the cannon by loosening the clamp handle assembly about a quarter of a turn and tightening the clamp handle assembly until he hears a metallic click.
- The cross-leveling locking knob is hand-tightened.
- The bipod support assembly is locked in the high or low position and the safety pin is installed.
- The turntable is in the locked position.
- The cargo hatches are open and locked.
- The BAD locking knob is hand-tightened.
- The firing pin is properly installed. If the mortar cannon is equipped with the old style of breech cap, the gunner ensures that the safety selector is in the FIRE position ("F" showing).
- The bipod is not forward of the vertical position.

MASK AND OVERHEAD CLEARANCE

5-47. To determine mask clearance, the gunner places an elevation of 0800 on the sight and lowers the cannon until the elevation bubble is level. He places his head against the breech cap and sights along the barrel to see if any obstructions are in front of the mortars. With the gunner's head still against the breech cap, the assistant gunner traverses the mortar through its full range of traverse (traversing extension) to ensure that no obstructions are in front of the mortar.

5-48. To determine overhead clearance, the gunner places an elevation of 1511 on the sight and raises the barrel until the elevation bubble is level. He places his head against the breech cap and sights along the barrel for any obstructions in front of the mortar. With the gunner's head still against the breech cap, the assistant gunner moves the mortar through its full range of traverse (traversing extension) to ensure that no obstructions are in front of the mortar.

> **NOTE:** If obstructions are found at any point in the full range of traverse or elevation, the mortar is not safe to fire. In a combat situation, however, it may be necessary to fire the mortar from that position. If this is the situation, traverse and/or elevate the mortar until it clears the obstruction and level the sight by using the elevation micrometer knob. Record the deflection and elevation where the mortar clears the obstruction and report this information to the FDC.

PERFORMING SMALL DEFLECTION AND ELEVATION CHANGES ON A CARRIER-MOUNTED 120-mm MORTAR

5-49. The gunner receives deflection and elevation changes from the FDC in the form of a fire command. If a deflection change is required, it precedes the elevation change.

> **NOTE:** Small deflection and elevation changes are greater than 20 mils but less than 60 mils for deflection and greater than 30 mils but less than 90 mils for elevation.

(1) The gunner sets the sight for deflection and elevation.
- He places the deflection on the sight by turning the deflection micrometer knob until the correct 100-mil deflection mark is indexed on the coarse deflection scale. He continues to turn the deflection micrometer knob until the remainder of the deflection is indexed on the deflection micrometer scale.
- The gunner places the elevation on the sight by turning the elevation micrometer knob until the correct 100-mil elevation mark is indexed on the coarse elevation scale. He continues to turn the elevation micrometer knob until the remainder of the elevation is indexed on the micrometer scale.

(2) The gunner lays the mortar for deflection.
- After the deflection and elevation are indexed on the sight, the gunner floats the elevation bubble.
- He turns the elevating hand crank to elevate or depress the mortar until the bubble in the elevation level vial starts to move. This initially rough lays the mortar for elevation.
- The gunner looks through the sight and traverses to realign on the aiming post. He traverses half the distance to the aiming post, then cross-levels.
- Once the vertical cross line is near the aiming posts (about 20 mils), the gunner checks the elevation vial. If required, he re-lays for elevation by elevating or depressing the elevating mechanism. He makes final adjustments using the traversing handwheel and cross-levels by traversing half the distance and cross-leveling.
- When the vertical cross line is within 2 mils of the aiming posts, all bubbles are leveled, and the sight is set on a given deflection, the mortar is laid.

> **NOTE:** If the given deflection exceeds left or right traverse, the gunner may choose to use the traversing extension assembly. This gives him an added number of mils in additional traverse to avoid moving the bipod. To use the traversing extension, the gunner pulls down on the traversing extension locking knob and shifts the cannon left or right. He re-locks the traversing extension locking knob by ensuring it is securely seated.

PERFORMING LARGE DEFLECTION AND ELEVATION CHANGES ON A CARRIER-MOUNTED 120-mm MORTAR

5-50. The squad receives a fire command requiring a deflection change of more than 200 mils but less than 300 mils and an elevation change of more than 100 mils but less than 200 mils. The gunner must announce, "Gun up," within 55 seconds. Time starts when the last digit of the elevation is given. The given deflection and elevation must be indexed, without error. The bubbles are centered (within the outer red lines). The vertical crossline is within ± 2 mils of an aligned or compensated sight picture. The traversing mechanism is within four turns of center traverse. The traversing extension is locked in the center position. The gunner receives deflection and elevation changes from the FDC in the form of a fire command. If a deflection change is required, it precedes the elevation change.

    (1) Upon receiving the fire command, the gunner places the data on the sight and elevates or depresses the mortar to "float" the elevation bubble.

- If in low range—
  - Gunner unlocks the low range support latch.
  - Gunner places his left hand on the bipod leg and his right hand on the traversing mechanism.
  - Assistant gunner places his right hand on the bipod leg and his left hand on the traversing mechanism
  - Gunner and assistant gunner together raise the bipod into high range
  - The assistant gunner inserts the high range-locking pin to secure the bipod.
- If in high range—
  - Gunner and assistant gunner together lower the bipod into low range
  - Gunner pulls the high range-locking pin
  - Gunner places his left hand on the bipod leg and his right hand on the traversing mechanism.
  - Assistant gunner places his right hand on the bipod leg and his left hand on the traversing mechanism.

    (2) The assistant gunner "dead levels" the elevation bubble.

    (3) The gunner looks into his sight to determine if he has to shift the mortar and announces to the assistant gunner to unlock the turntable.

    (4) The gunner looks through his sight and, with the assistant gunner's help, shifts the mortar until the vertical cross line is aligned with the aiming posts.

    (5) The gunner cross-levels the deflection bubble and looks into the sight. If the vertical cross line is still within 20 mils of the aiming posts, the gunner tells the assistant gunner to "lock it" (referring to locking the turntable).

    (6) With the turntable in the locked position, the gunner continues to traverse and cross-level until the vertical cross line is lined up with the aiming posts and the deflection bubble is level.

    (7) The gunner levels his elevation bubble and rechecks his data, bubbles and four turns of center.

    (8) Gunner announces "GUN UP."

REMOVAL OF A MISFIRE ON A CARRIER-MOUNTED 120-mm MORTAR

5-51. The following procedures are used when a misfire occurs while using the Mortar Fire Control System (MFCS).

NOTE: The squad leader immediately alerts the FDC of the misfire. The squad leader supervises the removal of the misfire using a printed copy of the current misfire procedures.

REMOVING A MISFIRED CARTRIDGE USING A CARTRIDGE EXTRACTOR

5-52. The following steps are taken from the time a misfire is identified to the time the misfired cartridge is extracted using a cartridge extractor.

    (1) When a misfire occurs, all crewmembers shout, "MISFIRE!"

> **WARNING**
>
> During peacetime live-fire training, all personnel, except the gunner, move 100 meters or farther to the rear of the mortar.

> **WARNING**
>
> Never change from high-range elevation to low-range elevation to perform misfire procedures.

(2) The gunner ensures that the cross-leveling locking knob is as tight as possible, stands to the left rear of the mortar, and kicks the barrel above the mounting bracket for the pointing device several times with the heel of his boot in an attempt to dislodge the round. If the round fires, the mortar is swabbed and re-laid on the aiming point, and the firing mission continues.

NOTE: Avoid kicking the pointing device and its mounting bracket.

> **WARNING**
>
> During peacetime live-fire training, if the round does not fire after the gunner kicks the barrel, the gunner joins the crew and waits one minute to avoid personal injury due to cookoff. After one minute, the gunner returns to the mortar.

(3) The gunner checks for heat by starting from just below the muzzle and working down to the base with his fingertips. If the cannon is too hot to be handled, he cools it with water or snow and checks it one minute later. If no water or snow is available, the cannon is air-cooled until it can be easily handled with bare hands.

NOTES: 1. Liquids must never be poured into the cannon.

2. During peacetime live-fire training, the gunner signals the squad to come forward once the cannon is cool. All crewmembers enter from the front of the vehicle.

(4) When the mortar is cool enough to handle, the gunner removes the firing pin by turning the firing pin wrench counterclockwise. If necessary to provide easier access to the firing pin, the gunner depresses the cannon until the firing pin can be completely removed from the breech cap. After removing the firing pin, the gunner shouts, "Firing pin is removed," and hands the firing pin to the squad leader. If the mortar cannon is equipped with the old style of breech cap, the gunner places the safety mechanism on SAFE ("S" showing) and shouts, "The safety mechanism is in the SAFE position." The squad leader confirms that the gunner has performed the correct actions, and misfire procedures continue. The gunner then depresses the barrel to its lowest elevation, leaving 0.25 inches (0.64 cm) of inner elevating sleeve showing. The squad leader confirms that the firing pin has been removed.

NOTE: Removing the firing pin ensures that the mortar will not fire should the round slip down the cannon during the subsequent drill.

Chapter 5

> **WARNING**
>
> When depressing the elevation using the elevating handwheel, ensure that no metal-to-metal contact is made.

> **WARNINGS**
>
> 1. Keep your head and body away from the front of the mortar when removing a misfire.
>
> 2. Do not stand directly behind the mortar when removing a misfire.
>
> 3. Do not open the buffer housing assembly when removing a misfire, except when using the barrel tip method.

(5) The gunner rotates the gunner's display until it is out of the way. If the carrier ramp is raised, the driver lowers it.

(6) When the M67 sight is installed, the gunner locks the deflection/elevation on the sightunit using the locking knobs and places the sightunit in the carrying case. The squad leader confirms that the gunner's actions were performed correctly.

(7) The assistant gunner removes the BAD from the barrel and stows it in a safe place. The squad leader confirms that the assistant gunner's actions were performed correctly.

(8) The assistant gunner inspects the catches on the cartridge extractor to ensure that they are in the latest configuration (as indicated by a 1/8 inch [0.32 cm] hole in the face of the catch) and to be sure that the catches are free of burrs, wear, rust, or corrosion that would impair proper function. Then, the assistant gunner tests each catch to ensure free operation and that each catch will snap into its original position. If the cartridge extractor fails to meet inspection standards, the cartridge extractor is not mission capable and must not be used. The squad leader confirms that the assistant gunner's actions were performed correctly.

NOTE: If the cartridge extractor is not mission capable and there are no other cartridge extractors available, or if there are no functional artillery cleaning staffs available, use the barrel tip method to remove the misfired cartridge (see paragraphs 5-33 to 5-35).

(9) The assistant gunner ensures that the artillery cleaning staff assembly section sleeves are fully extended and tightly locked so that the staff assembly will not extend or retract. If the section sleeves can't be tightly locked, the cartridge extractor assembly is not mission capable and must not be used. The squad leader confirms that the assistant gunner's actions were performed correctly.

(10) The assistant gunner attaches the cartridge extractor securely to the extended artillery cleaning staff assembly. The squad leader confirms that the assistant gunner's actions were performed correctly.

120-mm Mortars, M120 and M121

> **WARNINGS**
>
> 1. To avoid injury to the crew, if the bipod assembly is in the high range of bipod support, do not lower it to low range. If the bipod assembly is in the low range of bipod support, do not raise it to high range except when using the barrel tip method.
>
> 2. To avoid serious injury, do not stand in front of or behind the barrel when removing the cartridge.

(11) The assistant gunner rotates the artillery cleaning staff assembly until the cartridge extractor is secure against it. Then, he inserts the cartridge extractor into the barrel slowly (hand to hand) until it rests on the cartridge and will descend no further.

(12) The assistant gunner slowly rotates the cartridge extractor until it grasps the cartridge. As the cartridge extractor grasps the cartridge, the assistant gunner feels the spring-loaded catches click into the holes on the cartridge body. The assistant gunner continues to rotate the cartridge extractor an additional quarter of a turn.

NOTE: The assistant gunner should feel much more resistance to rotating the extractor. If not, the cartridge extractor has not successfully connected to the cartridge, and the assistant gunner must continue to slowly rotate the cartridge extractor until it grasps the cartridge. If the assistant gunner still feels no extra resistance, he removes the extractor assembly from the barrel and uses the barrel tip method to remove the misfired cartridge (see paragraphs 5-53 to 5-55).

(13) When the assistant gunner feels resistance, he slightly relaxes his grip on the artillery cleaning staff assembly for a moment. Then, without rotating, he slowly pulls the staff assembly slightly to ensure that the cartridge is grasped. If the assistant gunner feels resistance, he, with the gunner's hands at the muzzle, withdraws the cartridge from the barrel with the cartridge extractor, pulling the artillery staff with both hands (Figure 5-14). When the assistant gunner pulls the staff as high as he is able, he slides his lower hand down the staff and then slides his upper hand to his lower hand (hand to hand), maintaining positive control at all times. He repeats this action until the cartridge is exposed.

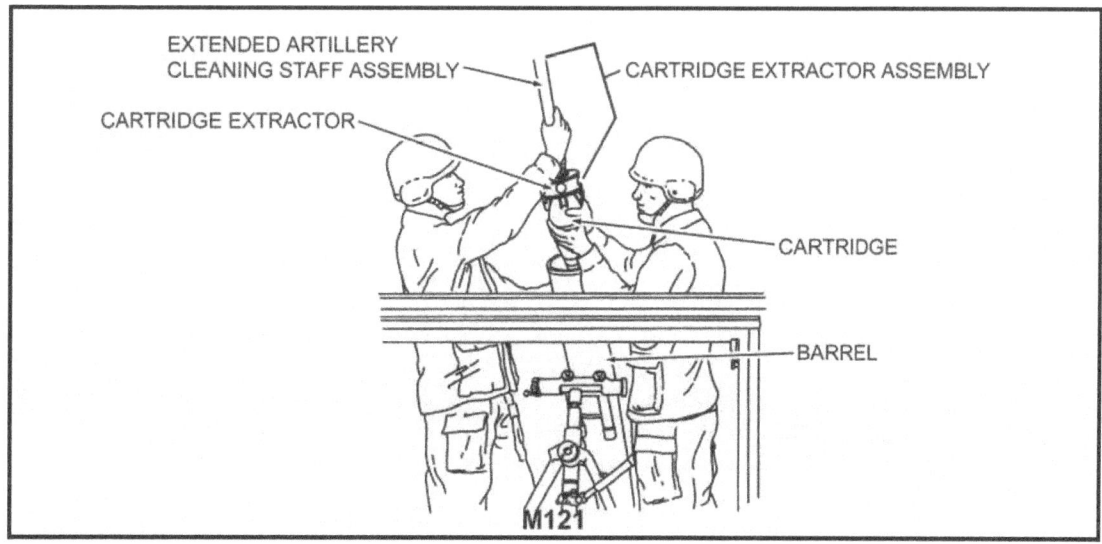

Figure 5-14. Withdrawing the cartridge from the barrel, M121.

Chapter 5

NOTE: If the assistant gunner can grasp the cartridge but can't withdraw it, he uses the barrel tip method (see paragraphs 5-53 to 5-55). If the assistant gunner feels no resistance, he removes the cartridge extractor assembly and uses the barrel tip method (see paragraphs 5-53 to 5-55).

WARNING

1. **When removing the cartridge, do not touch the primer, and do not stand directly in front of the barrel.**

2. **Care must be taken to ensure that the extractor catches are not depressed while removing the round from the barrel.**

NOTE: Due to the weight of the round, the gunner may assist during the entire extraction process.

(14) The gunner grasps the body of the cartridge as it comes out of the barrel. The gunner and assistant gunner secure the artillery staff, extractor, and misfired cartridge; move under the cross-member; and proceed down the carrier ramp to a location outside of the vehicle. The ammunition bearer assists in releasing the cartridge from the extractor by pressing all four extractor catches at the same time.

NOTE: Do not place the cartridge on the ground when removing the cartridge extractor.

(15) The gunner inspects the cartridge to see if the primer has been dented and gives the cartridge to the ammunition bearer. The ammunition bearer attempts to replace the safety wire (if applicable), places the cartridge in the dud pit, and tags the cartridge. The safety officer notifies EOD.

NOTE: If the primer has been dented, perform steps (4) through (8) of paragraph 5-55 before continuing with step (19). If the primer hasn't been dented, proceed to the next step. If two consecutive misfires occur and the primer has not been dented, perform steps (4) through (8) of paragraph 5-55.

(16) The assistant gunner swabs the barrel; the gunner inserts the firing pin into the breech cap and rotates the gunner's display into position; and the mission is continued.

REMOVING A MISFIRED CARTRIDGE USING THE BARREL TIP METHOD

5-53. The barrel tip method is used as circumstances dictate. Follow the steps outlined in paragraph 5-54 only if the extractor assembly is unserviceable or if extractor methods have failed. The steps outlined in paragraph 5-55 apply when the round is grasped by the cartridge extractor but can't be removed.

5-54. Follow these steps to remove a misfired cartridge if the extractor assembly is unserviceable or if extractor methods have failed.

(1) The squad leader shuts down the pointing device using the commander's interface (CI, refer to TM 9-1220-248-10). When "No Position Available" appears, turn off the pointing device and gunner's display at the power distribution assembly (PDA).

> **WARNING**
>
> Ensure that power is turned off at the PDA before disconnecting cables.

(2) The gunner disconnects the cables from the pointing device and the cable from the gunner's display. If the gunner's display is mounted on a bipod leg, remove the gunner's display from the bipod assembly and store it in a safe place.

(3) The gunner and assistant gunner carefully lower the bipod assembly into the lowest position in low range.

(4) The assistant gunner supports the bipod assembly, with his left hand grasping the left side of the traversing mechanism and his right hand grasping the right side of the traversing mechanism.

(5) The gunner cradles the barrel with his right arm near the muzzle and braces his left hand against the cross-member of the carrier for support.

(6) The ammunition bearer cradles the barrel with his right arm above the buffer housing assembly and unlocks the clamp handle assembly with his left hand, releasing the buffer housing assembly.

(7) The assistant gunner lowers the bipod assembly to the ramp. If the squad leader determines that the bipod assembly can interfere with the removal of the cannon, the bipod support pins are removed, and the bipod assembly is moved outside of the carrier.

> **WARNING**
>
> Once the cannon is horizontal, the rear of the cannon must not be lowered back down until the round is extracted. If the round slips down the cannon before extraction, it could ignite, causing death or personal injury.

(8) The gunner and ammunition bearer lift (approximately 60 degrees) and then rotate the barrel so that the white line is in the down position and the breech cap can be removed from the socket. Then, they remove the barrel from the breech cap socket and raise it to the horizontal position.

(9) Keeping the barrel horizontal and pointed in the direction of fire, the gunner, ammunition bearer, and crewmembers carry the barrel to a location outside of the carrier.

(10) The assistant gunner places the meaty portions of his thumbs over the edges of the muzzle, grasping the barrel with his fingers.

> **WARNING**
>
> 1. When removing the cartridge, do not touch the primer, and do not stand directly in front of the barrel.
>
> 2. Care must be taken to ensure that the extractor catches are not depressed while removing the round from the barrel.

(11) At the assistant gunner's command, crewmembers lift the cannon's breech cap assembly, causing the cartridge to slide down to the assistant gunner's hands.

Chapter 5

(12) The assistant gunner removes the cartridge, inspects the cartridge to see if the primer has been dented, attempts to replace the safety wire (if applicable), places the cartridge in the dud pit, and tags the cartridge. The safety officer notifies EOD.

NOTE: If the primer has been dented, perform steps (4) through (8) of 5-55 before continuing with step (13). If the primer has not been dented, proceed to the next step.

(13) The assistant gunner swabs the bore; the gunner inserts the firing pin into the breech cap; and the barrel is returned to action.

(14) The gunner installs the gunner's display, connects the cables to the gunner's display and pointing device, powers up, and checks the system. The mission is continued.

5-55. Follow these steps if the round is grasped by the cartridge extractor, but cannot be removed.

(1) The assistant gunner loosens the sleeve on the artillery cleaning staff assembly, depresses the ball-bearing spring lock, slowly lowers the upper staff section into the lower staff section, and tightens the sleeve.

(2) Perform steps (1) through (8) of paragraph 5-54.

NOTE: Prior to having EOD dispose of the barrel, notify field-level maintenance to remove the pointing device and its mounting bracket.

(3) Keeping the barrel horizontal and pointed in the direction of fire, the gunner, assistant gunner, and crewmembers carry the barrel from the carrier (with the extractor still attached to the round) to the dud pit. The safety officer notifies EOD.

NOTE: The following steps are part of the breech cap removal. These steps are used before firing, during firing, if two or more misfires occur, and after firing is completed. If the primer has been dented and time permits, perform steps (4) through (8).

(4) The assistant gunner and ammunition bearer place the barrel on two empty ammunition boxes and stabilize the barrel. The gunner inserts the breech cap removal tool into the cross bore of the breech cap and taps the end of the tool with a hammer to turn it clockwise. Then, he unscrews and removes the breech cap assembly from the barrel, and wipes away any debris from the inner part of the breech cap.

(5) The gunner inspects the firing pin housing on the breech cap assembly and removes any dirt or debris. The gunner replaces the firing pin by hand, ensuring that the threads are aligned; turns the firing pin clockwise until it is hand tight; and then tightens it further using the firing pin wrench.

(6) Using the firing pin gauge, the gunner measures the protrusion of the firing pin. The firing pin must protrude above the MIN and below the MAX cutouts. The gunner replaces the firing pin if it is defective.

(7) Using a wiping rag, the gunner cleans the inner threads of the breech cap assembly and outer threads of the barrel. The gunner applies a light coat of grease to the threads of the breech cap assembly and installs the breech cap on the barrel. He tightens the breech cap assembly using the breech cap removal tool found in the breech cap bore and backs off a quarter of a turn using a jerking motion to tighten.

(8) The barrel is returned to action.

NOTE: The white stripe on the barrel must align with the firing pin.

(9) Both the squad leader and the section leader confirm that the crew performed the actions correctly.

(10) The assistant gunner swabs the barrel; the gunner rotates the gunner's display into position; and the mission is continued.

TAKING THE MORTAR OUT OF ACTION (GROUND-MOUNTED TO M1064A3 CARRIER-MOUNTED)

5-56. The procedures for mounting the mortar on the carrier from a ground-mounted position are described herein.

(1) The squad leader commands, "OUT OF ACTION, PREPARE TO MARCH."

(2) If the mortar cannon is equipped with the old style of breech cap, the gunner places the weapon on SAFE ("S" showing).

(3) The gunner places a deflection of 3200 mils and an elevation of 0800 mils on the M67 sight.

(4) The gunner removes the sight from the dovetail slot. He places the sight in the sight case and secures it on the carrier.

(5) The gunner then replaces the sight mount cover and snaps it closed.

(6) The ammunition bearer retrieves the aiming posts and the aiming post lights and stows them on the carrier.

(7) The assistant gunner kneels down and secures the bipod and lowers it to its lowest elevation leaving about 1/4 of a turn of the shaft showing.

(8) The gunner opens the buffer housing assembly with his left hand while securing the cannon assembly with his right hand.

(9) The gunner and ammunition bearer rotate the cannon assembly until the white line is on the bottom. They raise the cannon to about a 60-degree angle and lift it from the baseplate socket. They carry the cannon assembly to the carrier.

(10) The gunner and ammunition bearer make sure the clamping support assembly is open. They ensure the bipod support assembly is in the low-range position and the turntable is centered and locked.

(11) The gunner and ammunition bearer position the mortar cannon assembly just above the clamping support assembly with the white line facing down.

(12) The gunner and ammunition bearer carefully insert the breech cap ball into the socket. They raise the mortar cannon assembly and rotate it until the white line is in the up position. They install the muzzle plug and lower the mortar cannon assembly onto the clamping support assembly and lock it.

(13) The assistant gunner takes the bipod assembly to the carrier.

(14) The assistant gunner rests the bipod assembly on the carrier ramp with the cross-leveling mechanism pointing toward the gunner's position.

(15) The assistant gunner removes the pins and leg extensions from the bipod legs and stows them in the step.

(16) The assistant gunner removes the handles from the bipod support assembly. He ensures the arrows are facing each other.

(17) The assistant gunner slides the bipod legs onto the attaching slot on the bipod support assembly.

(18) With the arrows facing each other and the holes matching, the assistant gunner inserts the handles through the bipod support assembly slot and legs. He turns the handles so that the arrows point toward the bipod legs and attaches the safety pins.

(19) The assistant gunner slides the buffer housing assembly up along the cannon assembly until it stops. He tightens the clamp handle assembly until he hears a metallic click.

(20) The assistant gunner and ammunition bearer carry the baseplate to the carrier,

(21) The assistant gunner and ammunition bearer make sure the pin is removed and the handle is up.

(22) The assistant gunner and ammunition bearer lift the baseplate with the spades away from the carrier and place the two corners without handles inside the rims of the lower brackets.

Chapter 5

(23) The assistant gunner and ammunition bearer tilt the baseplate toward the carrier and pull the handle down and lock it with the pin.

(24) The squad leader inspects to ensure all equipment is accounted for and properly secured.

(25) The squad leader announces, "Number two gun up and prepared for march."

RECIPROCALLY LAYING THE MORTAR CARRIER SECTION

5-57. To reciprocally lay the mortar, follow the procedures discussed below.

(1) The aiming circle operator lays the vertical cross line on the mortar sight and announces, "Aiming point this instrument."

(2) The gunner refers the sight (using the deflection micrometer knob) to the aiming circle with the vertical cross line splitting the lens of the aiming circle and announces, "Number (number of gun) gun, aiming point identified."

(3) The aiming circle operator turns the azimuth micrometer knob of the aiming circle until the vertical cross lines are laid on the center of the gun sight lens. He reads the deflection micrometer scales and announces, for example, "Number (number of gun) gun, deflection, two three one five (2315)."

(4) The gunner indexes the deflection announced by the aiming circle operator on the sight and re-lays on the center of the aiming circle lens by telling the driver to start the carrier. Looking through the sight, he tells the driver which direction to move the carrier (if necessary) and cross-levels. He ensures he has a correct sight picture (cross line splitting the aiming circle lens), and the elevation and deflection bubbles are leveled. Once he has accomplished this, he announces, "Number (number of gun) gun, ready for recheck."

(5) The aiming circle operator gives the new deflection, and the process is repeated until the gunner announces, "Number (number of gun) gun, zero mils (or one mil); mortar laid."

(6) As the squad announces, "Mortar laid," the aiming circle operator commands the squad to place the referred deflection (normally 2800) on the sight and to place out aiming posts.

(7) The gunner turns the deflection micrometer knob only and indexes a deflection of 2800 mils on his sight without disturbing the lay of the mortar.

(8) The ammunition bearer runs with the aiming posts along the referred deflection about 100 meters from the mortar, dropping one post halfway (about 50 meters).

(9) Once the far aiming post is placed out as described, the gunner uses arm-and-hand signals to guide the ammunition bearer to place out the near aiming post.

(10) Once the correct sight picture is obtained, the gunner announces, "Number (number of gun) gun, up."

SECTION VI. AMMUNITION

This section implements STANAG 2321.

The four types of ammunition for the 120-mm mortar are: HE, smoke, ILLUM, and training. All mortar cartridges, except training cartridges, are packaged as a complete round and have three major components—a fuze, body, and tail fin with propulsion system assembly.

CLASSIFICATION AND TYPES OF AMMUNITION

5-58. The ammunition used with M120 and M121 mortars can be classified by its use. The following cartridges can be fired from both M120 and M121 mortars, except for M91 ILLUM cartridges, which can only be fired from M120 and *ground-mounted* M121 mortars.

120-mm Mortars, M120 and M121

HIGH-EXPLOSIVE AMMUNITION

5-59. HE ammunition is used against enemy personnel and light materiel targets. Table 5-3 details the HE ammunition that can be used when firing M120 and M121 mortars.

Table 5-3. High-explosive ammunition for 120-mm mortars, M120 and M121.

CARTRIDGE/TYPE	MAXIMUM RANGE (METERS)	FUZE	CHARACTERISTICS AND LIMITATIONS
M934/M934A1 HE cartridges	7,200	M734 multioption (M934) fuze M734A1 multioption (M934A1) fuze	These cartridges' maximum range is achieved at charge 4. They weigh 30.17 pounds.
M933 HE cartridge	7,200	M745 PD fuze	This cartridge is identical to M934 and M934A1 HE cartridges, except for the fuze. It weighs 30.17 pounds.
NOTE: All rounds have a minimum range of 200 meters at charge 0 (charge may vary in firing table and whiz wheel).			

ILLUMINATION AMMUNITION

5-60. ILLUM ammunition is used for battlefield illumination and signaling. Table 5-4 details the ILLUM ammunition that can be used when firing M120 and M121 mortars.

Table 5-4. Illumination ammunition for 120-mm mortars, M120 and M121.

CARTRIDGE/TYPE	MAXIMUM RANGE (METERS)	FUZE	CHARACTERISTICS AND LIMITATIONS
M983 IR ILLUM cartridge	6,675	M776 MTSQ fuze	This cartridge is identical to the M930, except for the illuminating candle composition. It weighs 30.6 pounds. The candle provides IR illumination for 50 to 60 seconds and is used to support troops with night vision devices.
M930 and M930E1 ILLUM cartridges	6,675 (M930) 7,225 (M930E1)	M776 MTSQ fuze	These cartridges weigh 30.6 pounds. The filler in these cartridges provides an illumination of one million candlepower for 50 to 60 seconds.
M91 ILLUM cartridge	7,100	M776 MTSQ fuze	The filler provides an illumination of one million candlepower for 46 to 60 seconds. This cartridge can only be fired from M120 and ground-mounted M121 mortars. It weighs 27.01 pounds.
NOTE: All rounds have a minimum range of 200 meters at charge 0 (charge may vary in firing table and whiz wheel).			

Chapter 5

SMOKE, WHITE PHOSPHORUS CARTRIDGES

5-61. Smoke cartridges with WP filler are used for screening and spotting. Table 5-5 details the smoke WP ammunition that can be used when firing M120 and M121 mortars.

Table 5-5. Smoke, white phosphorus ammunition for 120-mm mortars, M120 and M121.

CARTRIDGE/TYPE	MAXIMUM RANGE (METERS)	FUZE	CHARACTERISTICS AND LIMITATIONS
M929/XM929 smoke WP cartridges	7,200	M734A1 multioption (M929) fuze M745 PD (XM929) fuze	These cartridges weigh 30 pounds. The current M929 utilizes the M734A1 multioption fuze and should be set to PROX for smoke curtains/screens. The XM929, a redesign of the original M929, utilizes the M745 PD fuze.
NOTE: All rounds have a minimum range of 200 meters at charge 0 (charge may vary in firing table and whiz wheel).			

WARNING

At temperatures exceeding 111.4 degrees Fahrenheit (44.1 degrees centigrade) (melting point of WP), store and transport WP rounds in a vertical position (nose up) to prevent voids in the WP.

TRAINING PRACTICE AMMUNITION

5-62. TP cartridges are used when training crews and indirect fire teams (less expensive than service ammunition) or when the use of service ammunition is restricted. Table 5-6 details the training ammunition that can be used when firing M120 and M121 mortars.

Table 5-6. Training practice ammunition for 120-mm mortars, M120 and M121.

CARTRIDGE/TYPE	MAXIMUM RANGE (METERS)	FUZE	CHARACTERISTICS AND LIMITATIONS
M931 FRTC	7,200	M781 PD fuze	This cartridge matches the M934 HE cartridge in size, shape, and weight, and has similar ballistic characteristics. It weighs 30.82 pounds. The M781 PD practice fuze produces a flash, a bang, and a smoke signature that provides audio-visual feedback to the mortar crew and FO.
NOTE: All rounds have a minimum range of 200 meters at charge 0 (charge may vary in firing table and whiz wheel).			

FUZES

5-63. The fuzes used with cartridges for the 120-mm mortar are described in the following paragraphs.

MECHANICAL TIME SUPERQUICK FUZE, M776

5-64. The M776 MTSQ fuze (Figure 5-15) has a mechanical arming and timing device, an IMP fuze, an expulsion charge, and a safety wire or pin.
- Functions: Air burst or impact.
- Settings: 6 to 52 seconds.

Setting the Fuze

5-65. Use the fuze setter to rotate the head of the fuze to the left (counterclockwise) until the inverted triangle or index line is aligned with the correct line and number of seconds of the time scale. If the desired setting is passed, continue counterclockwise to reach the desired setting. Try to index the desired setting in as few rotations as possible.

Resetting the Fuze

5-66. Rotate the head of the fuze (counterclockwise) until the safe line (S or inverted triangle of the time scale) is aligned with the index line of the fuze body. Replace the safety wire.

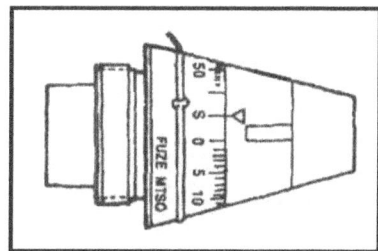

Figure 5-15. Mechanical time superquick fuze, M776.

POINT-DETONATING FUZE, M935

5-67. The M935 PD fuze (Figure 5-16) has a safety wire.
- Functions: Impact.
- Settings: SQ or 0.05-second delay action.

Setting the Fuze

5-68. These fuzes are shipped preset to function superquick on impact. Verify the setting before firing. The selector slot should be aligned with the SQ mark. To set the delay action, turn the selector slot in a clockwise direction until the slot is aligned with the DLY mark. Use a coin or a flat-tip screwdriver to change settings.

Resetting the Fuze

5-69. Align the selector slot with the SQ mark.

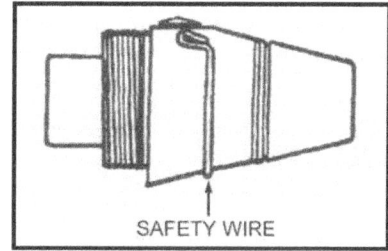

Figure 5-16. Point-detonating fuze, M935.

Chapter 5

MULTIOPTION FUZE, M734

5-70. This fuze (Figure 5-17) can be set by hand.
- Functions: PROX, NSB, IMP, or DLY.
- Settings: PRX, NSB, IMP, or DLY.

Setting the Fuze

5-71. The fuze can be set by rotating the fuze head (clockwise) until the correct marking (PRX, NSB, IMP, or DLY) is over the index line.
- PRX—The fuze comes set to PRX. Burst height is 3 to 13 feet (1 to 4 meters).
- NSB (nonjamming)—Burst height is 0 to 3 feet (0 to 1 meter).
- IMP (SQ).
- DLY (0.050 seconds).

Resetting the Fuze

5-72. Rotate the fuze head (counterclockwise) until the PRX marking is over the index line.

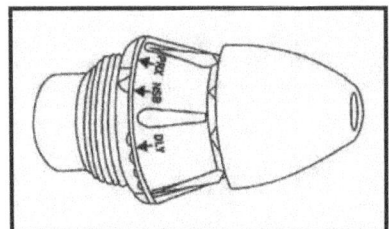

Figure 5-17. Multioption fuze, M734.

MULTIOPTION FUZE, M734A1

5-73. The air-powered M734A1 multioption fuze has four selectable functions:
- PRX 120.
- PRX 60/81.
- IMP.
- DLY.

5-74. In HE PROX mode, the HOB remains constant over all types of targets. The IMP mode causes the round to function on contact with the target and is the first backup function for either PROX setting. In the DLY mode, the fuze functions about 30 to 200 milliseconds after target contact. The DLY mode is the backup for the IMP and PRX modes. The IMP and DLY modes have not changed from the current M734 multioption fuze.

5-75. Radio frequency jamming can affect the functioning of PROX fuzes. Radio frequency jamming initiates a gradual desensitizing of the fuze electronics to prevent premature fuze function. Once the fuze is out of jamming range, the fuze electronics recover and function in the PROX mode if the designed HOB has not been passed. To limit the time of fuze radio frequency radiation, the proximity turn-on is controlled by an apex sensor that does not allow initiation of the fuze proximity electronics until after the apex of the ballistic trajectory has been passed.

5-76. In compliance with the safety requirements of military standard 1316C, the M734A1 uses ram air and setback to provide two independent environment sensors.

POINT-DETONATING FUZE, M745

5-77. The M745 fuze (Figure 5-18) functions on impact with SQ action only. Disregard the markings (PRX, NSB, IMP, and DLY) on the fuze head.
- Functions: Impact.
- Settings: None (no setting or resetting is required).

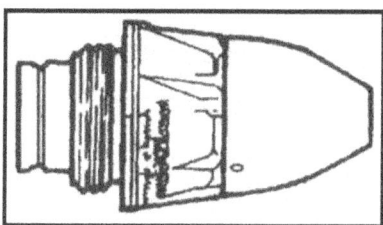

Figure 5-18. Point-detonating fuze, M745.

CARTRIDGE PREPARATION

5-78. Cartridges are prepared as close to the actual firing period as possible.

UNPACKING

5-79. Remove the protective bag and any desiccants secured to or protecting the fin assemblies. Also remove the plastic shell/insert assembly (if any) surrounding the propelling charges.

INSPECTION OF THE CARTRIDGE

5-80. Examine the cartridge fin assembly for any visible damage or looseness. Examine the fuze and propelling charges for any visible damage. Tighten loose fin assemblies by hand. Cartridges with damaged (bent) fin assemblies, fuzes, or propelling charges are turned in to the ASP as unserviceable.

> **WARNING**
>
> The M781 fuze on the M931 FRPC may be armed if the packing clip is missing or the red band on the striker is protruding from the nose cap. Any force applied to the nose of an armed fuze can result in ignition of the propelling charge. Do not attempt to fire the cartridge. Remove it to the dud pit without striking the nose of the fuze.

PREPARATION TO FIRE

5-81. When the ammunition bearer receives a fire command, he prepares the ammunition for firing. The number of cartridges, type of cartridge, fuze setting, and charge are all included in the fire command. To apply the data, the ammunition bearer selects the proper cartridge, sets the fuze, and adjusts (removes or replaces) the number of propelling charges on the quantity of cartridges called for in the fire command. He inspects each cartridge for cleanliness and serviceability. He also ensures that all packing material is removed and that there is no glue or other foreign substance adhering to the cartridge, especially around the obturator band. Any safety wires are pulled just before firing.

Chapter 5

> **WARNINGS**
>
> Always check the mask and overhead clearance before firing.
>
> PD and PROX fuzes may prematurely function when fired through extremely heavy rainfall.
>
> Do not fire ammunition in temperatures above +146 degrees Fahrenheit (+63 degrees centigrade) or below -28 degrees Fahrenheit (-3 degrees centigrade).
>
> Short rounds and misfires can occur if an excessive amount of oil or water is in the cannon during firing.
>
> Before loading the cartridge, ensure the cannon and cartridge are free of sand, mud, moisture, snow, wax, or other foreign matter.
>
> Ensure that all packing materials (packing stops, supports, and plastic bags) are removed from the cartridge.
>
> Remove any glue or other foreign substance adhering to the cartridge, particularly at or near the obturator band. If the substance cannot be removed, the cartridge should not be fired.

NOTE: The ammunition bearer adjusts the number of charges **immediately** in a FFE mission. In an adjust-fire mission, he prepares the cartridge but delays adjusting the charges until the FFE is entered because the charge may change during the subsequent adjustment of fire.

UNFIRED CARTRIDGES

5-82. If a cartridge is not fired, replace the safety wire/clip if it was removed from the fuze. Re-install the propellant increments so that the cartridge has a full charge in the proper order. Install the packing stop, and repack the cartridge in its original packaging.

NOTES: 1. If the safety pins cannot be fully inserted into the fuze, notify EOD.

2. M68 and M91 cartridges should have seven propellant charges. The correct order is one brown, two blue, and four white. Use the original increments and do not mix propelling charge models or lots.

CARE AND HANDLING OF CARTRIDGES

5-83. To properly care for and handle cartridges, the squad uses the following procedures.
- Do not throw or drop live ammunition.
- Do not break the moisture resistant seal of the ammunition container until the cartridges are to be fired.
- Protect cartridges when removed from the ammunition container. Protect ammunition from rain and snow. Do not remove the plastic shell or insert assembly around the propelling charge until the propelling charge is to be adjusted. If protective bags were packed with the cartridge, cover the fin assembly and propelling charge to prevent moisture contamination.

Stack cartridges on top of empty ammunition boxes or on 4 to 6 inches (10 to 15 centimeters) of dunnage. Cover cartridges with the plastic sheets provided.

- Do not expose cartridges to direct sunlight, extreme temperatures, flame, or other sources of heat.
- Shield cartridges from small-arms fire.
- Store WP-loaded cartridges at temperatures below 111.4 degrees Fahrenheit (44.1 degrees centigrade) to prevent melting of the WP filler. If this is not possible, WP-loaded cartridges must be stored fuze-end up so that WP will resolidify with the void space in the nose end of the cartridge (after temperature returns below 111.4 degrees Fahrenheit [44.1 degrees centigrade]). Failure to observe this precaution could result in rounds with erratic flight.
- Store WP-loaded ammunition separately from other types of ammunition.
- Notify EOD of leaking WP cartridges. Avoid contact with any cartridges that leak.
- Protect the primer of cartridges during handling.
- Do not handle duds other than when performing misfire procedures.

This page intentionally left blank.

Chapter 6
Mortar Fire Control System

The Mortar Fire Control System (MFCS) provides a complete, fully-integrated, digital, onboard fire control system for the carrier-mounted 120-mm mortar. It provides a "shoot and scoot" capability to the carrier-mounted M121 mortar. With the MFCS, the mortar FDC continues to compute fire commands to execute fire missions and controls its gun tracks. The carrier-mounted MFCS components work together to compute targeting solutions, direct the movement of vehicles into firing positions, allow real-time orientation, and present gun orders to the gunner.

DESCRIPTION

6-1. The MFCS is an automated fire control system designed to improve the command and control of mortar fires and the speed of employment, accuracy, and survivability of mortars. The commander's interface (CI) microprocessor, controlled by a software operating system, manages computer activities, performs computations, and controls the interface with peripheral and external devices. The CI operator enters data at the keypad and composes messages using the liquid crystal display (LCD). Completed messages are then transmitted digitally or by radio. Should the FDC become disabled, each mortar crew can compute its own fire missions if the FDC is configured as a gun/FDC. System accuracy is increased through the use of a Global Positioning System (GPS), an onboard azimuth reference for the gun, and digital meteorological data (MET) updates. The MFCS (Figure 6-1) provides the capability to have self-surveying mortars, digital CFF exchange, and automated ballistic solutions.

> **NOTE:** For more information about the MFCS, see TM 9-1220-248-10 and FM 3-22.91.

Chapter 6

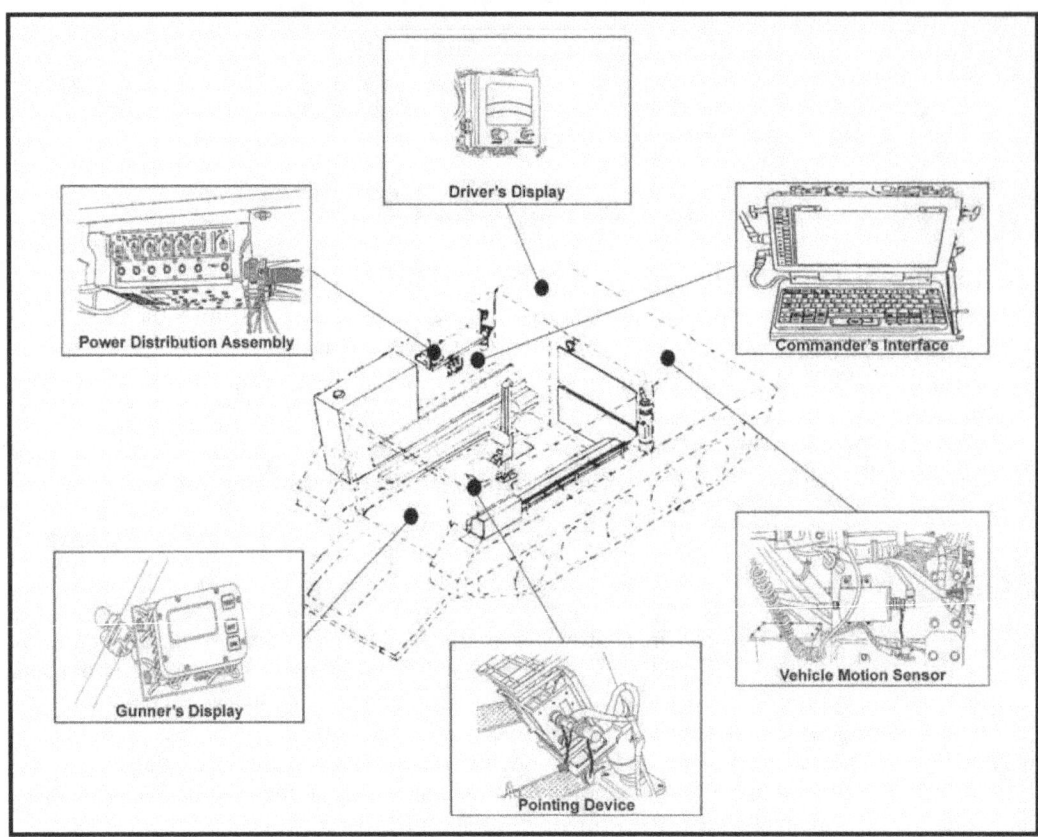

Figure 6-1. Mortar Fire Control System.

COMMANDER'S INTERFACE

6-2. The commander's interface (CI, shown in Figure 6-2) is a computer terminal that provides high-speed data processing, an LCD, and a QWERTY-style keyboard similar to a common personal computer. It has a built-in modem allowing a wide variety of data exchange requirements. The operator works with the CI's graphic user interface (GUI) to operate the system by using the built-in mouse or the keyboard and pointing and clicking buttons and tabs. Data are presented on screens designed for specific missions and operations. The CI's keys and their functions are as follows.

F1-F12 Function Keys

6-3. Although the primary method of operating the CI is the built-in mouse, the operator can use the F1 through F10 function keys (1, Figure 6-2) to make selections in the software (F11 and F12 are not used). Table 6-1 identifies the assignment of keys.

Table 6-1. Function keys.

FUNCTION KEY	ASSIGNMENT
F1	This Menu
F2	Use All
F3	Undo Changes
F4	Fire Support Coordination Measures (FSCMs)
F5	Hipshoot
F6	Final Protective Fire (FPF)
F7	Boresight
F8	Safety Fan
F9	Checkfire
F10	To Be Determined
F11	Not Used
F12	Not Used

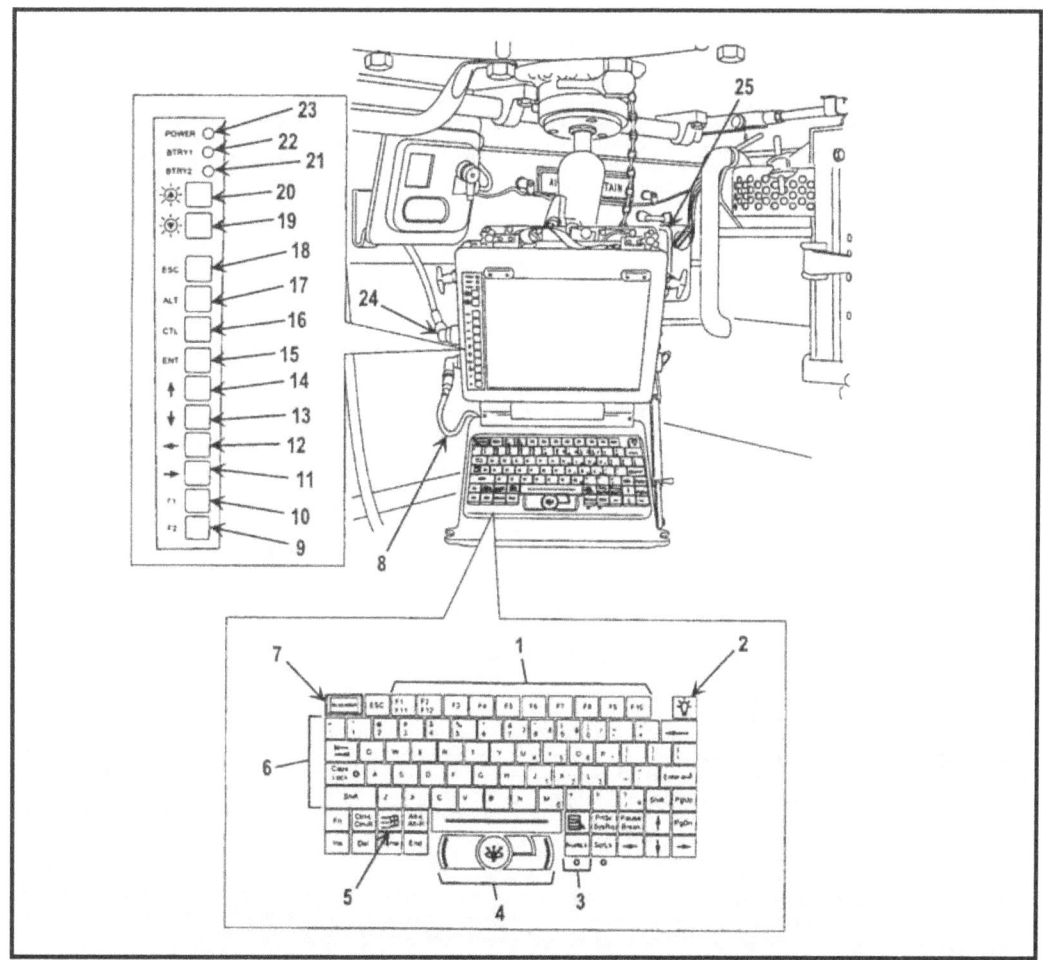

Figure 6-2. Commander's interface.

Chapter 6

Keyboard Backlighting Control

6-4. This control (2, Figure 6-2) adjusts the intensity of the light (with off, low, and high settings).

Number Lock Key and Indicator

6-5. Keys with blue numerals and arithmetic functions can be used as a number pad. Inadvertent use of the number lock (NUM LK) key (3, Figure 6-2) may result in the inability to perform other desired functions. When activated, the indicator light is illuminated.

Mouse

6-6. The mouse (4, Figure 6-2) allows the user to move the cursor on the screen.

Windows Key

6-7. The Windows key (5, Figure 6-2) is not used by the operator.

Alphabetic, Numeric, and Special Character Keys

6-8. These alphabetic, numeric, and special character keys (6, Figure 6-2) function as a standard keyboard to compose messages, and are another method to enter data into the system.

Blackout Key

6-9. The blackout key (BLACKOUT, as shown in 7, Figure 6-2) blacks out the screen to guard against enemy detection in a tactical environment.

F1 and F2 Function Keys

6-10. The F1 and F2 function keys (9 and 10, Figure 6-2) make a selection in the software. Mouse or keyboard use is recommended.

Right, Left, Down, and Up Direction Arrow Keys

6-11. The right, left, down, and up direction arrow keys (11, 12, 13, and 14, Figure 6-2) make a selection in the software. Mouse or keyboard use is recommended.

Enter Key

6-12. The enter (ENT) key (15, Figure 6-2) brings up a menu of function keys.

Control, Alternate, and Escape Keys

6-13. The control (CTL), alternate (ALT), and escape (ESC) keys (16, 17, and 18, Figure 6-2) are not used in this application.

Screen Brightness Intensity Buttons

6-14. The screen brightness intensity buttons (19 and 20, Figure 6-2) decrease and increase the brightness of the LCD screen.

Battery 1 and Battery 2 Indicators

6-15. The battery 1 (BTRY1) and battery 2 (BTRY2) indicators (21 and 22, Figure 6-2) illuminate with a green light when the capacity of the respective battery is 50 to 100 percent of power, with amber light when capacity is 25 to 50 percent, and with no illumination when capacity drops below 25 percent.

Power Indicator

6-16. The power indicator (23, Figure 6-2) illuminates with green light when the computer is being powered with external power (PDA, AC/DC adapter) and with amber light when battery power only is used.

POWER DISTRIBUTION ASSEMBLY

6-17. The power distribution assembly (PDA, shown in Figure 6-3) accepts direct current (DC) and alternating current (AC) to power the MFCS components. It filters vehicle power through a DC-to-DC power system that isolates the MFCS components from fluctuations in vehicle power, including vehicle starting. It also provides protection to the MFCS components against reverse polarity and power surges. The PDA also provides a 115-220v AC interface for connection to available line power that supports classroom training as well as nonfield usage, which preserves the battery charge level.

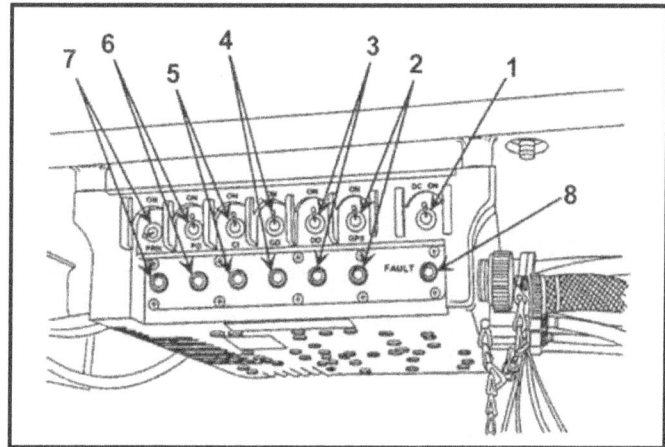

Figure 6-3. Power distribution assembly.

Power ON Switch

6-18. The power ON switch (1, Figure 6-3) turns the PDA on and off for DC power only.

Switch and LED Indicator for Global Positioning System

6-19. This switch (2, Figure 6-3) distributes power to the precision lightweight GPS receiver (PLGR) on the M1064 only. The LED turns green when the switch is ON.

Switch and LED Indicator for Driver's Display

6-20. This switch (3, Figure 6-3) distributes power to the driver's display (DD) on the M1064 only. The LED turns green when the switch is ON.

Switch and LED Indicator for Gunner's Display

6-21. This switch (4, Figure 6-3) distributes power to the gunner's display (GD) on the M1064 only. The LED turns green when the switch is ON.

Switch and LED Indicator for Commander's Interface

6-22. This switch (5, Figure 6-3) distributes power to the CI. The LED turns green when the switch is ON.

Chapter 6

Switch and LED Indicator for Pointing Device

6-23. This switch (6, Figure 6-3) distributes power to the pointing device on the M1064 only. The LED turns green when the switch is ON.

Switch and LED Indicator for Printer

6-24. This switch (7, Figure 6-3) distributes power to the printer (PRN) or, in an emergency, the PRN port can be used to power the pointing device. The LED turns green when the switch is ON. This switch is not used in MFCS application except in an emergency situation.

FAULT LED Indicator

6-25. The FAULT LED indicator (8, Figure 6-3) illuminates with amber light and, possibly, flickers if the PDA malfunctions.

POINTING DEVICE

6-26. The pointing device (PD, shown in Figure 6-4) is mounted in the M1064 mortar carrier and aligns the M121 mortar. It can maintain alignment and accuracy within 3 mils of azimuth and 1 mil of elevation in all conditions. It provides pointing and positional performance at an operational range of 80 degrees south to 84 degrees north latitude. An inertial measurement unit (IMU) provides the weapon with absolute knowledge of vehicle position and mortar barrel azimuth and elevation. The IMU can determine the orientation of the mortar barrel without the need for survey control points, aiming circles, or aiming posts, which allows the mortar platoon to emplace at any location and at any time. To maintain a high degree of accuracy, the IMU incorporates information from the GPS and a vehicle motion sensor (VMS). The design of the pointing device allows for the loss of the GPS, the VMS, or both devices without substantial degradation of overall performance.

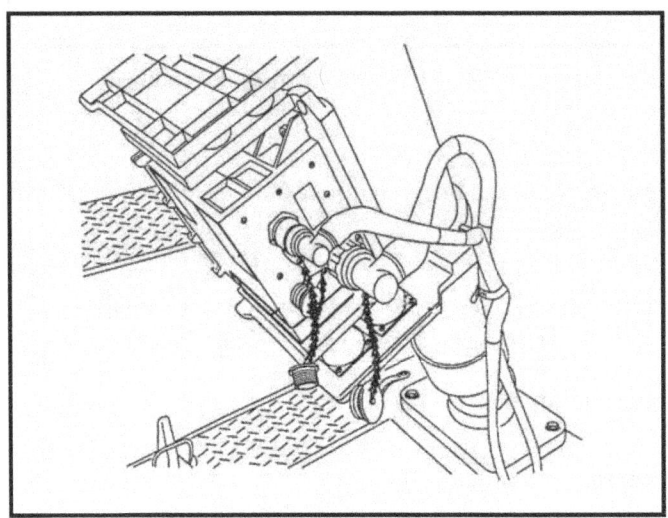

Figure 6-4. Pointing device.

GUNNER'S DISPLAY

6-27. The gunner's display (GD, shown in Figure 6-5) provides the gunner with the necessary information to aim and fire the mortar by displaying deflection and elevation commands. It also displays "check fire" and CFF commands. It is mounted to the left center bipod leg. Function keys are used to start various displays that cover the gunner's functional needs for information, status, and reporting.

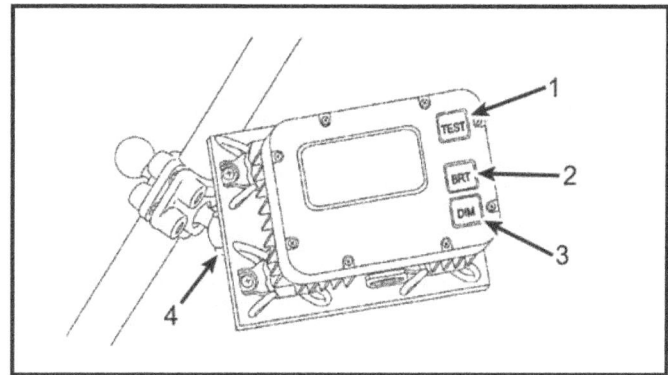

Figure 6-5. Gunner's display.

TEST Key

6-28. The TEST key (1, Figure 6-5) initiates the internal built-in test (BIT) display.

BRT and DIM Keys

6-29. The BRT and DIM keys (2 and 3, Figure 6-5) increase and decrease the brightness of the LCD.

Locking Clamp

6-30. The locking clamp (4, Figure 6-5) is used to allow the gunner to adjust the position of the display.

DRIVER'S DISPLAY

6-31. Located within the driver's vision, the driver's display (DD, shown in Figure 6-6) provides the driver with the steering directions and compass orientation to move the vehicle to the next firing location or waypoint. It also provides information to correctly orient the vehicle at the next emplacement. Information is graphically displayed for steering directions and compass orientation, and numerically for distance and heading.

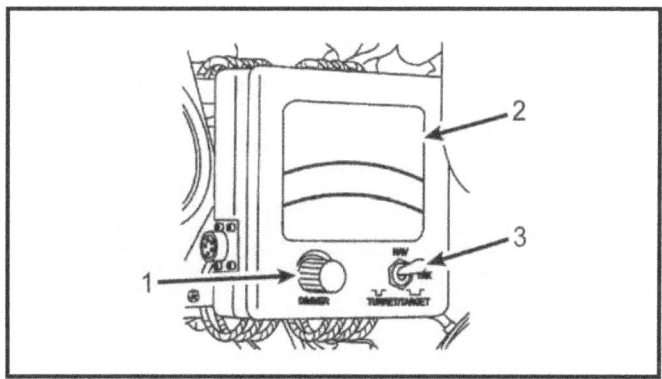

Figure 6-6. Driver's display.

Chapter 6

Dimmer Knob

6-32. The dimmer knob (1, Figure 6-6) controls the brightness of the LCD.

Liquid Crystal Display

6-33. The liquid crystal display (LCD, shown in 2, Figure 6-6) displays directions to the next position. Blinking of the LCD indicates a CFF.

Toggle Switch

6-34. With the toggle switch (3, Figure 6-6) turned to NAV, the LCD displays directional instructions. The TRK and TURRET/TARGET toggle positions are not used.

Vehicle Motion Sensor

6-35. The vehicle motion sensor (VMS, Figure 6-7) provides the pointing device with vehicle velocity data to reduce the vertical position error and improve location accuracy. Rotation of the drive wheels, and, thus, motion of the vehicle, creates a pulse, which represents forward or backward motion of the vehicle. The VMS is mounted inside the engine compartment in front of the driver's cab on the M1064 carrier.

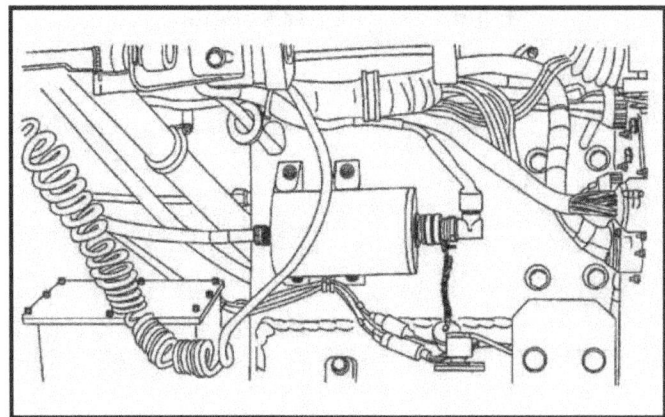

Figure 6-7. Vehicle motion sensor.

CAPABILITIES

6-36. The computer design permits the system to be upgraded to match expanding demands on field terminal equipment. Currently, the MFCS is capable of storing—
- Eighteen gun positions.
- Three gun sections.
- Fifty known points.
- Fifty known targets.
- Sixteen registration points.
- Three FPFs.
- Twelve FO locations.

COMMANDER'S INTERFACE PROCESSING CAPABILITIES

6-37. The CI can simultaneously process up to six active fire missions consisting of one to six guns in a mission as long as no mission consists of guns from different sections and no gun is assigned to more than one mission concurrently. It can also—
- Accept up to 100 digital messages.
- Handle a full range of mortar ammunition per type.
- Conduct registration missions and apply all registration corrections automatically.
- Receive, compute, and automatically apply all MET corrections.
- Store a maximum of 10 safety fans.
- Provide a 13-digit grid to impact for all rounds.
- Provide azimuth and range from gun to impact for each round.
- Connect to radio or wire for digital communications.

6-38. The CI can be powered with dual nickel metal hydride (NiMh) rechargeable lithium batteries. Power can also be supplied through power adapter options of 12- to 32-volt DC, or 110- or 220-volt AC power. The CI also contains power saving circuitry.

6-39. The CI accepts digital messages from the Forward Observer System (FOS), the Advanced Field Artillery Tactical Data System (AFATDS), and other digital messaging devices to support the following FSCMs, maneuver control measures, and airspace coordinating measures. It has an integral speaker and software to control its output and provides the following safety geometry:
- No-fire areas (NFAs).
- Coordinating fire lines (CFLs).
- Restrictive fire areas (RFAs).
- Restrictive fire lines (RFLs).
- Lateral boundaries (LBs).
- Airspace coordination areas (ACAs).
- Forward line of own troops (FLOT).

COMMANDER'S INTERFACE BATTERY SPECIFICATIONS

6-40. The CI holds two rechargeable NiMh batteries, which are for backup power purposes only. The CI can operate without batteries. Features include—
- Single batteries can be replaced while the system is in operation.
- Batteries cannot be installed incorrectly.
- Battery power will last approximately four hours total at ambient conditions.

SOLDIER GRAPHIC USER INTERFACE

6-41. A Soldier GUI is a display that allows the operator to use menus, windows, and icons rather than type in complicated commands (Figure 6-8).

SCREEN AREA

6-42. The screen area is located on the screen of the laptop computer. The screen will display an image with buttons, tabs, fields, and menus. It displays the classification of the information on the current screen in the upper center of the screen.

Chapter 6

Control Button Area

6-43. The control button area displays the buttons to access all available functions of the MFCS for the gun and is located on the right side of the GUI. The buttons include—
- SETUP
- POINTING DEVICE (PD)
- AMMO/STATUS
- NAVIGATION/EMPLACEMENT (NAV/EMPLACE)
- FIRE COMMAND (FIRE CMD)
- CHECK FIRE
- PLAIN TEXT MESSAGES (PTM)
- ALERTS.

NOTE: The control button area displays only those buttons available for the function currently being used. For example, the setup display above does not include the fire mission functions of FPF, FSCM, MET, and so on.

Tabs

6-44. Tabs are similar to buttons and are displayed on the top of the screen after one of the control buttons has been selected. When a tab is selected, the appropriate screen appears and the operator can review and enter information and execute actions.

Work Area

6-45. The work area is located in the center of the screen. This is where information can be viewed, selected, or modified.

Work Area Buttons

6-46. Work area buttons are located in the center and at the bottom of the screen. They are used to accept, process, or refuse data presented in the work area.

Position Location

6-47. Position location is in the upper right corner of the screen. It can be the actual position, a series of question marks if the position is not established, or a note such as "POSITION NOT AVAILABLE."

Unit Name and Long Name

6-48. These are located just below the position location.

COMMON ACTIONS

6-49. When using the MFCS, the operator repeatedly uses the same commands. These are usually done by clicking a button in the working button area of the screen or in response to a query in a message box. These commands include: operationally acknowledge (OpACK), message to observer (MTO) accept or deny, process or delete, selecting the guns to fire the mission, confirming gun orders, and accepting or modifying data.

Operationally Acknowledge

6-50. Clicking the operationally acknowledge (OpACK) button acknowledges receipt of a message and, if turned on, deactivates the audio alarm. The phrase "Click the OpACK button," is used throughout the chapter to indicate this action.

Recording Data

6-51. Data is recorded throughout the process to maintain a record and to preserve data should the MFCS fail. Data is transcribed onto DA Form 2399 (Computer's Record) or DA Form 2188-R (Data Sheet). The phrase "Record data," is used throughout the chapter to indicate this action.

MTO ACCEPT or MTO DENY

6-52. The MFCS uses the terms MTO ACCEPT to accept the mission and MTO DENY to refuse the mission. If the operator selects MTO DENY, the DELETE button becomes visible and, when clicked, deletes the mission. If an error is beyond FDC control, the only choice is MTO DENY. The phrases "Click MTO ACCEPT" to accept the mission or "Click MTO DENY" to refuse the mission are used throughout the chapter to indicate this action.

Process or Delete

6-53. The MFCS usually presents a choice to process or delete the fire mission. The operator can click PROCESS to continue with the mission or DELETE to stop the mission. The terms "PROCESS" or "DELETE" are used throughout the chapter to indicate this action.

Selecting the Guns to Fire the Mission

6-54. Guns for the mission are preselected by the software and are checked in the SEL box in GUN SELECT. The operator has the option to modify these selections by clicking in the SEL box to select or deselect any operationally ready (OpRDY) gun. The operator select can also select all or none of the guns listed by clicking ALL or NONE. The term "Select guns" is used throughout the chapter to indicate this action.

Accepting or Modifying Data

6-55. If required, the operator makes adjustments to the mission data. To undo any changes, he clicks UNDO CHANGES and the data fields display the original data. To accept all the changes made, click USE ALL. If no changes are made, click USE ALL. The phrase "Modify data if necessary" is used throughout the chapter to indicate this action.

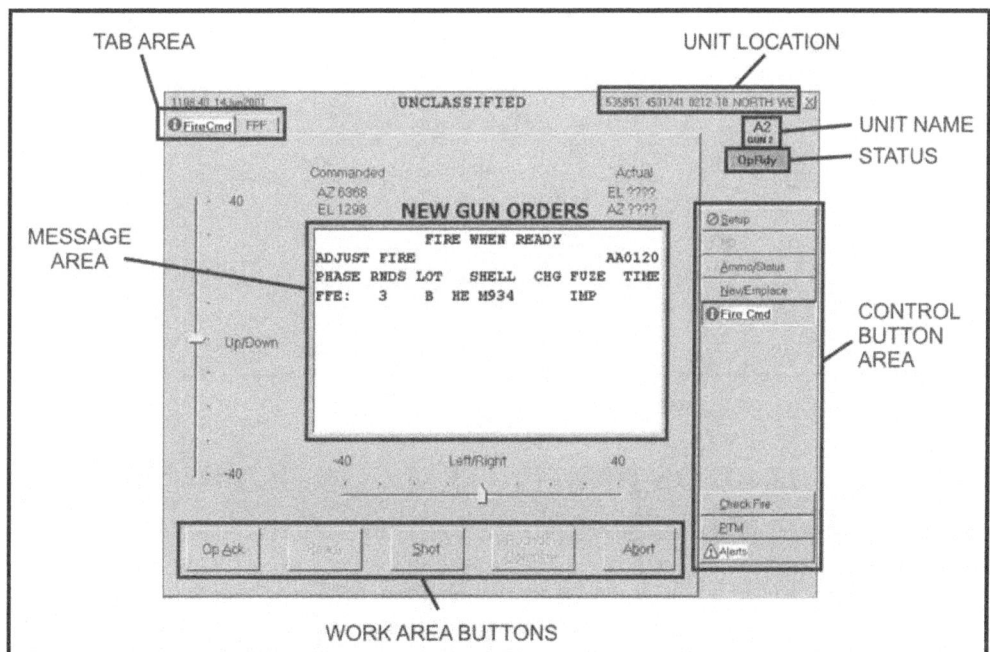

Figure 6-8. Graphic user interface.

Chapter 6

STARTUP

6-56. Starting the MFCS on the M1064 must be done in sequence to ensure the system performs properly. Follow these steps to start the MFCS and its ancillary equipment. (See TM 9-1220-248-10 for more detailed information).

(1) Ensure that the radios are in the stand-by mode and turn on the vehicular or AC power. Check to ensure sufficient vehicular or auxiliary power is available by checking the power gauge meter on the driver's instrument panel.

(2) Turn on the PDA. The PDA distributes power to the MFCS components and the PLGR. It has a toggle switch to turn on power for each.

(3) Turn on the PLGR. The PLGR self-test is successfully performed when the PLGR screen states, "NO FAULTS FOUND." For all MFCS operations, the PLGR must have current communications security (COMSEC) codes installed.

(4) Turn on the pointing device. The PLGR screen displays "REMOTE CONTROL only ZEROIZED key activated."

(5) Turn on the CI. Check to see if the batteries are installed in the battery compartment and the power cable is connected to the computer. Turn the switch on the PDA to the ON position for the CI. If the POWER indicator illuminates with amber light, the CI is not receiving external power. If the POWER indicator illuminates with a green light, the CI has external power.

(6) Establish continuous mode for the PLGR. The upper left portion of the PLGR screen should display "CONT."

(7) Turn on the radios.

(8) Turn on the driver's display.

(9) Turn on the gunner's display.

LOG-IN PROCEDURES

6-57. The log-in screen is displayed after system startup is complete (Figure 6-9). The log-in screen ensures only authorized personnel have access to the system. The user enters his name and password. The operator then clicks the ENTER button.

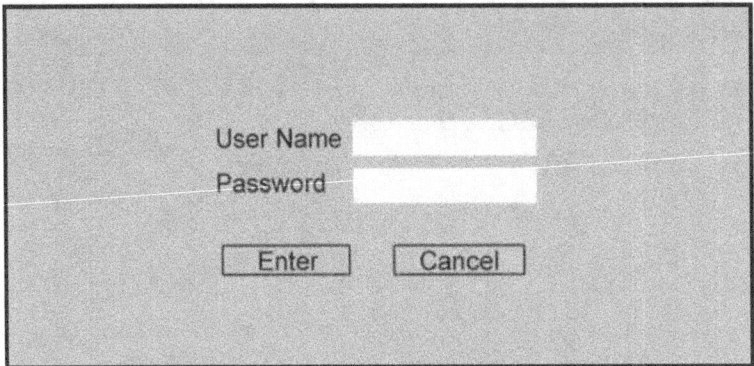

Figure 6-9. Log-in screen.

DATA INITIALIZATION AND SYSTEM CONFIGURATION

6-58. CI initialization and configuration is performed after the startup is completed. The user inputs or selects information for fields to provide the system with the necessary information to function correctly. The user configures the system to the platform using the mouse or keyboard. Various symbols or icons may appear on the screen while using the CI. These symbols assist the operator by highlighting actions that must be completed such as status of operations and changes to data. (See TM 9-1220-248-10 for a list of these icons.)

6-59. The following screens appear as the operator initializes and configures the MFCS. The user must enter, modify, or confirm the information on each of these screens.

- The "Unit List" screen lists all of the units assigned or used by the FDC. It allows the operator to view, enter, edit, or delete a unit name, unit reference number (URN), long (explanatory) name, devices, and observer number (OBSNUM).
- The "Configuration" screen provides the means to enter or reconfigure information on the operator's vehicle by selecting one of the options on the drop-down menus for—
    - Platform.
    - Role.
    - Mode.
    - Security.
    - Unit.
    - Controlling FDC fires cell (FC).
- The "Data" screen provides the means to verify, view, or set the system date/time, audio alarm, target number, and splash offset.
- The "Geographical Reference" screen provides the means to set the geographical positioning system by choosing the ELLIPSOID, DATUM, and MAP MOD fields.
- The "Position" screen provides the means to view or update the operator's current position and other units' position.
- The "Channel A" and "Channel B" screens provide the means to enter, view, or update communication net parameters for Channel A or B subscribers.

6-60. Until the system is configured, only the SETUP and ALERT buttons appear. The SETUP button is located at the top of the control button area. Once the operator clicks the SETUP button, the "Unit List" screen appears. The UNIT LIST and CONFIG tabs are the only tabs shown. After the operator enters the required information for these screens, all other setup tabs display. To complete the setup, data must be entered and accepted on the following screens:

- Unit List.
- Configuration.
- Data.
- Geographic Reference.
- Position.
- Channels A and B.

"Unit List" Screen—Setup

6-61. The "Unit List" screen (Figure 6-10) is the default screen; therefore, it appears after the operator clicks the SETUP button. If previously configured, the system defaults to the data already entered. There is a limit of 100 unit entries. A lock icon displays in the index box at the beginning of a row when that platform has already been assigned a unit name. The unit name must be two to four characters long, and the first character must be a letter. The URN must be unique to the unit, and the long name must be less than 30 characters long. The operator accesses the drop-down menu in the DEVICE field to select FDC, GUN, FO, FSE, or OTHER. In the last field, OBSNUM, the operator assigns a number to each FO. Once complete, the operator accepts the existing or modified data.

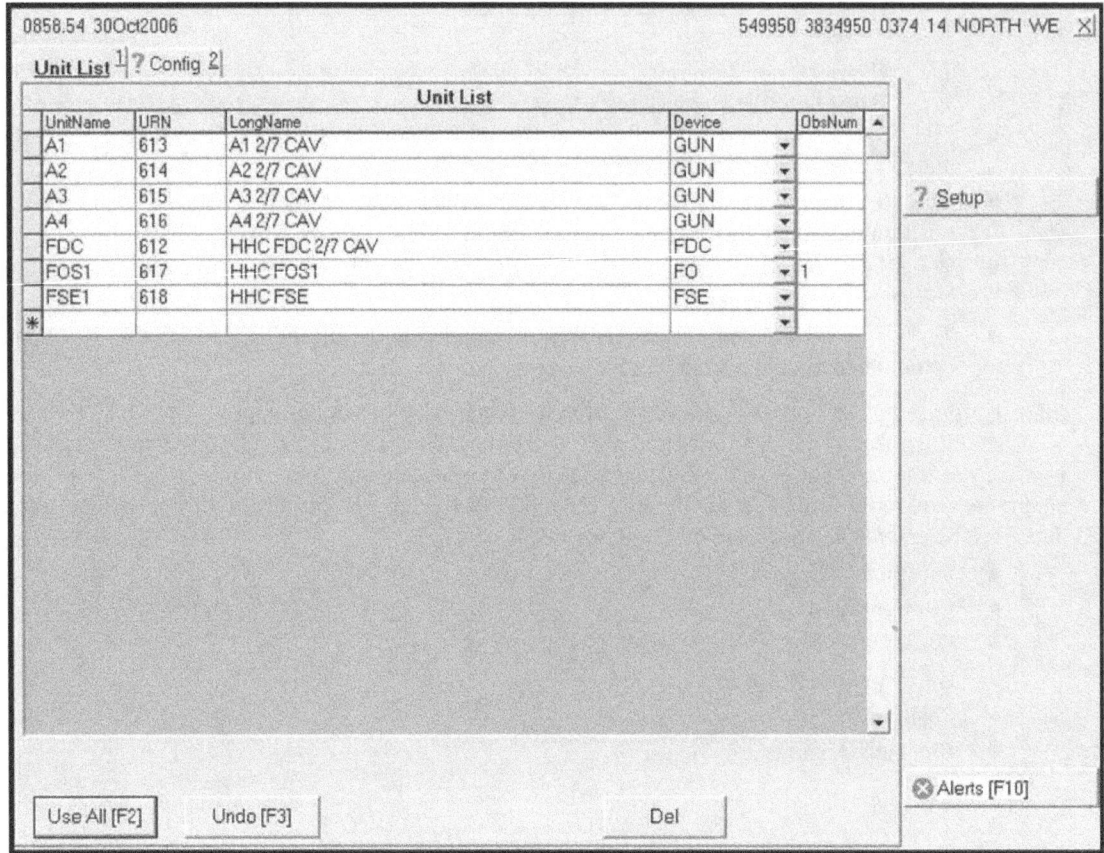

Figure 6-10. "Unit List" screen.

"CONFIGURATION" SCREEN—SETUP

6-62. The operator selects the USE ALL working area button on the "Unit List" screen to display the "Configuration" screen (Figure 6-11). This screen allows the operator to enter or reconfigure information by selecting one of the options on the drop-down menus. If previously configured, the system defaults to the data already entered. A red question mark displays on the CONFIG tab until the operator clicks USE ALL.

NOTE: GUN/FDC is the role used for units that do not have an independent FDC.

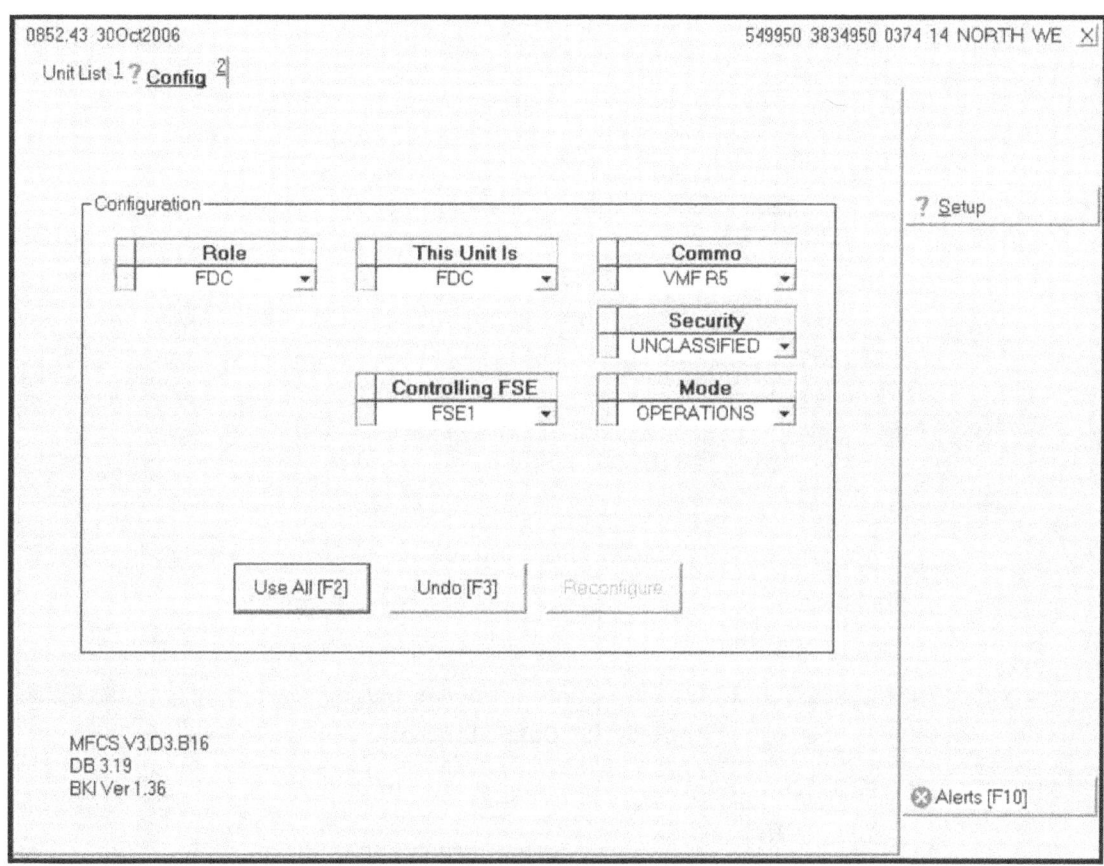

Figure 6-11. "Configuration" screen.

Chapter 6

"DATA" SCREEN—SETUP

6-63. The operator selects the DATA tab to display the "Data" screen (Figure 6-12). This screen allows the operator to verify, view, or set the system date/time, audio alarm, target number, and splash offset.

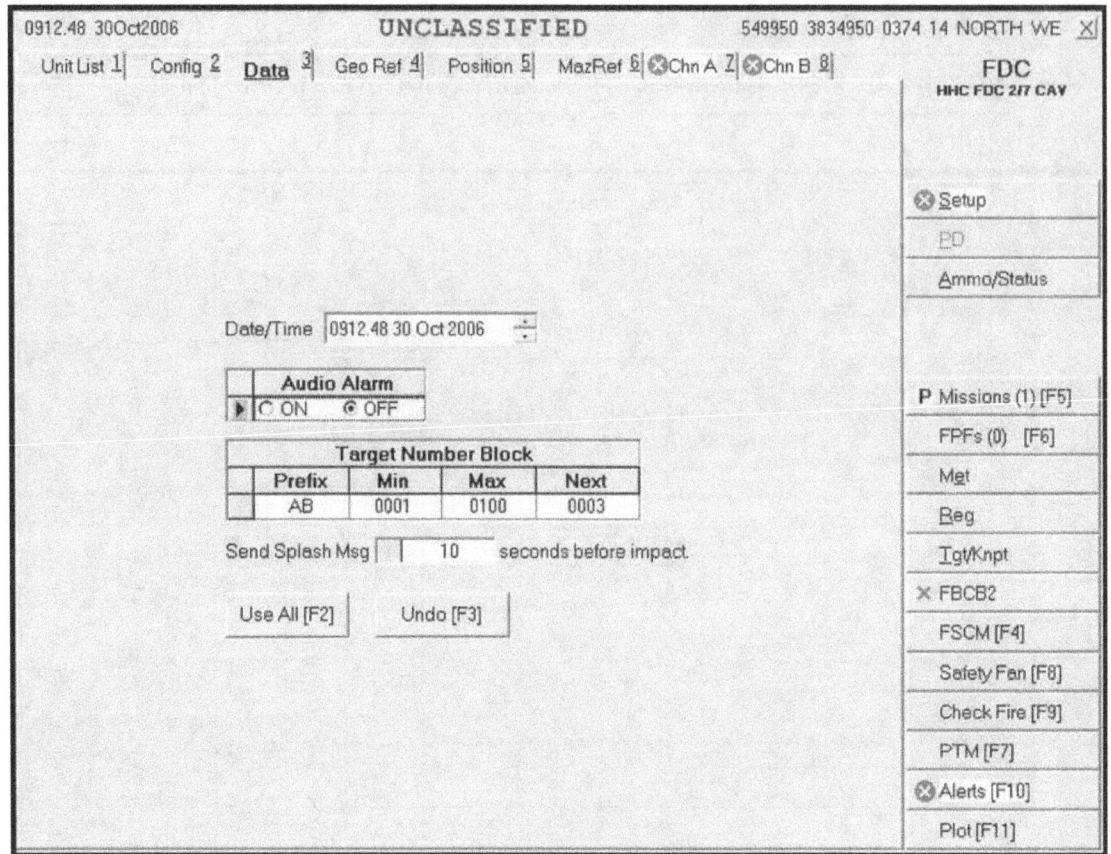

Figure 6-12. "Data" screen.

"Geographic Reference" Screen—Setup

6-64. This screen (Figure 6-13) allows the operator to set the geographic positioning system by filling the ELLIPSOID, DATUM, and MAP MOD fields.

ELLIPSOID and DATUM fields

6-65. The operator clicks the down arrow in the ELLIPSOID field, and then selects the appropriate choice. The corresponding default datum for the ellipsoid automatically displays in the DATUM field. The operator can change the DATUM field by choosing another option. The screen value is updated, and a pencil icon displays in the index box at the beginning of each row. To undo the changes, the operator clicks UNDO CHANGES; any unsaved changes (indicated by a pencil icon) are refreshed with the original data.

MAP MOD field

6-66. The MAX EASTING and MAX NORTHING fields are "read only" and automatically filled when the operator enters the MIN EASTING or MIN NORTHING fields. The GRID ZONE and HEMISPHERE fields in the MAXIMUM field row are also "read only" and automatically filled and updated when the operator enters the MINIMUM values. The MIN EASTING and MIN NORTHING fields are disabled if there are any active missions.

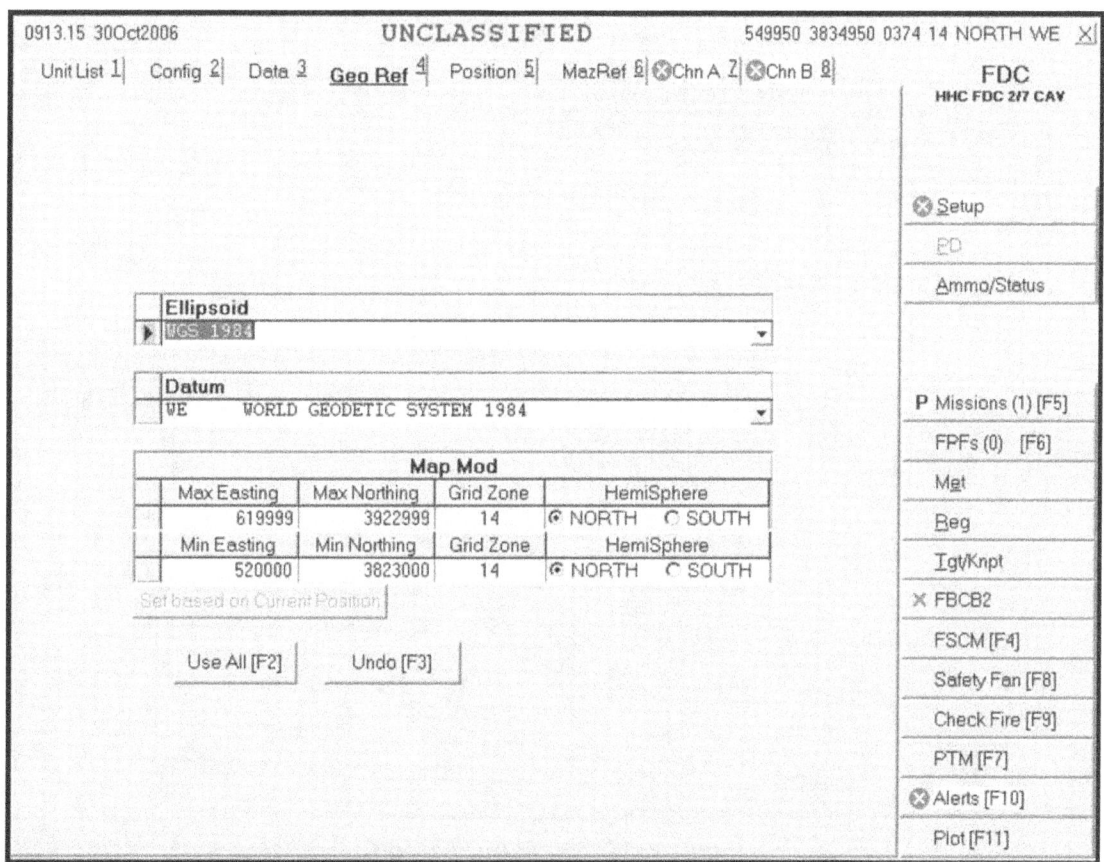

Figure 6-13. "Geographic Reference" screen.

Chapter 6

"POSITION" SCREEN—SETUP

6-67. The operator selects the POSITION tab to display the "Position" screen (Figure 6-14). This screen allows the operator to view or update his and other units' current positions. The operator must update the position if a red circle with an X through it displays on the POSITION tab. The UNITNAME field is "read only" and automatically filled with the unit names selected in the "Configuration" screen. The ZONE and DATUM fields are also "read only" and automatically filled when the operator enters the MAP MOD field. The operator can also manually update or modify the fields as follows:

- The EASTING field with six digits.
- The NORTHING field with seven digits.
- The ALT field with a range of values from -400 to 9999 meters.
- The HEMISPHERE field with the appropriate hemisphere.

NOTE: If the pointing device is available and operational, the operator must ensure that the position data source is set to PD or GPS. The CI cannot process digital input from the pointing device when the position data source is set to FBCB2.

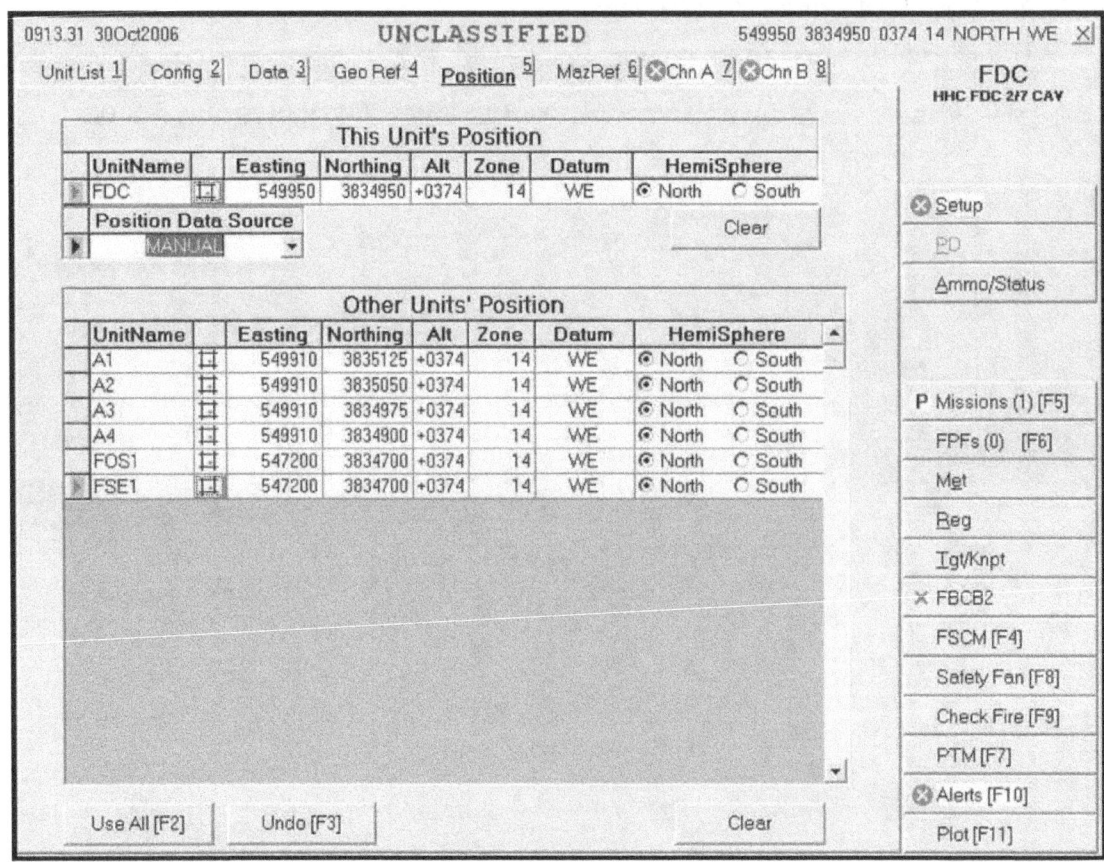

Figure 6-14. "Position" screen.

Mortar Fire Control System

"Mounting Azimuth and Reference" Screen—Setup

6-68. This screen (Figure 6-15) is utilized for degraded mode setup. If the pointing device is not operational or unavailable, solutions in deflection are necessary.

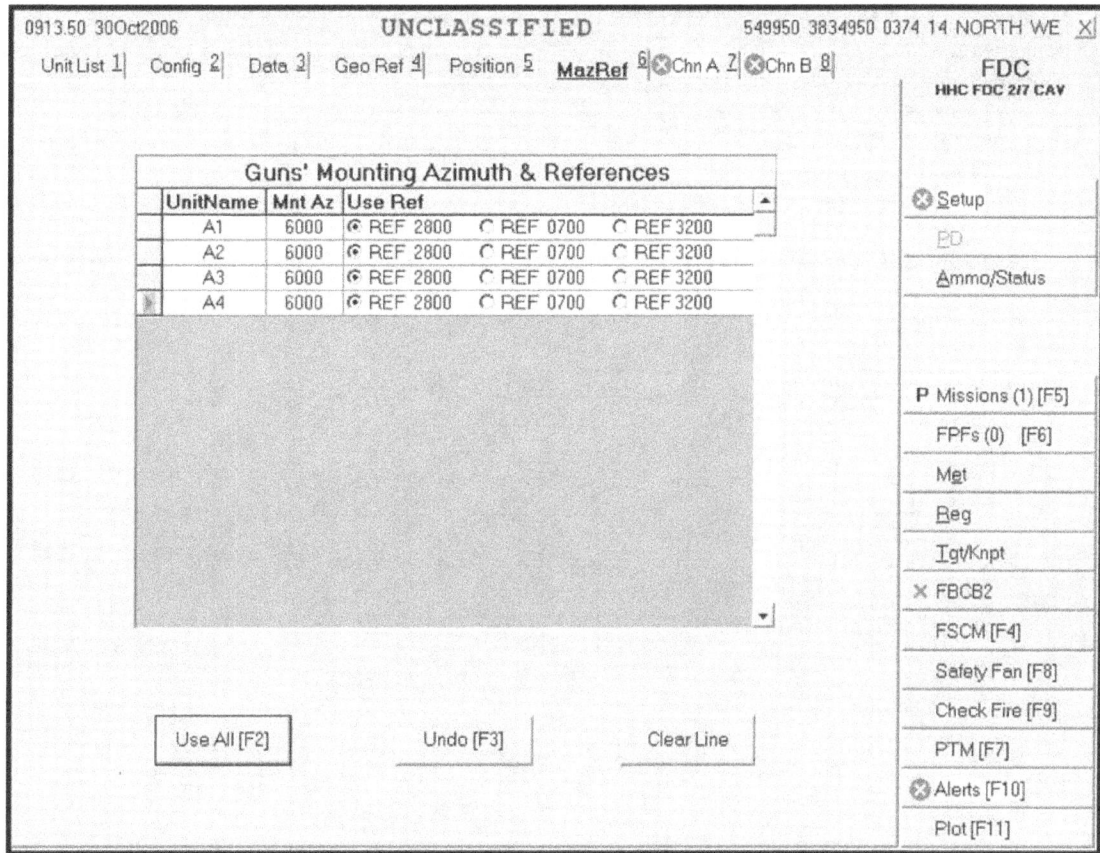

Figure 6-15. "Mounting Azimuth and Reference" screen.

Chapter 6

"CHANNEL A" AND "CHANNEL B" SCREENS—SETUP

6-69. These screens (Figure 6-16) allow the operator to enter, view, or update net parameters and Channel A or B subscribers. Both channels share the same setup procedures and can be set up for Single Channel Ground and Airborne Radio System (SINCGARS), but only Channel A can be set up for wire. Using CHN A for SINCGARS and digital communication between the FDC and the gun tracks is recommended. If a red circle with an X through it displays on the CHN A or CHN B tab, the parameters and subscribers must be set up.

NOTE: When using SINCGARS FSK188-C, ensure that SINCGARS is indicated as the device type.

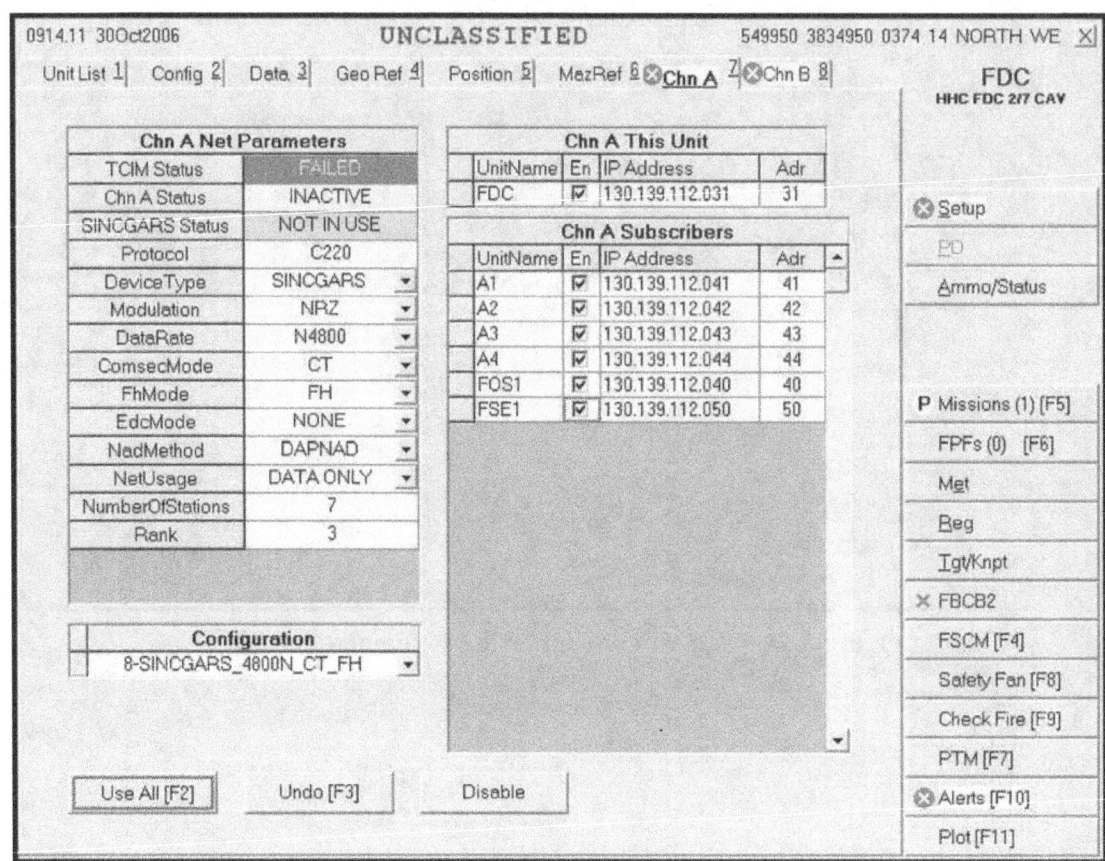

Figure 6-16. "Channel A" screen.

CHANNEL A and CHANNEL B NET PARAMETERS Fields

6-70. Tactical communication interface modem (TCIM) STATUS, CHN A STATUS, CHN B STATUS, and PROTOCOL are "read only" and automatically filled. TCIM STATUS and CHN A/CHN B STATUS reflect the current status. The operator sets the radio or wire defaults or manually selects the parameters to set the net parameters. To determine the parameters, he coordinates with the communications officer to receive the number of stations and rank, and enters this information.

SET RADIO DEFAULTS Button

6-71. To set the radio defaults, the user clicks the SET RADIO DEFAULTS working area button for the device type, modulation, data rate (baud), COMSEC mode, frequency hopping (FH) mode, error detection and correction (EDC) mode, net access delay (NAD) method, and net usage.

SET WIRE DEFAULTS Button

6-72. To set the wire defaults (for Channel A only), the user selects SET WIRE DEFAULTS. To manually set the net parameters, he clicks the down arrows in each field and selects the appropriate choice. He also enters the number of stations and rank, as supplied by the communications officer.

CHANNEL A and CHANNEL B THIS UNIT Fields

6-73. The UNITNAME field is "read only" and automatically filled with the unit name listed in the "Unit List" screen. To update or enter the Internet protocol (IP) address, the user types the correct address.

CHANNEL A and CHANNEL B SUBSCRIBER Fields

6-74. The UNITNAME field is "read only" and automatically filled with the unit names listed in the "Unit List" screen. In the IP ADDRESS field, the user enters the last two digits of the IP address for each unit name. The rest of the address (ADR) is automatically filled when he moves to another field. Then, he clicks the EN box for each unit name to enable communication with that subscriber. To undo any changes, the operator clicks UNDO CHANGES; any unsaved changes (indicated by a pencil icon at the beginning of the row) are refreshed with the original data. The user clicks the USE ALL working area button to save all screen information and activate the channel status, which will cause CHN A STATUS to go from inactive to loading to active, and TCIM STATUS to go from operating to downloading to operating. If disabling communication becomes necessary at any time, the user clicks DISABLE.

ADDITIONAL FUNCTIONS

6-75. Additional functions and procedures of the MFCS are listed below:
- Determining ammunition status.
- Transmitting check fires.
- Sending and receiving PTMs.
- Using the Alerts function.
- Using the Plot function.

6-76. This paragraph displays the sequences used to review, modify, and use data during FDC operations. As previously discussed, the MFCS operator uses the GUI or keystrokes to navigate and use the system.

AMMO/STATUS FUNCTION

6-77. Ammunition information about the unit, lot, shell, lot number, and quantity is transmitted digitally from the gun tracks. This function allows the operator to enter, modify, delete, and sort these data. This button also includes procedures for checking, entering, or changing ammunition status. When the operator presses the AMMO/STATUS button, three tabs display: AMMO FIRE UNIT, AMMO ROLL UP, and STATUS FIRE UNIT.

Chapter 6

"Ammo Fire Unit" Screen

6-78. The operator clicks the AMMO/STATUS button in the control button area, and then clicks the AMMO FIRE UNIT tab to display the "Ammo Fire Unit" screen (Figure 6-17). An exclamation point (!) indicator in the index box at the beginning of a row indicates that the FDC has updated ammunition. Ammunition information about the unit, lot, shell, fuze, lot number, and quantity is received digitally from the gun. A green sunburst icon on the AMMO/STATUS button indicates that the ammunition has been changed. The system automatically arranges ammunition alphabetically by lot, but the operator can click another column heading to sort data by another field.

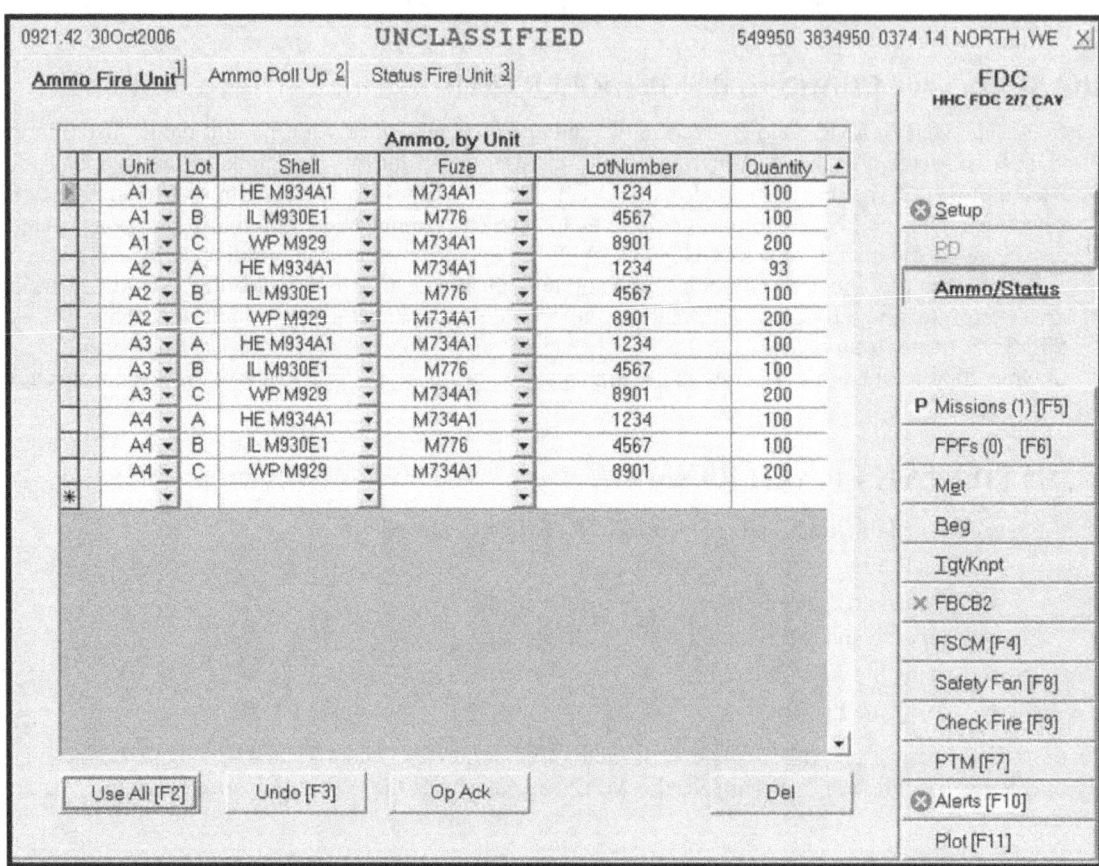

Figure 6-17. "Ammo Fire Unit" screen.

"Ammo Roll-Up" Screen

6-79. This screen (Figure 6-18) displays all of the ammunition carried by the gun carrier. Ammunition information is "read only" and automatically filled, and includes lot, shell, lot number, and total number of rounds.

NOTE: Fire unit ammunition will roll up by lot.

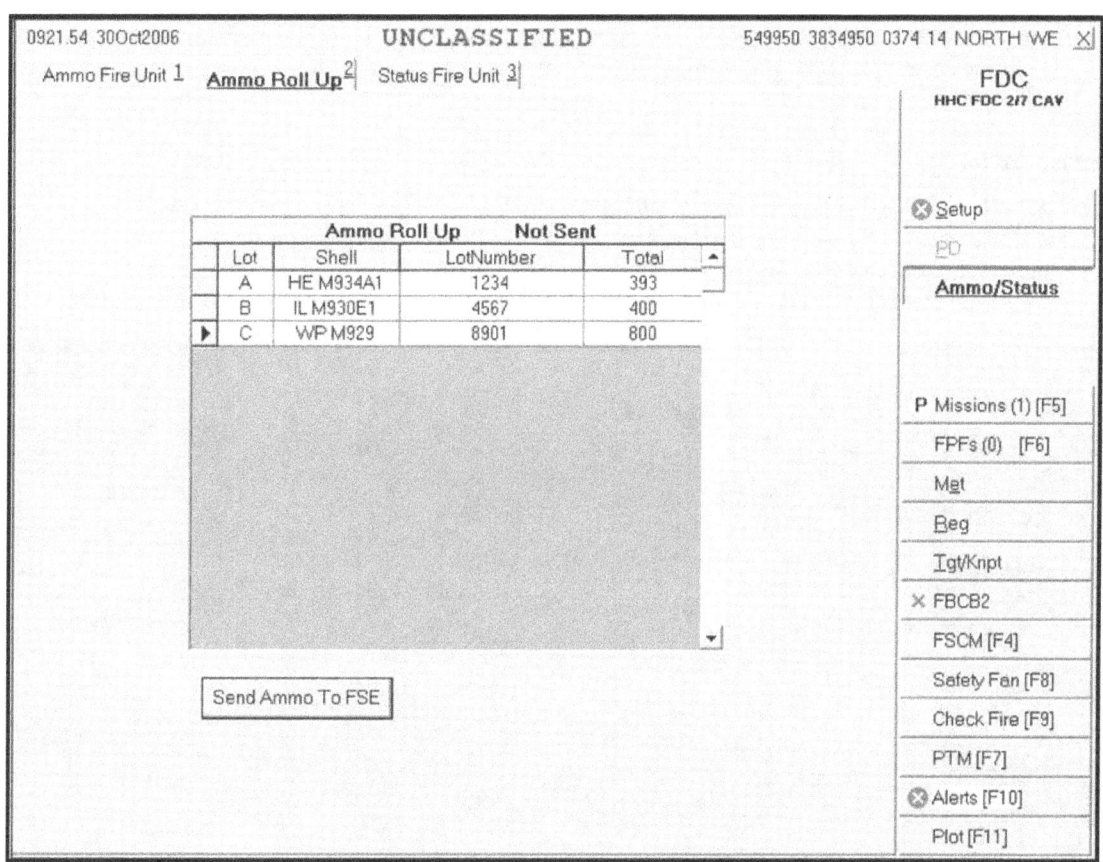

Figure 6-18. "Ammo Roll-up" screen.

Chapter 6

"Status Fire Unit" Screen

6-80. This screen (Figure 6-19) contains two parts. The upper section displays the status by gun, and the lower section displays the status of the fire unit.

Status, By Gun

6-81. Status is received digitally from the guns, and the information is automatically filled. Meteorological data (MET) is automatically sent to the FDC Alerts messages when a gun reports "OpRdy." The GUN, IN MSN, IN FPF, and LOCATION fields are "read only." The operator can modify the OPSTATUS, WEAPON, and MOUNT fields using the drop-down menu. The operator also checks the MAN box if the gun is operating on manual position data. Temperature can be changed, but must override a warning.

Status Fire Unit

6-82. This section displays information about the guns as a fire unit. To enter or change the number of guns, the user clicks the GET # GUNS button, and the field automatically updates according to the number of guns enabled. To get the geographical center of the fire unit, the user clicks GET FU CENTER, which displays the Easting, Northing, and Alt to the center point of all available guns.

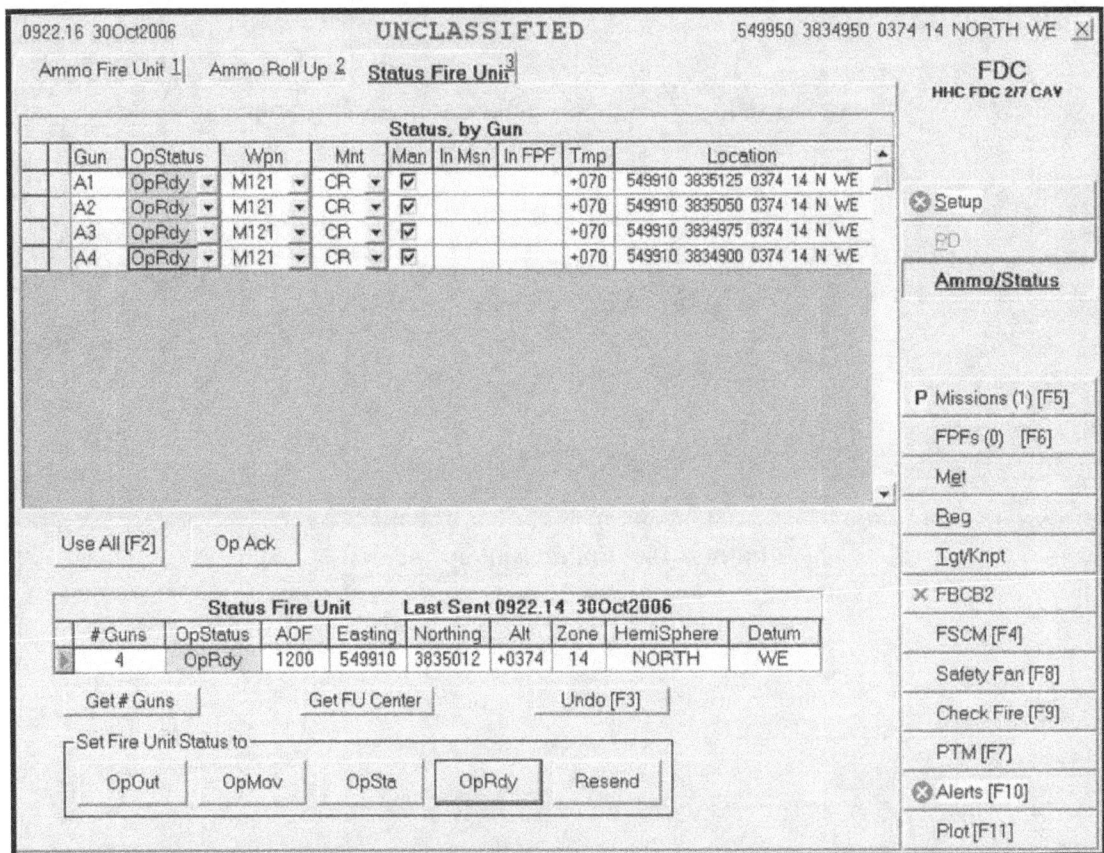

Figure 6-19. "Status Fire Unit" screen.

CHECK FIRE FUNCTION

6-83. The CHECK FIRE button lets the FDC send a message to the guns to immediately cease fire and preserve data. When a Check Fire command is received, all active fire missions come to a halt, and inactive fire missions cannot be activated. When the "Check Fire By Target Number" command is received, the specified fire mission is halted. A "Check Fire" message from the FC or FO is sent as an alert.

6-84. When the operator receives a "Check Fire" command from the FC, a red "CF" message displays in the control tab area in front of MISSIONS and CHECK FIRE. He clicks the CHECK FIRE button to display the "Check Fire" screen (Figure 6-20).

6-85. To initiate or transmit a "Check Fire" message to all guns in the fire unit, the operator clicks the CHECK FIRE ALL button on the "Check Fire" screen. This sends a "Check Fire" banner message to all guns and halts the mission until a valid "Cancel Check Fire" command is transmitted.

6-86. To initiate a "Check Fire Target" message, the operator enters the target number and clicks CHECK FIRE TARGET. This transmits a "Check Fire" banner to the guns engaged in a mission with the associated target.

6-87. To cancel the "Check Fire" command, the operator clicks the indicator for the appropriate "Check Fire" message and clicks CANCEL CHECK FIRE. This forwards the "Cancel Check Fire" message to all units in the fire unit. Fire missions may then continue.

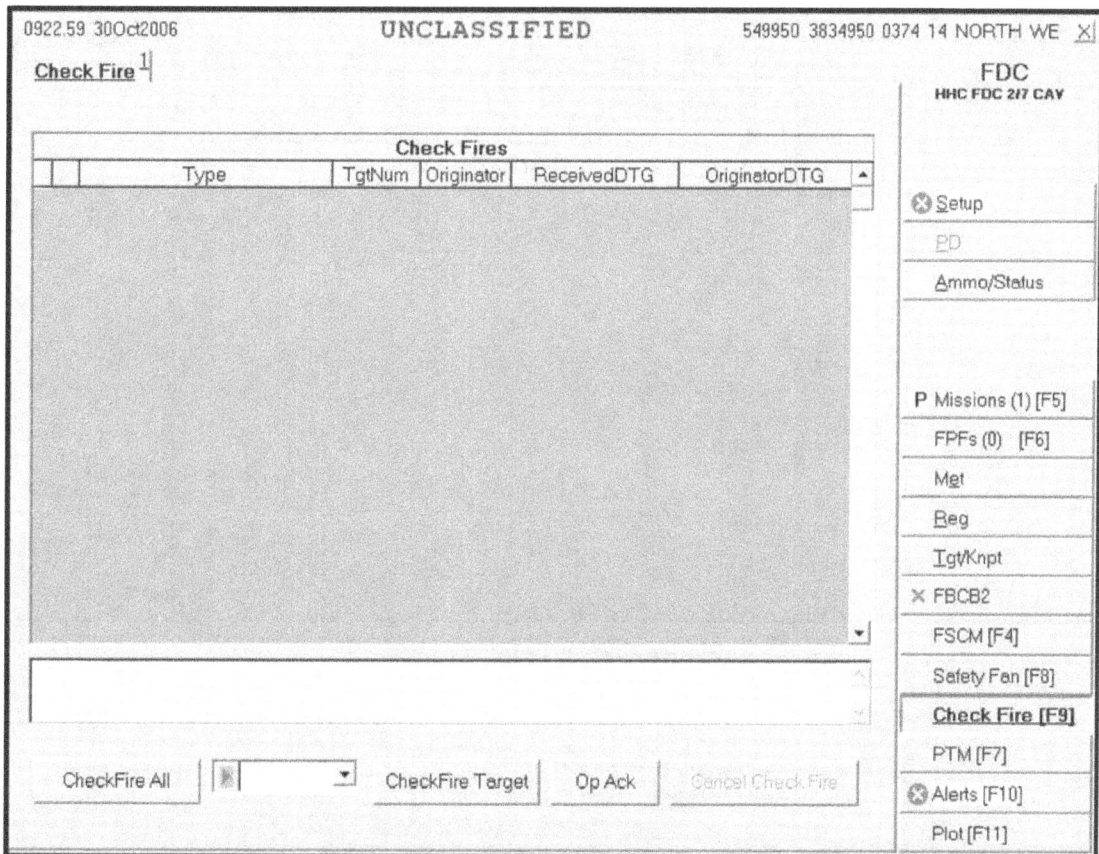

Figure 6-20. "Check Fire" screen.

Chapter 6

PLAIN TEXT MESSAGES FUNCTION

6-88. The PTM function allows the FDC to send and receive messages concerning supply, administration, or other subjects.

"Read" Screen

6-89. Upon receipt of a PTM, an indicator displays on the PTM button. The operator clicks the PTM button to display the "Read" screen (Figure 6-21). The four message categories are—
- Flash (F).
- Immediate (I).
- Priority (P).
- Routine (R).

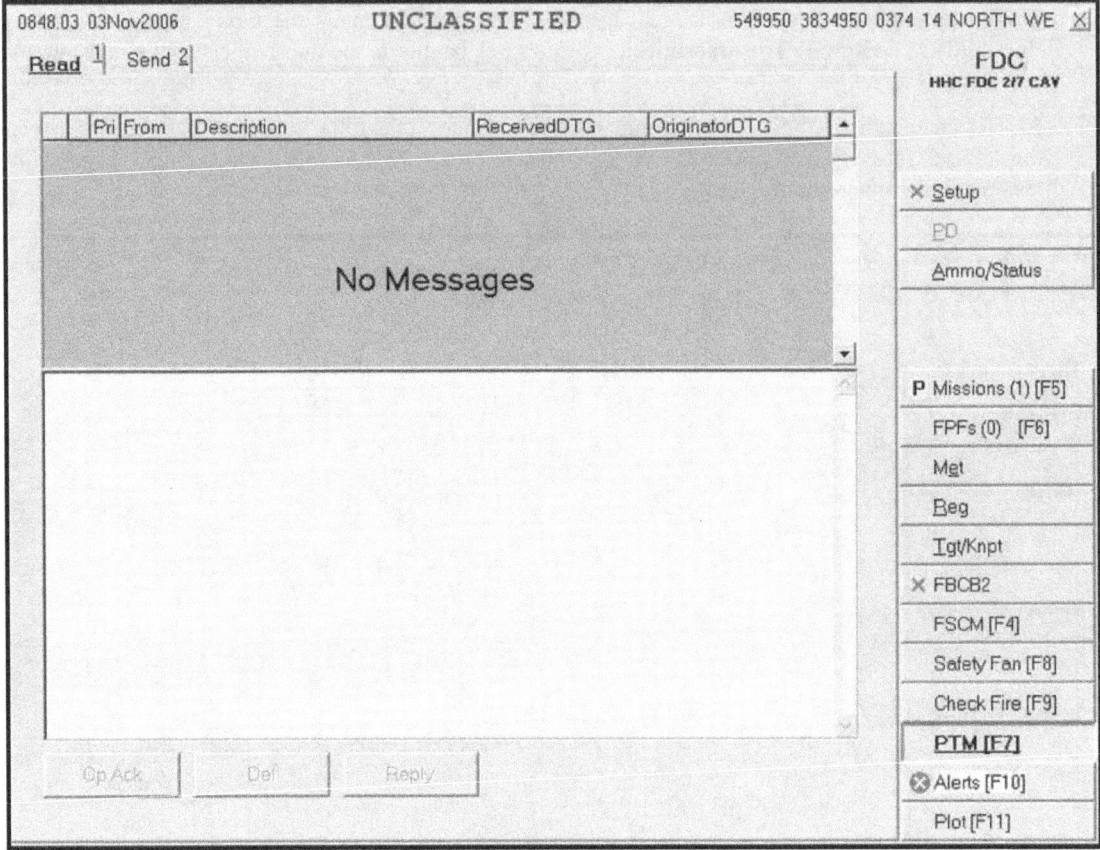

Figure 6-21. "Read" screen.

"Send" Screen

6-90. The operator selects the SEND tab to display the "Send" screen (Figure 6-22). Then, he selects the units to which he will send the message by clicking SEL in DESTINATIONS, types the message using the keyboard, and clicks SEND.

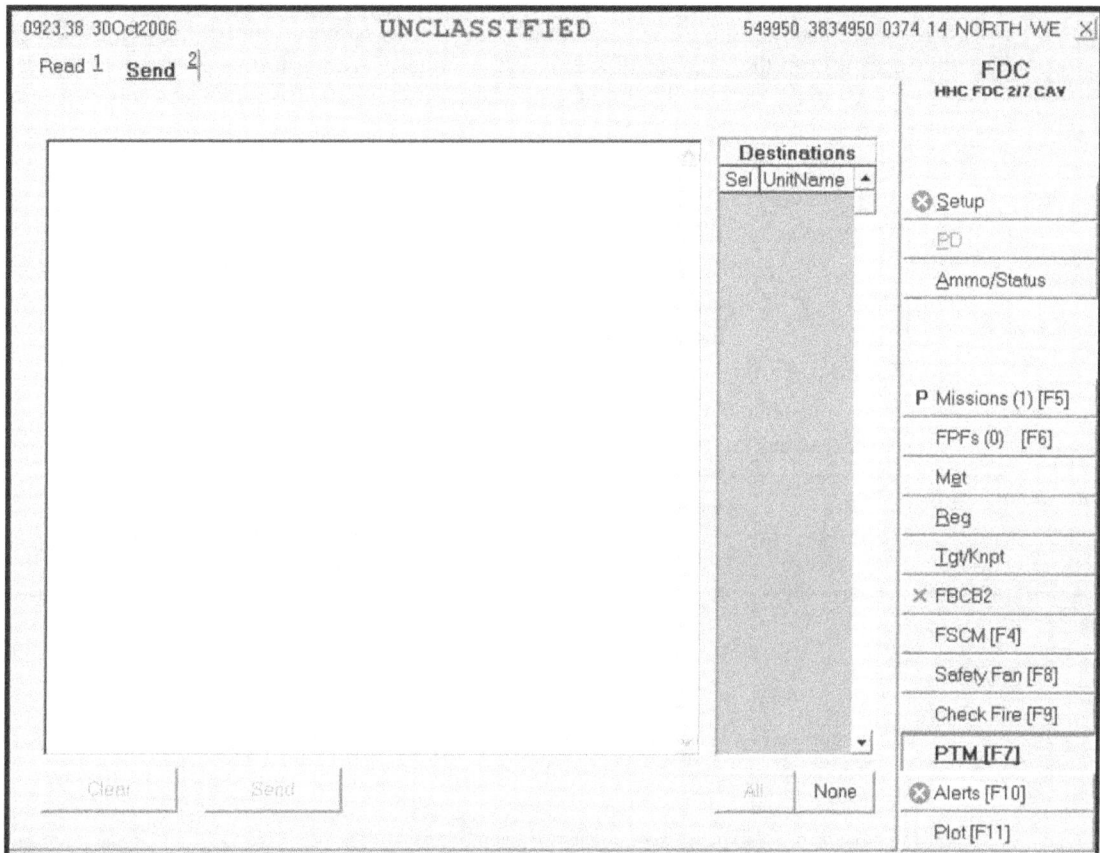

Figure 6-22. "Send" screen.

Chapter 6

ALERTS FUNCTION

6-91. The Alerts function allows the operator to receive automated system messages (Figure 6-23), such as information, error, and warning alerts.

- Information messages provide the operator with general types of information, such as "Ammo Updated for Gun A1." No audio alert sounds.
- Error messages display when operator data entries are performed incorrectly. These messages inform the operator of specific errors and provide information about the correct input. They must be cleared before the operator can continue. Error messages are generated for—
  - Incorrect data entry/combinations/duplications.
  - Ballistic computation failure.
  - Hardware and software failures.
- Warning messages display when events or conditions produce an unsafe condition, a potential loss of data, or system degradation. They also signify that—
  - Digital messages have not been received.
  - The message queue is full.
  - Stored digital messages have been deleted.
  - An FSCM has failed.

NOTE: The operator must respond to a warning message.

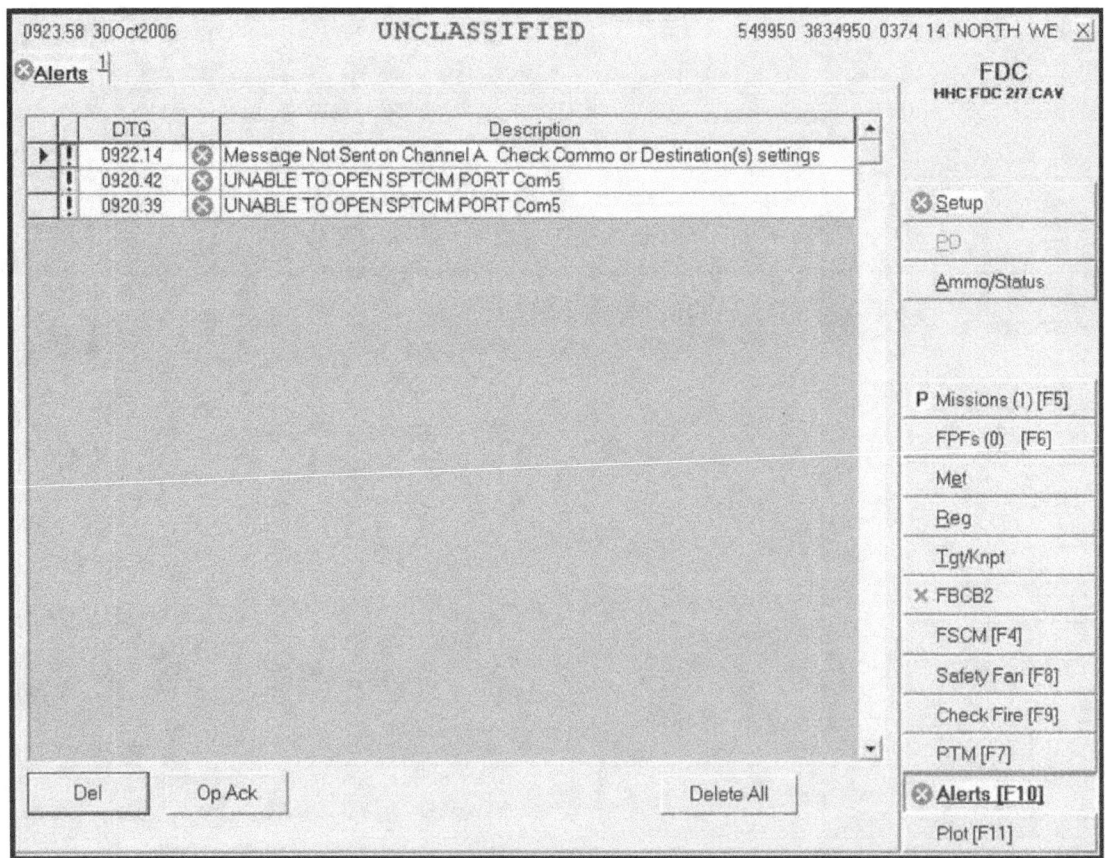

Figure 6-23. "Alerts" screen.

PLOT FUNCTION

6-92. The Plot function provides a graphic representation of the firing mission.

6-93. When the operator accesses this function, a menu of selectable items displays on the right side of the screen (Figure 6-24). Then, the operator can individually select or deselect items by clicking the item's checkbox; the checkbox will indicate an item's selection status. To obtain specific information about an object, the operator places the cursor over the object to display a label with the location or name of the object.

6-94. The Plot function also allows the operator to access information crucial to reading this representation.

- The operator can access a legend of the selectable items by clicking the LEGEND button (Figure 6-25).
- The slider allows the operator to zoom in/out on the "Plot" screen and can center on a given location when the user double-clicks (only available in Map Mod).

NOTE: Items on the "Plot" screen are not directional. For instance, regardless of the direction of fire, all mortars appear to be oriented north.

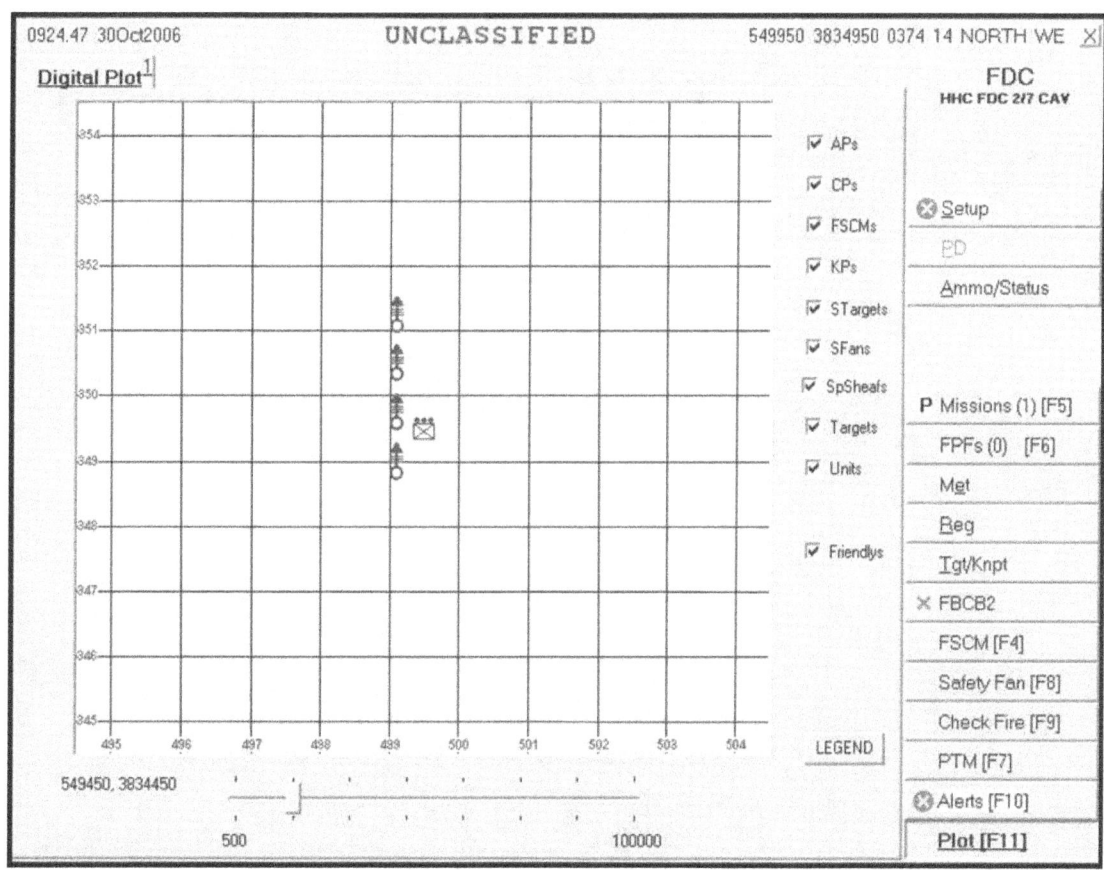

Figure 6-24. "Plot" screen.

Chapter 6

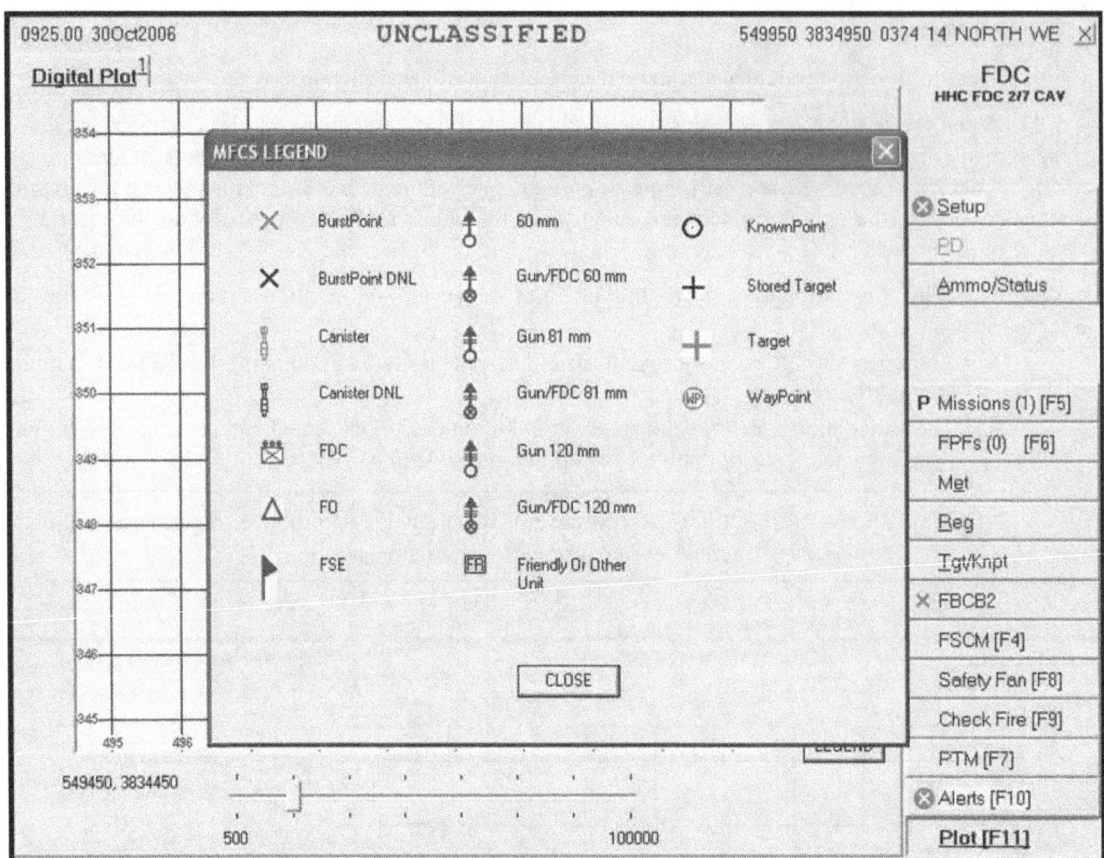

Figure 6-25. "Plot" screen legend.

Chapter 7
Conduct Fire Missions Using the Mortar Fire Control System

The purpose of the MFCS is to bring fires quickly and accurately in a fire mission. This chapter explains how the system assists the gun crew in moving to designated points, rapidly laying the mortar, and firing the mission. It describes how the following MFCS functions work: the pointing device, gun navigation and emplacement, fire missions, and FPFs. (More detailed instructions for these and the other functions are found in TM 9-1220-248-10.)

POINTING DEVICE

7-1. The pointing device is mounted on the M121 mortar tube in the M1064A3 mortar carrier and aligns the M121 mortar. It determines the correct gun alignment without survey control points, aiming circles, or aiming posts so that the gunner can quickly adjust the mortar to the proper elevation and azimuth to accurately fire the cannon. Throughout the vehicle movement and gun emplacement, the MFCS receives, calculates, and displays instructions to the driver, gunner, and squad leader. It can maintain alignment and accuracy within 3 mils of azimuth and 1 mil of elevation in all conditions and provides pointing and positional performance throughout most of the world. The system uses information from the GPS and a vehicle motion sensor (VMS). The design of the pointing device allows for the loss of the GPS, the VMS, or both devices without substantial degradation of overall performance. Prior to a fire mission, the mortar crew ensures that the pointing device is functioning and that the mortar is properly boresighted.

Chapter 7

POINTING DEVICE "STATUS" SCREEN

7-2. This screen (Figure 7-1) shows the status of the pointing device, PLGR, and VMS. It also displays the pointing device update position and permits the manual input of position data.

Pointing Device Status

7-3. The "Status" screen displays errors, warnings, and status messages for the pointing device. An icon with the word "NEW" appears in front of a new message or a message that has changed.

> **CAUTION**
> Errors displayed on the "Status" screen require immediate attention to continue the operation.

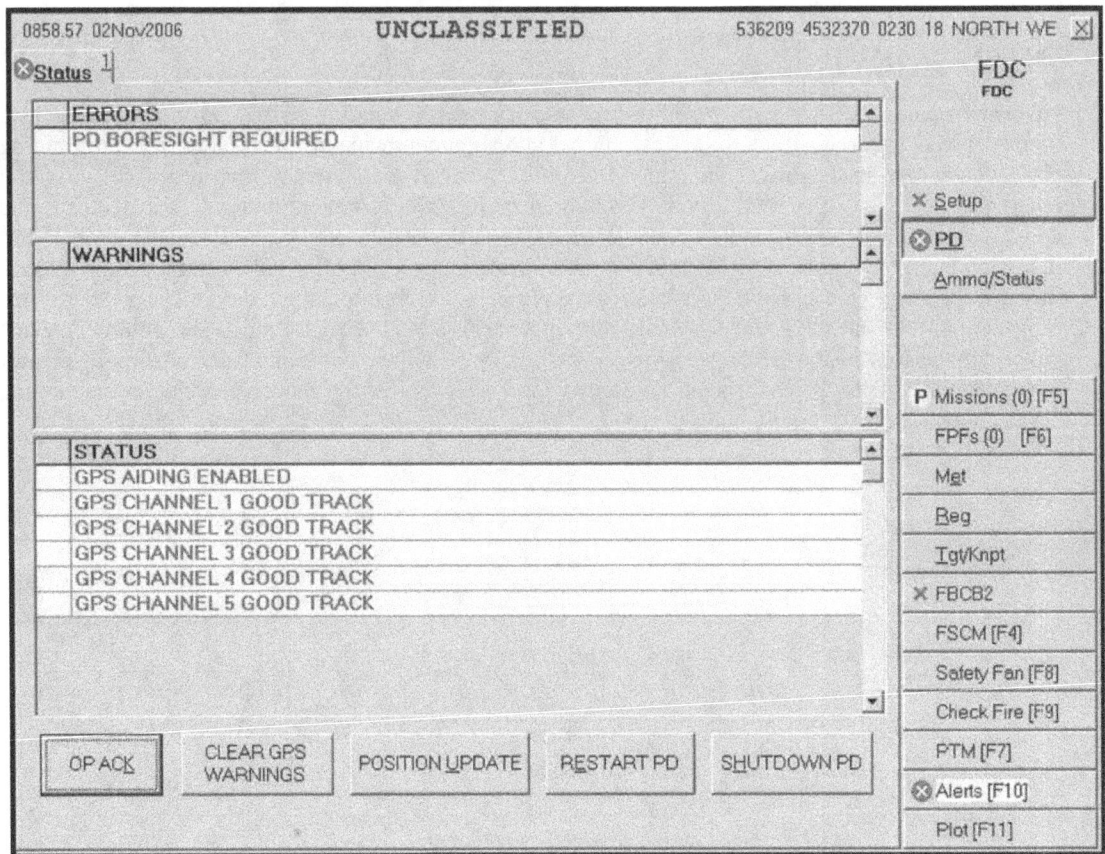

Figure 7-1. Pointing device "Status" screen.

Manually Update Pointing Device Position Fields

7-4. When the pointing device does not automatically align using PLGR data, it must be manually updated. If the pointing device has not automatically aligned, the words "PD ALIGNING WAIT 600" remain in the upper right corner of the "CI" screen, and the numbers will not automatically count down.

Conduct Fire Missions Using the Mortar Fire Control System

> **CAUTION**
> To prevent alignment errors, keep movements inside the vehicle to a minimum.

(1) Click POSITION UPDATE. Highlight the information to be changed, or click CLEAR POSITION to clear all information in the POSITION UPDATE fields.

(2) To manually update the pointing device position information, enter the correct data into the EASTING, NORTHING, ALT, ZONE, DATUM, and HEMISPHERE in PD UPDATE POSITION fields. To display or use the last recorded pointing device position, click the SETUP button and select the POSITION tab.

(3) Click SEND UPDATE. The upper right corner of the screen changes to either "ALIGNING:___SECONDS REMAIN" or "POS UPDATE IN PROGRESS," depending on whether the pointing device is in need of alignment.

Pointing Device Data Precedence Rules

7-5. The pointing device has a set of rules to determine whether to use the manually input locations or locations generated by the GPS.

During Alignment

7-6. If the pointing device is aligning using data from a manual position update and the PLGR starts to acquire accurate satellite data, the pointing device will begin to use the PLGR data to complete alignment and, within 2 minutes, the *GPS Aiding Off* box becomes unchecked. When the pointing device alignment is complete, the pointing device's current position is displayed in the upper right-hand corner of the "CI" screen and is automatically sent to the FDC.

After Alignment

7-7. If the PLGR acquires accurate satellite data after the pointing device has been aligned using data from a manual position update, the *GPS Aiding Off* box becomes unchecked and the current coordinates displayed in the upper right-hand corner of the "CI" screen will begin to drift toward PLGR data. When the drifting is stabilized, the position shown is not automatically sent to the FDC or updated in *This Unit's Position* or *Status*. The operator must click SEND POSITION UPDATE TO PD. If the two positions do not agree, click SEND POSITION UPDATE TO PD again.

Restart Pointing Device

7-8. This button restarts the system after it is shut down. It takes approximately 10 minutes to restart.

Shut Down Pointing Device

7-9. Following the proper pointing device shutdown procedures will ensure a quicker alignment at the next startup, if the vehicle/system has not been moved. To shut down the pointing device without shutting down the computer, press PERFORM PD SHUTDOWN. This allows the system to shut down and save its position. It also enables the system to boot up in approximately 2 minutes.

Chapter 7

POINTING DEVICE "BORESIGHT" SCREEN

7-10. The pointing device "Boresight" screen (Figure 7-2) allows the operator to compensate for mechanical alignment errors and must be completed before using the pointing device. Before beginning, ensure that the power is on and that the pointing device is aligned.

- The fields at the top left corner display the azimuth, elevation, and roll of the current azimuth correction.
- Step 1 of the boresighting procedures is to boresight the mortar. (See Chapter 2 for more details.)
- In step 2, the gunner verifies that the mortar is laid on the DAP at an azimuth of 800 mils, and then clicks the ON DAP AT 800 MILS button.
- The gunner completes steps 3 and 4.
- The ZERO button is only used if boresighting with a new pointing device for the first time, when the pointing device bracket has been adjusted/retorqued, or when troubleshooting.

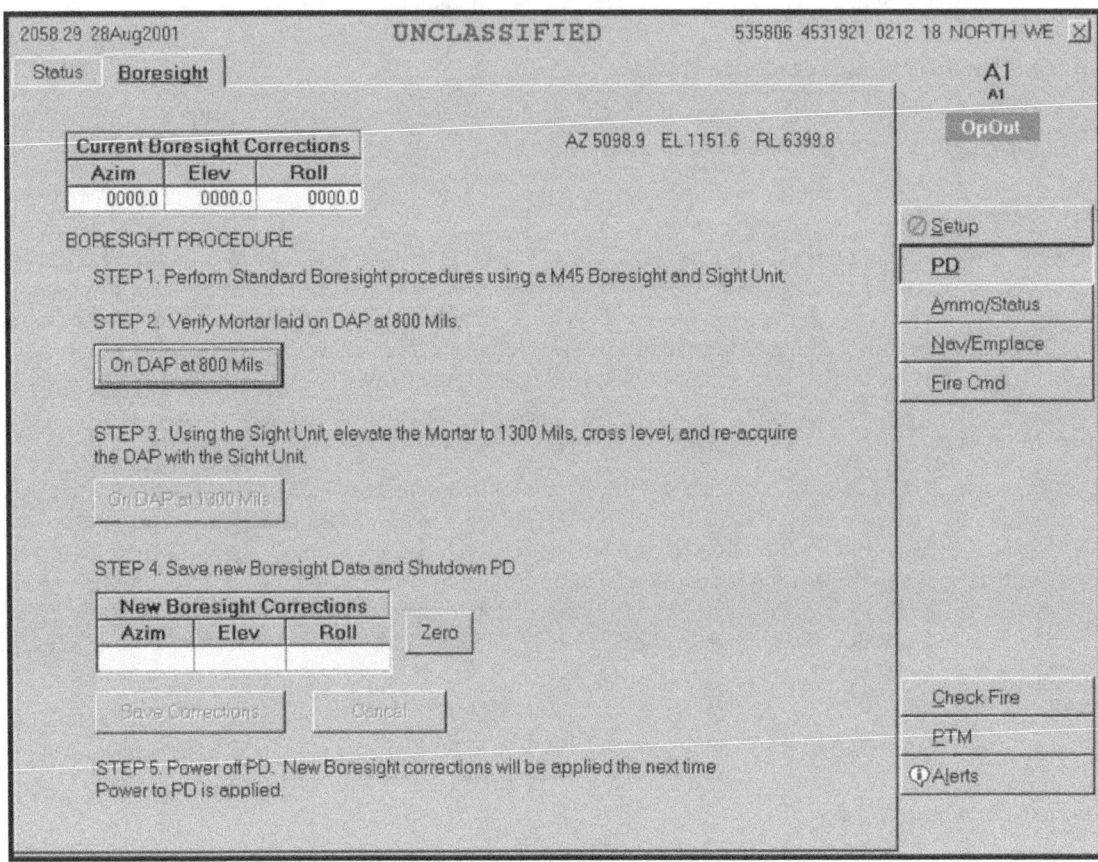

Figure 7-2. "Boresight" screen.

Conduct Fire Missions Using the Mortar Fire Control System

NAVIGATION AND EMPLACEMENT

7-11. The squad leader receives movement orders from the FDC via voice radio or PTM and inputs the destination into his CI. The MFCS then translates the destination into directions displayed on the driver's display and the CI. These procedures describe how to use waypoints and fire areas for navigation. A waypoint is any nonfiring destination the gun track is required to go (for example, refueling). An FP is any location planned as a position to fire the mortar. (TM 9-1220-248-10 includes detailed procedures for the emplacement of the M1064A3 gun track using the MFCS.)

NAVIGATION TO WAYPOINT

7-12. Upon receipt of a movement order from the FDC, the squad leader selects the NAV/EMPLACE button in the control button area and clicks the NAV tab to display the "Navigation/Emplacement" screen (Figure 7-3).

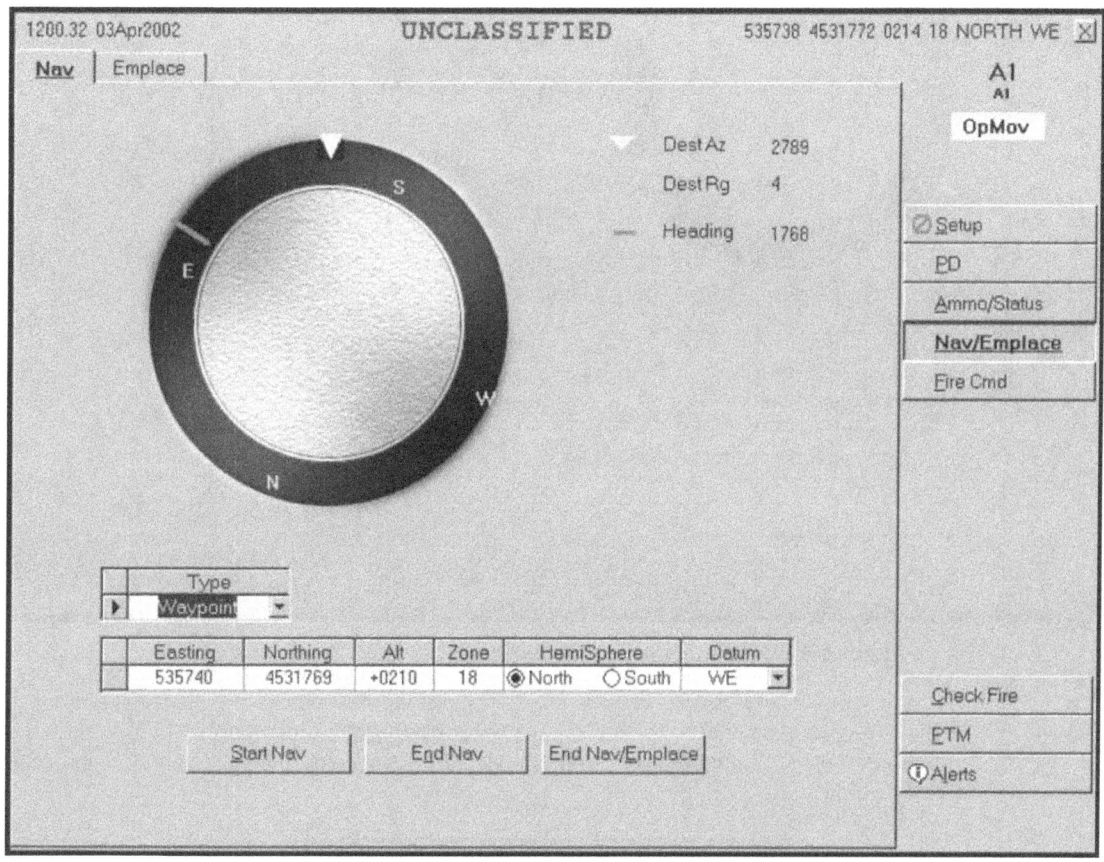

Figure 7-3. "Navigation/Emplacement" screen.

Chapter 7

(1) The squad leader enters TYPE by clicking the down arrow and choosing WAYPOINT from choices of FIRE AREA or WAYPOINT. He also enters or selects data for the following fields: EASTING, NORTHING, ALTITUDE (ALT), ZONE, HEMISPHERE, and DATUM. The squad leader then clicks START NAV, which automatically transmits the status "Operationally Moving" (OpMov) to the FDC. It also activates the driver's display. The screen shown in Figure 7-4 is then displayed.

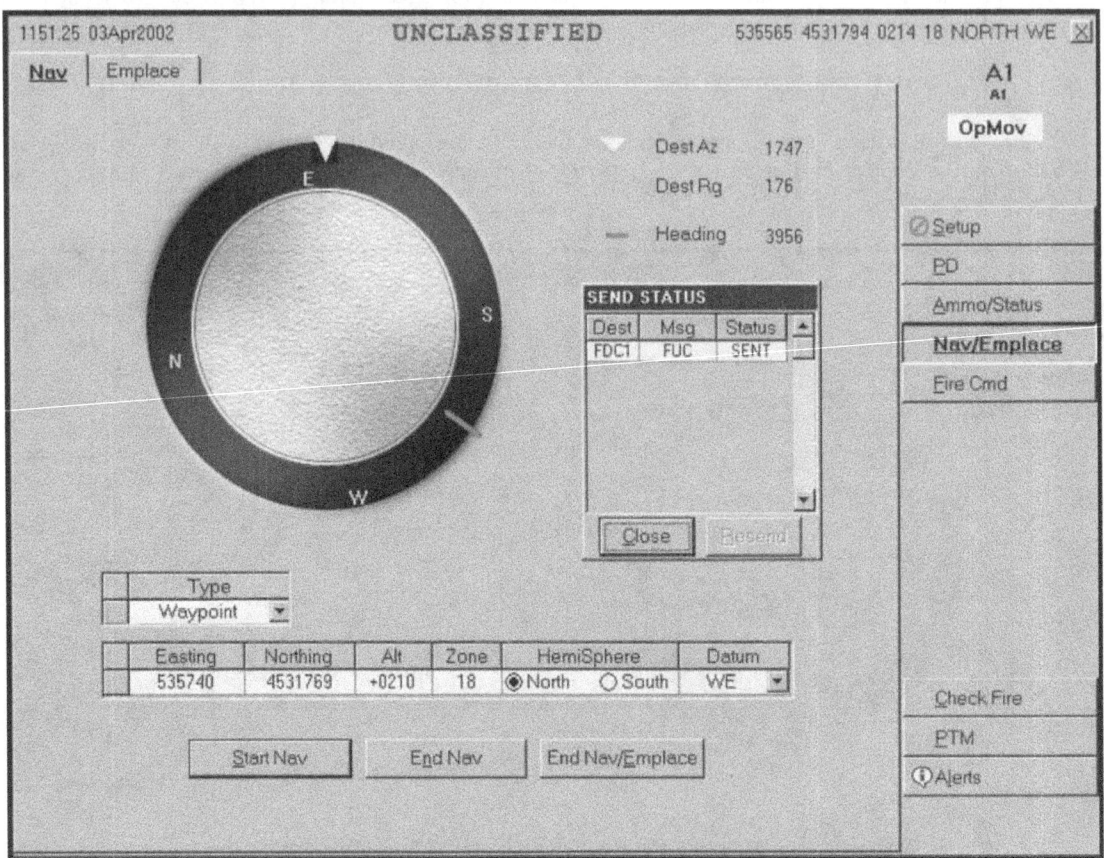

Figure 7-4. "Navigation/Emplacement" screen: Send status.

(2) After activation, the driver's display shows the driving direction and distance (Figure 7-5). The driver's display shows "Steer To" arrows indicating the direction in which to turn. The destination range and current position is updated continuously until the vehicle approaches the specified area.

Figure 7-5. Driver's display showing steering directions, distance, and position.

Conduct Fire Missions Using the Mortar Fire Control System

(3) The squad leader verifies that the gun status is changed to "Operationally Moving" (OpMov). The *Send Status* box also appears, specifying that an "OpMov" status was sent to the FDC. The CI now includes the destination azimuth (DestAZ), destination range (DestRg) and heading (Figure 7-6). If the pointing device is aligned, a position update will be sent every 1,000 meters.

(4) When the vehicle/gun is within 30 meters or less of the waypoint, the driver's display indicator will display "Arrived" (Figure 7-7). The CI however, may not display "Arrived" if the CI has been set for a radius other than 30 meters, in which case the squad leader determines when the vehicle/gun has arrived by the destination range of 30 meters or less.

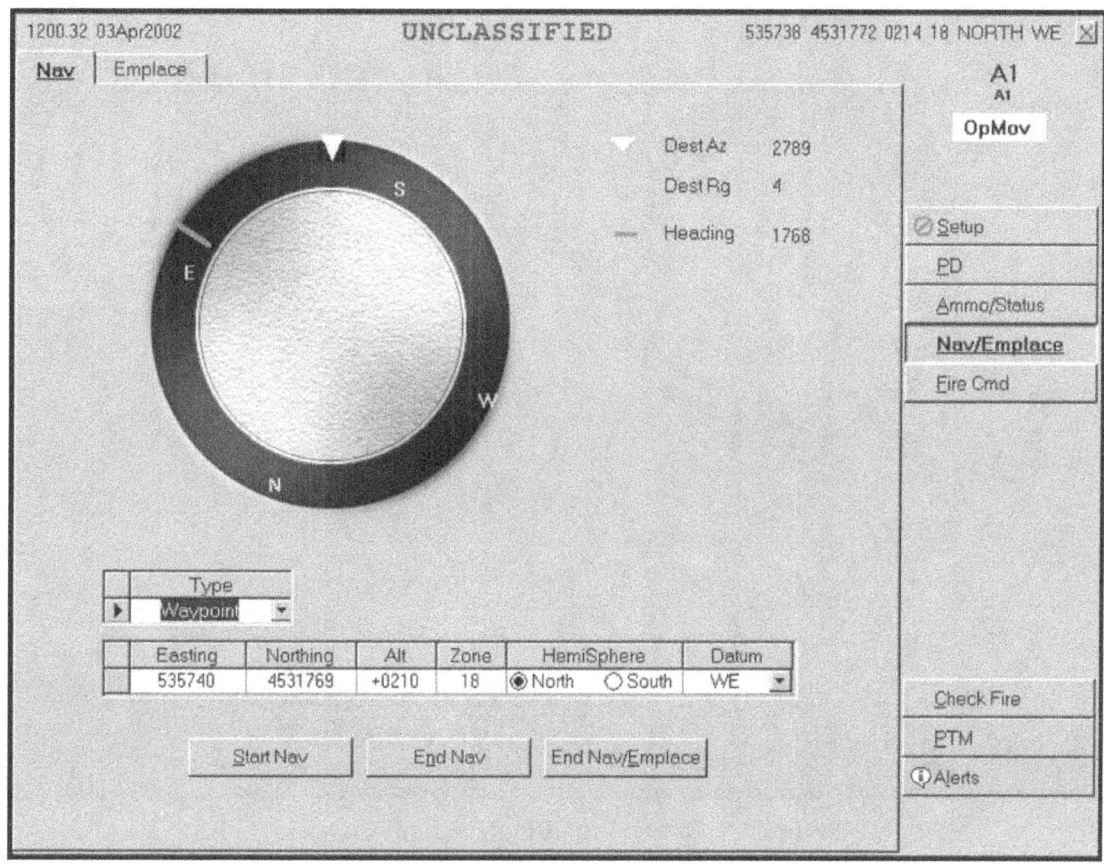

Figure 7-6. "Navigation/Emplacement" screen: destination azimuth, destination range, and heading.

Figure 7-7. Driver's display showing arrival at the waypoint.

Chapter 7

(5) The driver continues to maneuver the vehicle left or right until the driver's display indicator is within a minimum of 20 mils and a maximum of 100 mils of the waypoint.

(6) When the destination is reached, the squad leader clicks END NAV, which sends the FDC an "Operationally Stationary" (OpSta) status and the gun's current location. The *Send Status* box appears, specifying that the message was transmitted and received by the FDC (Figure 7-8). At the waypoint, the squad leader accomplishes the task given or waits for further orders.

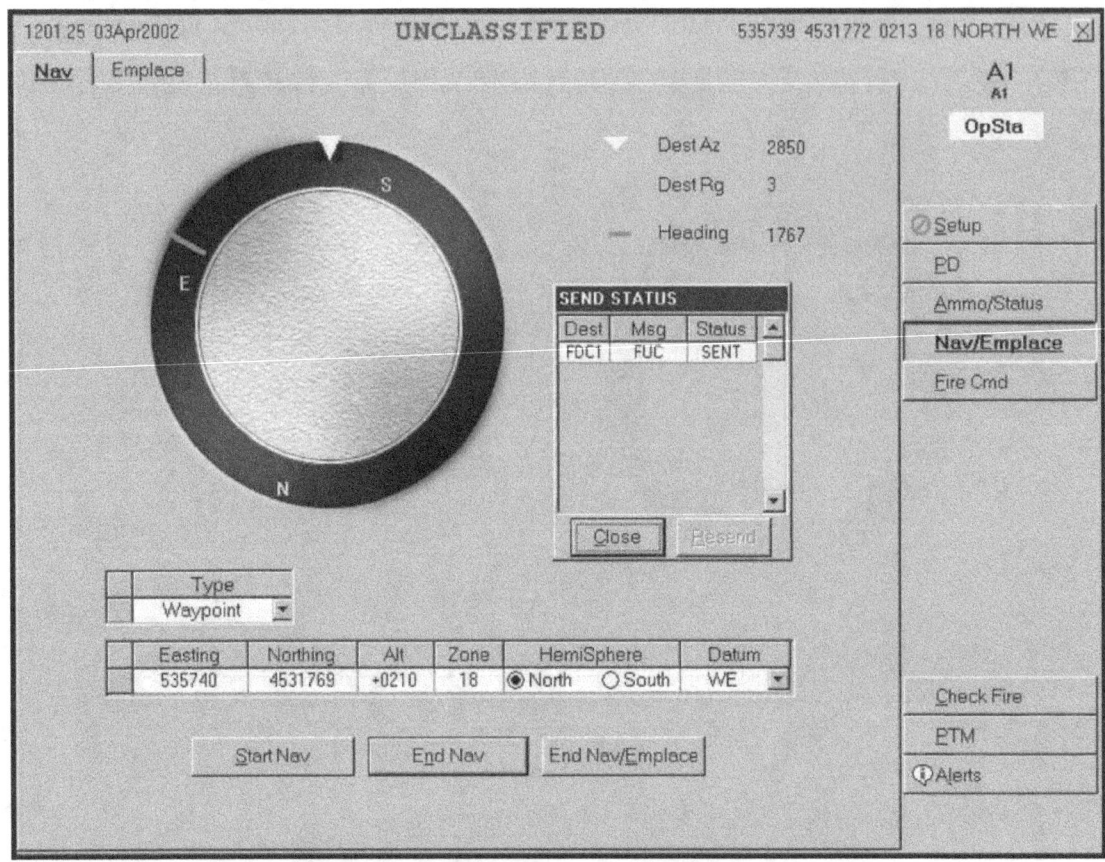

Figure 7-8. "Navigation/Emplacement" screen: message transmitted and received by the FDC.

NOTE: The current position, which is shown in yellow on the "Position" screen in Setup and which has also been sent to the FDC, is the locked-in current position. The position shown in the upper right-hand corner of the CI may change slightly due to PLGR drift and may differ from the locked-in position.

Conduct Fire Missions Using the Mortar Fire Control System

NAVIGATION TO FIRE AREA

7-13. On receipt of a movement order from the FDC, transmitted through PTM or voice, the squad leader selects NAV/EMPLACE button in the control button area and clicks the *Nav* tab. The "CI" screen shown in Figure 7-9 is displayed. In addition to the fields in the "Waypoint" screen, this screen has: FIRE AREA RADIUS, AZIMUTH OF FIRE, and ELEVATION fields.

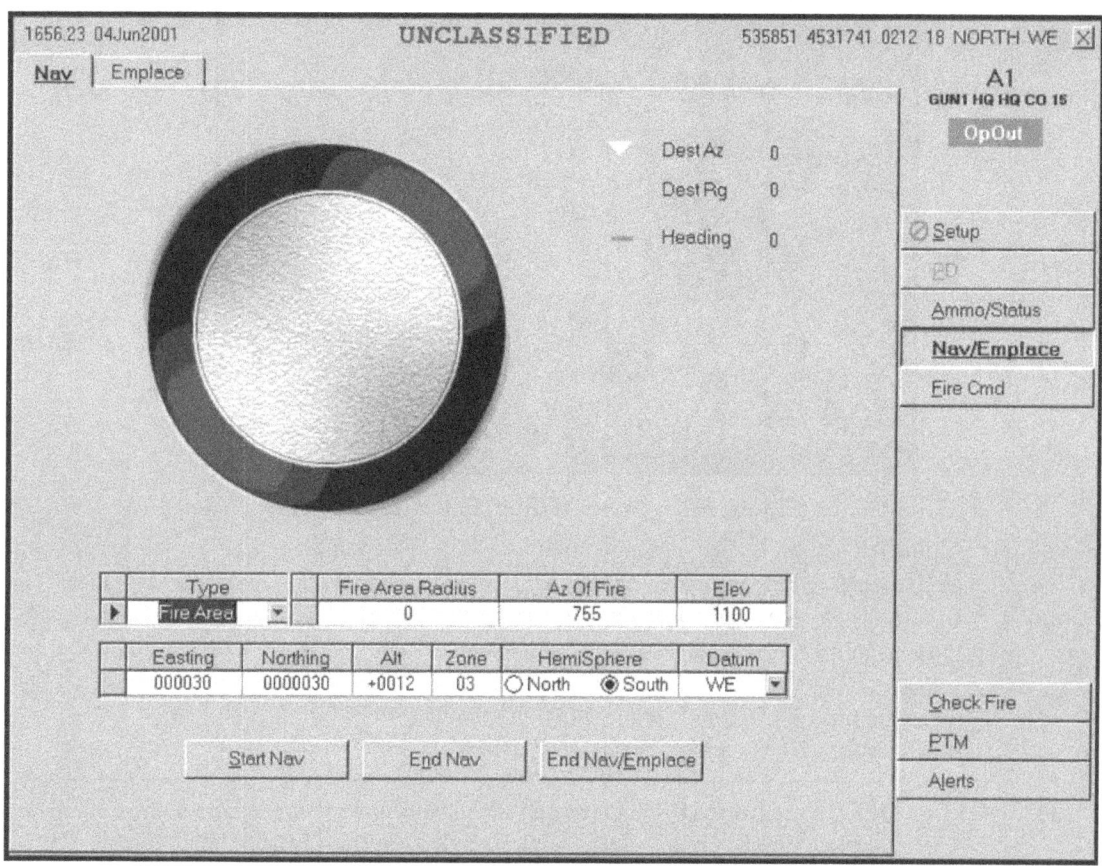

Figure 7-9. "Navigation/Emplacement" screen: fire area.

(1) The squad leader enters TYPE by clicking the down arrow and choosing FIRE AREA from choices of FIRE AREA or WAYPOINT, and then enters the received FIRE AREA RADIUS, AZIMUTH OF FIRE (AZ of Fire), and ELEVATION (Elev). He also enters the received EASTING, NORTHING, ALTITUDE (Alt), ZONE, HEMISPHERE, and DATUM data into the applicable fields.

Chapter 7

(2) The squad leader clicks START NAV. The driver's display is activated and the driver begins to navigate to the firing position using the steering arrows and distance displayed on the screen (Figure 7-10).

Figure 7-10. Driver's display activated.

(3) On the CI, the gun status is changed to "Operationally Moving" (OpMov). The *Send Status* box appears, specifying that an "OpMov" status was sent to and received by the FDC. The CI now displays destination azimuth (DestAZ), destination range (DestRg) and heading.

(4) If the pointing device is aligned, a position update will be sent every 1,000 meters.

(5) When the vehicle/gun is within 30 meters or less of the fire area, the driver's display indicator shows "Arrived." If the FIRE AREA RADIUS of 30 meters was entered, the squad leader's CI will also indicate "ARRIVED."

NOTE: If FIRE AREA RADIUS entered in the CI was more than 30 meters, the CI will indicate "ARRIVED" before the driver's display does, because the radius is not changeable in the driver's display.

FIRE COMMANDS

7-14. The FDC uses fire commands to give the mortar sections the information necessary to start, conduct, and cease fire. The fire commands are used for all types of shell and fuze combinations. The three phases of the fire command are the initial fire command, subsequent fire command, and the end of mission.

Conduct Fire Missions Using the Mortar Fire Control System

INITIAL FIRE COMMAND

7-15. The squad uses the following procedures when they receive an initial fire command.

 (1) The squad leader receives and carries out a movement order from the FDC to proceed to his waypoint/fire area. The gunner ensures that the turntable and traversing extension are both centered, and that the traversing mechanism is center of traverse. The squad leader updates and verifies AMMO and STATUS.

 (2) When gun orders are sent from the FDC, the action is indicated by an "I" (immediate) in a red circle beside the FIRE COMMAND button (Figure 7-11). The "Fire Command" screen is automatically displayed and the words "NEW GUN ORDERS" appear at the top of the gun order.

 (3) The gun orders include: method of control (MOC), method of fire (MOF), target number, phase, rounds, lot, shell, charge, fuze, and time (fuze time). The squad leader reads the gun order and acknowledges receipt by clicking the OpACK button. The gunner's display and driver's display are now activated.

 (4) If necessary, the driver maneuvers the vehicle so that the mortar tube is pointing a minimum of 20 mils and a maximum of 100 mils of the azimuth of fire. The gunner and assistant gunner manipulate the gun system to bring it to the correct azimuth and elevation displayed on the gunner's display. The gun system is ready to fire when the word "LAID" appears in both AZ and EL of the gunner's display. After firing, the squad leader clicks SHOT and, if more than one round is shot, ROUNDS COMPLETE.

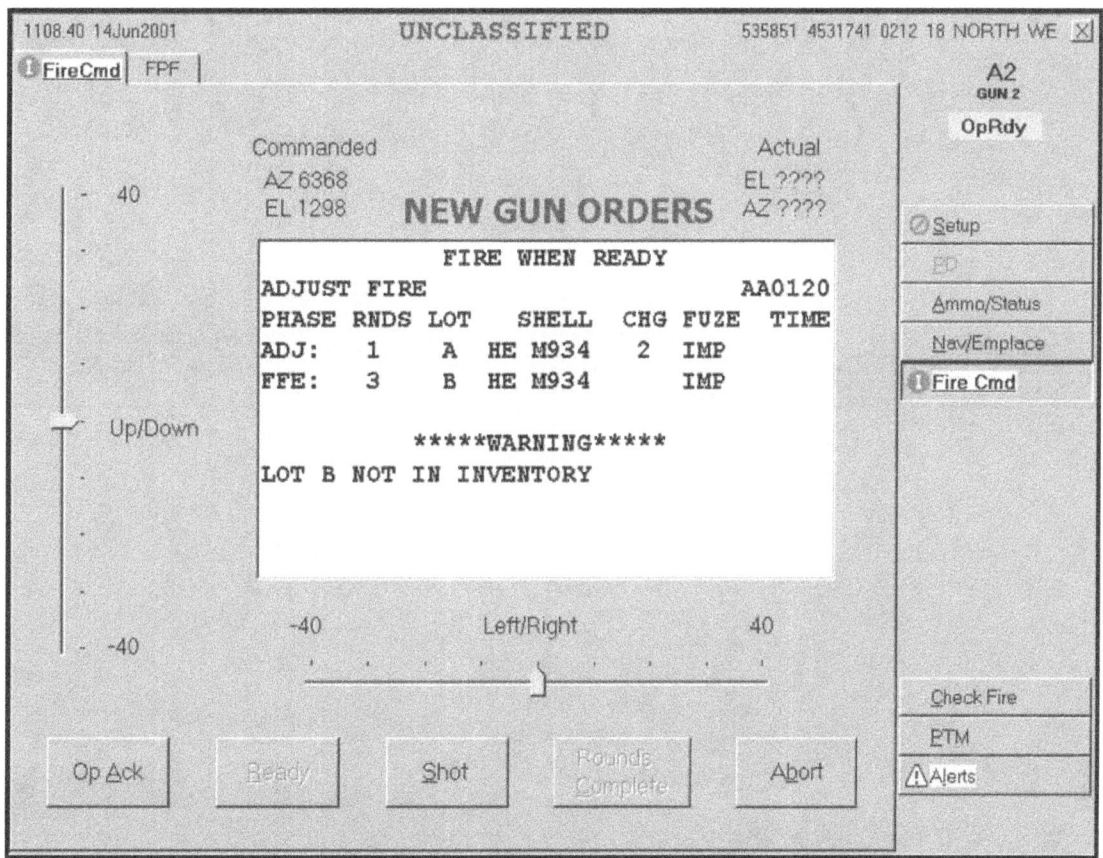

Figure 7-11. "Fire Command" screen.

Chapter 7

SUBSEQUENT FIRE COMMAND

7-16. The operator uses subsequent "Fire Command" screen (Figure 7-12) to receive further commands. The words "MORTARS LAID" in large blue letters are also displayed. If there is a SUBSEQUENT ADJUST, the squad leader receives another gun order. The MFCS calculates solutions, and the gun crew performs the same duties as in the CFF. The crew continues to respond to commands until the gun is adjusted. Once adjusted, the FDC sends an FFE gun order.

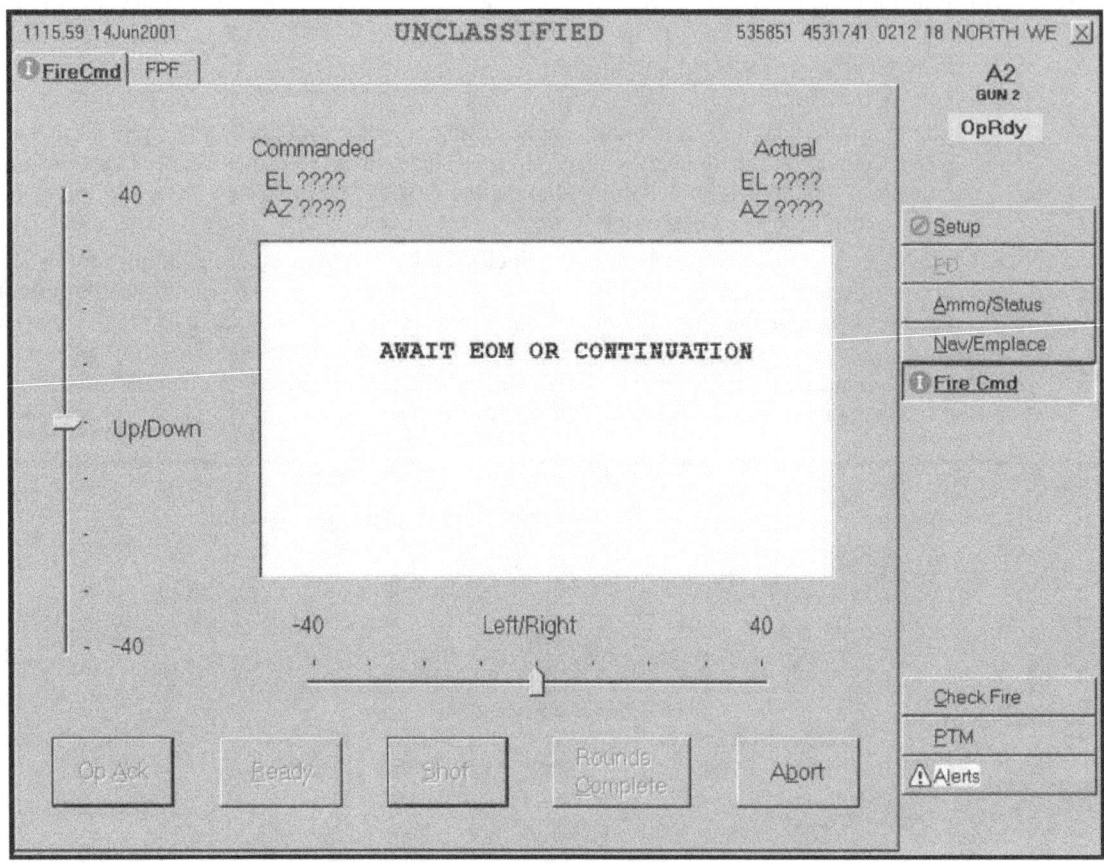

Figure 7-12. Subsequent "Fire Command" screen.

Conduct Fire Missions Using the Mortar Fire Control System

END OF MISSION

7-17. Upon receipt of the end of mission command, the "End of Mission" screen displays (Figure 7-13), and the operator confirms the amount of ammunition expended. If correct, he clicks YES. If incorrect, he clicks NO. (When NO is clicked, the "Ammo/Status" screen is displayed and allows a change to the ammunition status.) When YES is clicked, the "Not In Mission" screen is displayed (Figure 7-14).

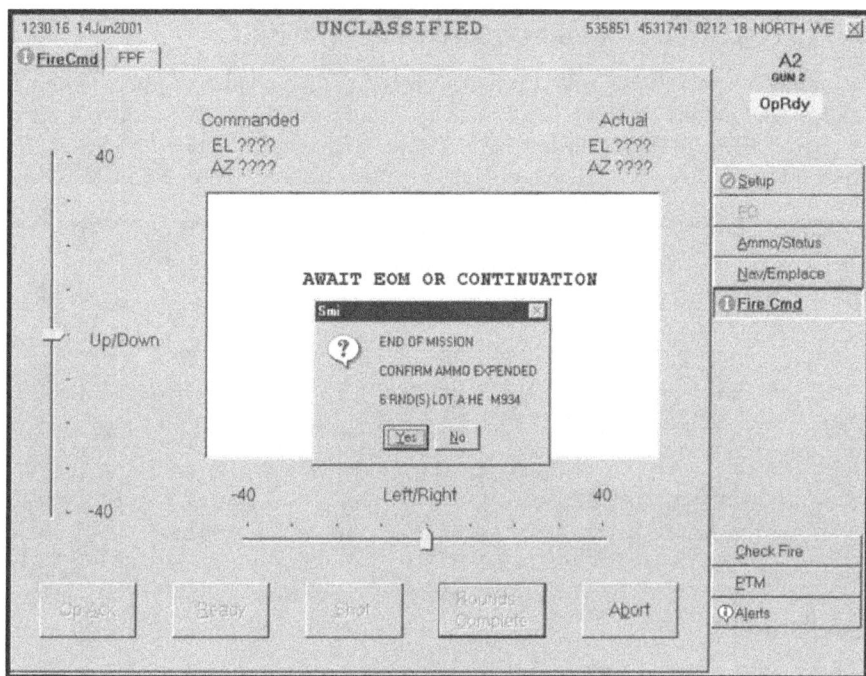

Figure 7-13. "End of Mission" screen.

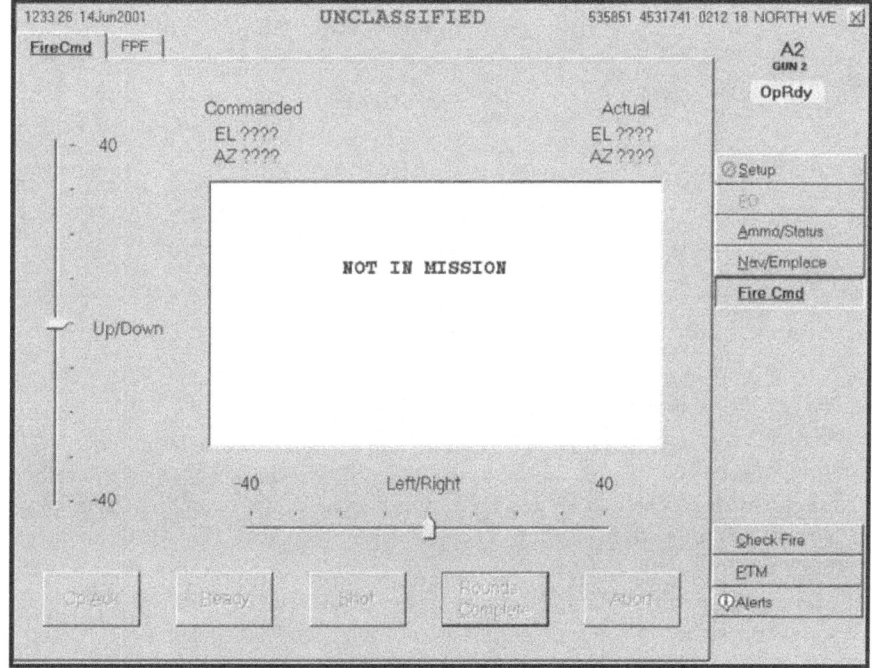

Figure 7-14. "Not in Mission" screen.

Chapter 7

FINAL PROTECTIVE FIRES

7-18. These are the procedures for receiving and carrying out a final protective fire (FPF) mission. The MFCS differentiates between an assigned FPF, where the guns are adjusted before an FFE, and a stored FPF, where the guns fire the FPF when ready. The gun MFCS can store only one FPF at a time. If an active mission is in progress and the order to fire the FPF is received, the FPF mission takes precedence.

FIRE COMMAND FOR AN ASSIGNED FINAL PROTECTIVE FIRE

7-19. Gun orders for an FPF mission are indicated by an "I" (immediate) in a red circle beside the FIRE COMMAND button. The "Fire Command" screen is automatically displayed (Figure 7-15) and the words "NEW GUN ORDERS" appear at the top of the gun order. The method of control is AT MY COMMAND, method of fire is ADJUST FIRE and method of attack is DANGER CLOSE.

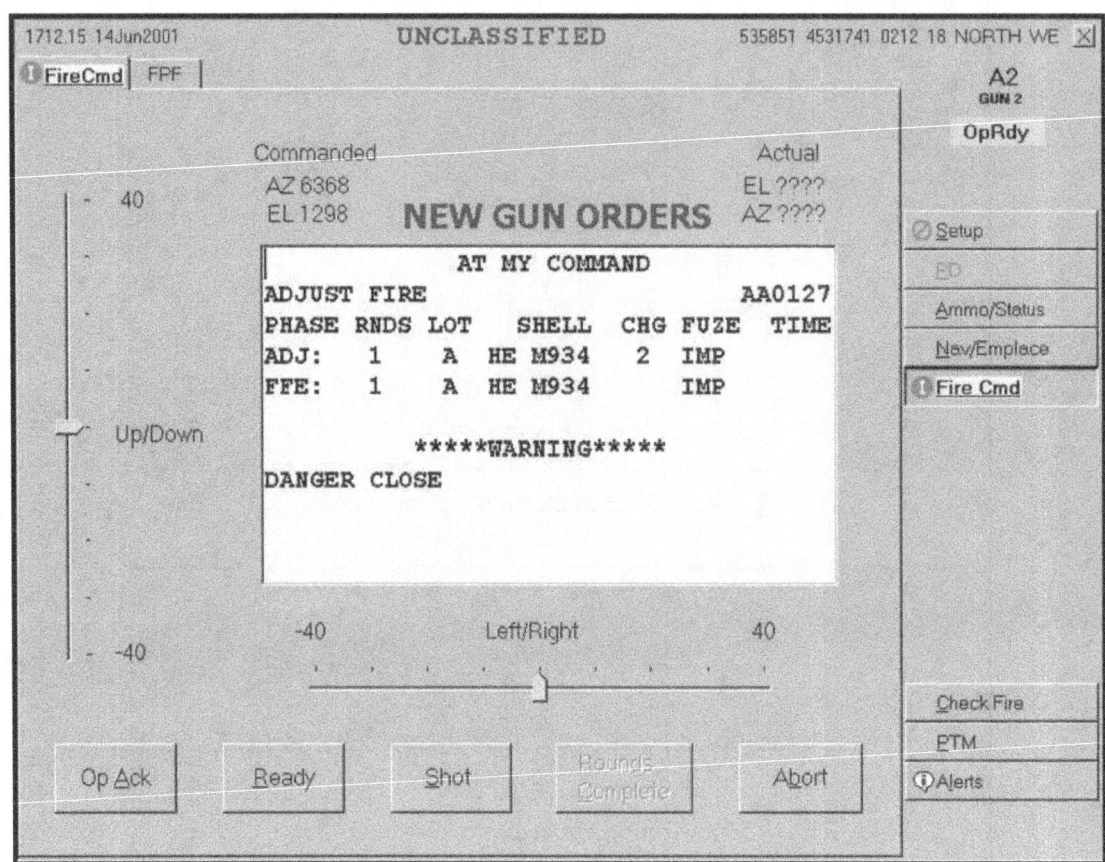

Figure 7-15. Fire command for an assigned FPF.

(1) The squad leader reads the gun order and acknowledges receipt. The gunner's display and driver's display are now activated.

(2) If necessary, the driver maneuvers the vehicle to within a minimum of 20 mils and a maximum of 100 mils of the azimuth of fire. The gunner and assistant gunner manipulate the gun system. The gun system is ready to fire when the word "LAID" appears in both AZ and EL of the gunner's display. The squad leader sends a "READY" message to the FDC and waits for a reply to fire.

(3) After firing the first round, the squad leader sends SHOT. The *Send Status* box is displayed and then a screen message displays "Await EOM" or "Continuation."

(4) If there is a SUBSEQUENT ADJUST, the squad leader receives another gun order and acknowledges the order. The gun squad then follows the same procedures previously described.

(5) When the gun is adjusted, the FDC sends an FFE gun order. The method of control is AT MY COMMAND.

(6) Upon receipt of the end of mission command (Figure 7-16), the operator confirms the amount of ammunition expended. (He can manually change the amount of ammunition expended, if necessary.) When YES is clicked, the message "NOT IN MISSION" is displayed.

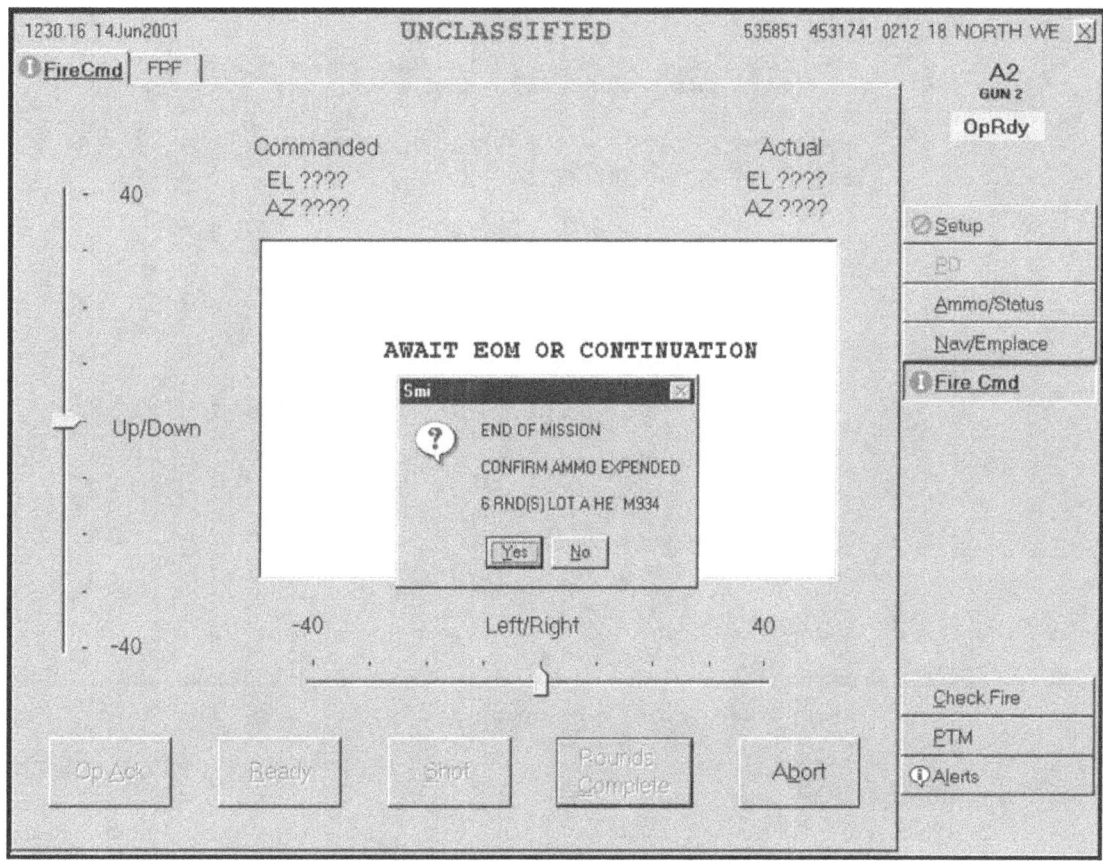

Figure 7-16. End of mission.

(7) The FPF is stored in the FPF mission buffer. The operator can click on the FPF tab to verify that the FPF has been stored.

Chapter 7

FIRE A STORED FINAL PROTECTIVE FIRE MISSION

7-20. When the FDC sends the fire the FPF command, the FPF tab is enabled and the "FPF" screen is displayed (Figure 7-17).

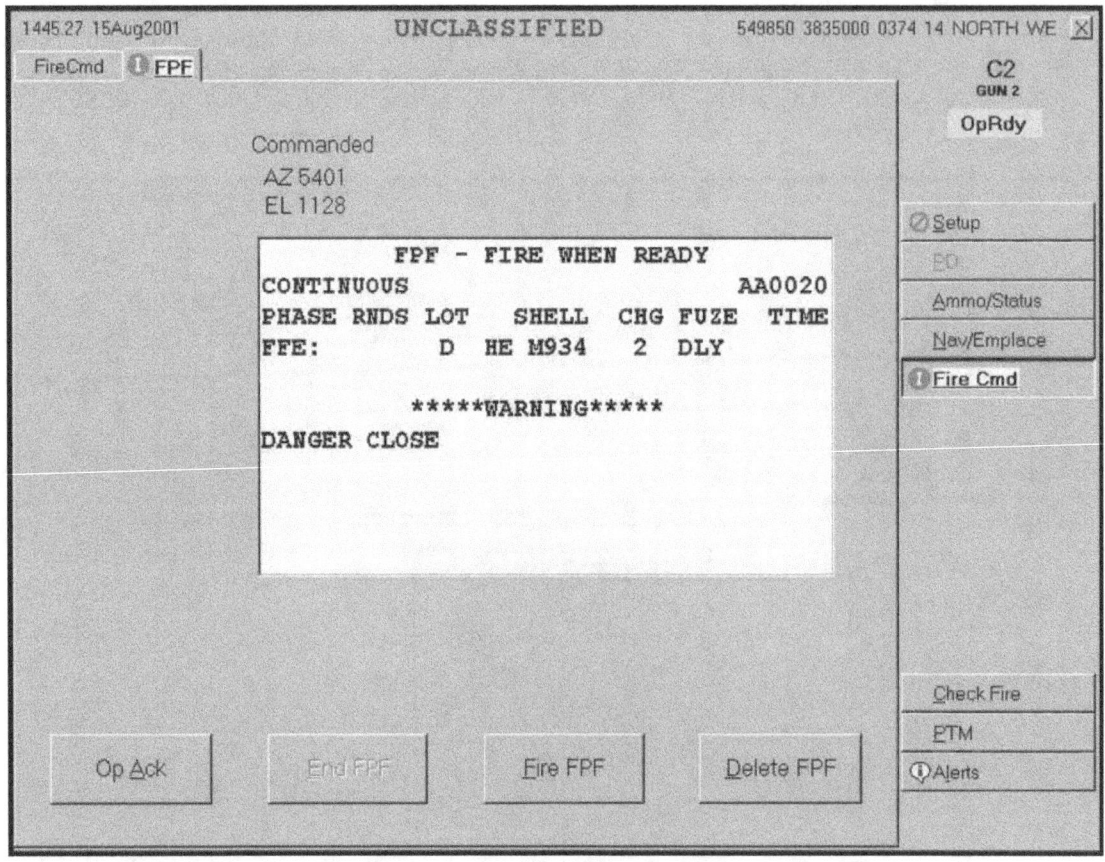

Figure 7-17. Fire the stored FPF.

(1) The squad leader reads the FPF and clicks OpACK to acknowledge receipt. The gunner's display and driver's display are activated.

(2) If necessary, the gun squad manipulates the gun as previously described until the word "LAID" appears in both AZ and EL of the gunner's display.

(3) Begin firing and continue until the "END FPF" message is sent. An "END OF MISSION" message is also sent.

Chapter 8
Fire Without a Fire Direction Center

The use of the target-grid method of fire control may not always be possible or desirable for placing fire on a target. Communications failure, casualties from enemy fire, lack of equipment, or the tactical situation may require that one or more mortars be employed without an FDC. Firing without an FDC, however, greatly increases the vulnerability of the crew and substantially reduces the capabilities of the weapon.

SECTION I. FIRE PROCEDURES

When the squad is under squad control, the gunner or an FO from the squad observes the target area. Adjustments and fire commands are sent directly to the mortar crew. The two methods used to fire the mortar without an FDC are *direct lay* and *direct alignment*. In the direct-lay method, the gunner sees the target through the mortar sight. No directional posts, aiming posts, FO, or FDC are used. In the direct-alignment method, an FO (usually the squad leader) observes the target near the mortar or the gun-target (GT) line. He observes the fall of the rounds, makes corrections relative to the GT line, and gives gun commands directly to the mortar squad.

NOTE: Employment of mortars without an FDC is only temporary—the FDC should be established as soon as possible.

ADVANTAGES AND DISADVANTAGES

8-1. Advantages of operating without an FDC include—
- Speed in engaging a target.
- Better response to commanders.
- Fewer requirements for personnel and equipment.

8-2. Disadvantages of operating without an FDC include—
- Limited movement capability of the FO.
- Increased vulnerability to direct and indirect fire.
- Difficulty of massing or shifting fires on all targets within the range of the mortar.
- Necessity of locating the mortar position too far forward where it is subject to enemy fire delivered on the friendly frontlines.
- Greater ammunition resupply problems.

FIRING DATA

8-3. The direct-alignment or direct-lay methods can be used to lay the mortar for direction. Initial range can be determined by—
- Estimation using the appearance of objects or 100-meter unit of measure methods.
- Map, photographic map, or aerial photograph.
- Mil-relation formula.

Chapter 8

OBSERVER CORRECTIONS

8-4. In direct alignment, the FO makes corrections differently than when operating with an FDC. He makes all his deviation corrections with respect to the GT line rather than with respect to the observer-target (OT) line. All deviation corrections are sent in mils or turns of the traversing handwheel (the FO should consistently use one or the other method and not mix the two).

FORWARD OBSERVER WITHIN 100 METERS OF MORTAR POSITION

8-5. The best FO location for rapid fire adjustment on the mortar is at the mortar position.

8-6. The tactical employment of the mortar usually requires the FO to be in a position other than at the mortar. However, if the FO is located within 100 meters of the mortar position, the deviation error he reads in his binoculars can be applied directly to the sight without computation. This is true because the angle between the observer-burst line and OT line is close to the angle between the mortar-burst line and GT line. The inherent dispersion of the weapon and the bursting area of the round compensate for any slight difference between these two angles.

8-7. For example, if the FO from a position within 100 meters of the mortar location observes the burst to the left of the target and reads that it is 40 mils left on the mil scale of his binoculars, he orders a correction of RIGHT FOUR ZERO. This correction is sent to the mortar in mils and is not converted to meters. The gunner applies this correction directly to the previous deflection setting using the LARS (left add, right subtract) rule.

FORWARD OBSERVER MORE THAN 100 METERS FROM MORTAR POSITION

8-8. The FO cannot always be located within 100 meters of the mortar position.

8-9. When the FO cannot locate within 100 meters of the mortar position, he must be within 100 meters of the GT line (Figure 8-1). If he is not, then making the correct adjustments may be difficult. If the FO was attacking targets over a wide frontage, he would also be required to move often.

8-10. Since the angle that exists between the mortar-burst line and GT line is not equal to the angle that exists between the observer-burst line and OT line, computations must be made to correct these differences. Because the FO is also the FDC, he must use a correction factor to figure the adjustments from the gun and not from himself.

8-11. The correction factor is a fraction, the numerator of which is the OT distance and the denominator is the GT distance (OT distance/GT distance). For example, if the FO is halfway between the mortar and target, the correction to be made on the sight is one-half his deviation spotting; if the mortar is half-way between the FO and the target, the correction is twice his deviation spotting. As other distances give other ratios, a correction factor must be applied to the number of mils spotted before ordering a deflection change. In applying this factor, simplicity and speed are important. The distances used should be to the nearest 100 meters.

EXAMPLE

If the distance from the FO to the target is 1,000 meters, the GT distance is 1,200 meters, and the deviation of the burst from the target, as read by the FO, is 60 mils (Figure 8-1), the correction is—

    1,000 ÷ 1,200 = 5/6
    5/6 x 60 mils = 50 mils

Fire Without a Fire Direction Center

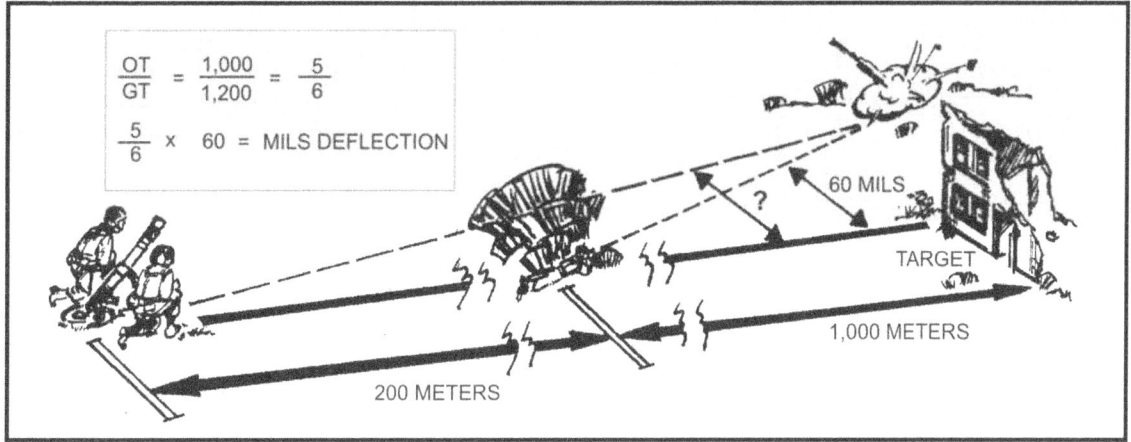

Figure 8-1. Observer more than 100 meters from mortar but within 100 meters of GT line.

INITIAL FIRE COMMANDS

8-12. Initial fire commands contain the necessary data to lay the mortars and fire the first round. The sequence for transmission of the initial fire command is as follows:

- Mortars to follow.
- Type of projectile and fuze.
- Mortars to fire.
- Method of fire.
- Deflection.
- Charge.
- Time setting.
- Elevation.

FIRE COMMANDS

8-13. When mortars are employed without an FDC, fire commands originate with the squad leader at the observation post.

NORMAL FIRE COMMANDS

8-14. The commands used in observed fire procedures without an FDC follow the procedure outlined above with the following exceptions.

Direction

8-15. When operating without an FDC, the FO gives the direction as a shift from a known point (normally the registration point) in the initial fire command. In subsequent commands, he gives the deflection correction from the last round fired. For example, during an adjustment the FO requests moving the next burst right 50 mils. Regardless of the sight setting, his command for deflection is RIGHT FIVE ZERO. The gunner applies the LARS rule to obtain the deflection to be placed on the sight. The sight has a deflection setting of 20 mils; therefore, the gunner subtracts 50 mils from 20 mils and obtains the new sight setting of 3170 mils. Normally, when the registration has been completed, the aiming posts are placed out on a referred deflection.

Chapter 8

Elevation (Range)

8-16. The FO may refer to a firing table, determine the charge and elevation (in mils) corresponding to the GT range, and announce this charge and elevation in his fire command. He may, however, announce the range in meters and have the gunner refer to a firing table to determine the charge and elevation.

Gunner's Correction

8-17. The gunner applies the deflection correction to the previous deflection setting, determines the charge and elevation for the given range on the firing table (if required by the FO), announces the charge to the ammunition bearer, sets the elevation on the sight, and then lays for elevation and deflection.

MODIFIED FIRE COMMANDS

8-18. Modified fire commands differ from normal fire commands only in that the deflection and elevation changes in subsequent commands are given as turns of the traversing handwheel and elevating crank. The advantages of modified fire commands are speed and simplicity of execution by the gunner. One turn of the traversing handwheel is equal to about 10 mils of deflection for the 60- and 81-mm and 5 mils for the 120-mm. The unabridged firing tables have a column for the number of turns of the elevation crank to change the range 100 meters.

   (1) When using modified fire commands, deflection and elevation changes are computed to the nearest quarter turn. When the FO anticipates using modified fire commands involving turns of the elevating crank, he must announce the range element of the initial fire command as a charge and elevation. This ensures that the FO and gunner are working in the same charge zone, since the number of turns required to move the burst of a round a given distance on the ground can vary considerably between two charge zones.

   (2) The gunner lays the mortar for direction and elevation as given in the initial fire command; he does not need to refer to a firing table. Following the initial fire command, the gunner makes no attempt to align the sight on the aiming point or to level the elevation bubble. He makes the corrections by taking the turns given in the subsequent commands and keeping the cross-level bubble centered. If the gunner can no longer traverse in the desired direction, he should be able to align the sight on the aiming point, center the traversing bearing, re-lay on the aiming point, and resume traverse.

   (3) In computing the number of turns of the elevation crank between two elevations in mils (taken from the firing table), the FO subtracts the smaller elevation from the larger and divides by 10 (one turn of the elevation crank being equal to about 10 mils of elevation for the 60- and 81-mm mortars) or by 5 (for the 120-mm mortars). In the ladder and bracketing methods of adjustment for range, once the FO has obtained a bracket on the target, he does not need the firing table. He continues the adjustment by halving the number of turns of the elevating crank that established the preceding bracket. In the following example, the first round burst between the FO and target.

Fire Without a Fire Direction Center

EXAMPLE (for a 60- or 81-mm mortar)

MODIFIED FIRE COMMANDS:	RIGHT FOUR TURNS
	DOWN NINE TURNS
	LEFT TWO TURNS
	UP FOUR AND ONE-HALF TURNS
	RIGHT ONE TURN
	DOWN TWO TURNS
	THREE ROUNDS
	UP ONE TURN
NORMAL FIRE COMMANDS:	RIGHT FOUR ZERO
	ONE SEVEN ZERO ZERO
	LEFT TWO ZERO
	ONE SIX ZERO ZERO
	RIGHT ONE ZERO
	ONE SIX FIVE ZERO
	THREE ROUNDS
	ONE SIX TWO FIVE

In a different example, assume that the first round was fired at a range of 900 meters and burst beyond the target. The FO wants to DROP ONE ZERO ZERO (100) for the next round and gives a modified fire command in turns of the elevating crank. Using charge 1, the elevation for the first round at 900 meters is 1275 mils. The elevation for a range of 800 meters is 1316 mils. Subtracting 1275 mils from 1316 mils gives a difference in elevation for the two ranges of 41 mils, or 4 turns. Therefore, the subsequent command to fire the second round is UP FOUR TURNS. The second round now bursts short of the target, establishing a bracket. The FO wants to split the bracket and commands, DOWN TWO TURNS, or one-half the number of turns that he previously gave to bracket the target. With this command, the FO is splitting a 100-meter bracket and could specify an FFE if he was engaging a tactical area target.

FIRE CONTROL

8-19. The FO controls the fire from an observation post, issuing fire commands directly to the mortar crew. He may select an observation post close enough to the mortar so that he can give his fire commands orally to the mortar crew. When the observation post is not close to the mortar position, the FO uses a telephone or radio to transmit fire commands.

MOVEMENT TO ALTERNATE AND SUPPLEMENTARY POSITIONS

8-20. When time or the situation dictates, the mortar may be moved to both alternate and supplementary positions, and registered on the registration point, FPF (in the defense), and as many targets as possible.

SQUAD CONDUCT OF FIRE

8-21. Conduct of fire includes all operations in placing effective fire on a target. Examples include: the FO's ability to open and adjust fires, the ability to distribute fire on the target, shifting fire from one target to another, and regulating the type and the amount of ammunition to be expended. Quick actions and teamwork are required to efficiently conduct fires. Training increases the squad's ability to fire without an FDC.

Chapter 8

8-22. The normal sequence of instruction begins on the training shell range using practice ammunition, and progresses to field training exercises (FTXs) with practice or combat ammunition. Training ammunition allocation is outlined in DA Pam 350-38, Standards in Training Commission (STRAC).

8-23. To ensure maximum efficiency, each squad member is acquainted with the principles of technique of fire for each type of adjustment and FFE. Frequent rotation of duty helps squad members to better understand this technique of fire. The designated FO (squad member) is trained in all methods and techniques used in bringing effective fire on a target as quickly as possible.

REFERENCE LINE

8-24. The normal method of establishing the initial direction when operating without an FDC is the direct-alignment method. After the initial direction has been established, the FO should conduct a registration on his registration point using only the direction stake as a reference point. After registration is completed, a reference line should be established by placing out aiming posts on a referred deflection, which then becomes the registration point or base deflection.

SQUAD USE OF SMOKE AND ILLUMINATION

8-25. Smoke is used to obscure the enemy's vision for short periods. Illumination is designed to assist friendly forces with light for night operations.

USE OF SMOKE

8-26. The squad leader must be authorized to employ smoke. The authority to fire smoke rests with the battlefield commander that the screen will affect. After careful evaluation of the terrain and weather, the FO locates a point on the ground where he wishes to place the screen. If necessary, the FO adjusts fire to determine the correct location of this point. For a screening mission, splitting a 100-meter bracket is normally sufficient.

USE OF ILLUMINATION

8-27. The battalion commander exercises control over the use of infantry mortar illuminating rounds after coordination with adjacent units through the next higher headquarters. The correct relative position of the flare to the target depends upon the wind and terrain. The point of burst is placed so as to give the most effective illumination on the target and to make sure that the final travel of the flare is not between the FO and target. Adjusting the round directly over the target is not necessary due to the wide area of illumination.

8-28. If there is a strong wind, the point of burst must be placed some distance upwind from the target so the flare drifts to the target location. The flare should be slightly to one flank of the target and at about the same range. When the target is on the forward slope, the flare is placed on the flank and at a slightly shorter range.

8-29. For adjustment on a prominent target, better visibility is obtained by placing the flare beyond the target to silhouette it and to prevent adjustment on the target's shadow. When firing continuous illumination, a strong wind can decrease the time interval between rounds. For maximum illumination, the flare is adjusted to burn out shortly before reaching the ground.

ATTACK OF WIDE TARGETS

8-30. To attack wide targets, the FO must use distributed FFE. In distributed FFE, the gunner fires a specified number of rounds but manipulates the mortar for range or deflection between each round. Distributed fire on wide targets is called *traversing fire*. To place traversing fire on a target, the FO must adjust fire on one end of the target, normally the end nearest to a known point.

Fire Without a Fire Direction Center

(1) After adjustment, the FO determines the width of the target in mils by using the mil scale in the binoculars, or by reading an azimuth to each end and subtracting the smaller from the larger. He then divides the mil width by 10 for the 60- and 81-mm, and 5 for the 120-mm (10 and 5 are the number of mils that one turn of the traversing handwheel moves the mortar). This determines the number of turns needed to traverse across the target (computed to the nearest one-half turn).

(2) In computing the number of rounds, the FO divides the width of the target by the bursting area of the round. He then divides the total number of turns by the number of intervals between the rounds to be fired to determine the number of turns between rounds (computed to the nearest one-half turn). There will always be one less interval than the number of rounds fired in the FFE phase.

(3) After adjustment and before issuing the subsequent fire command, the FO must tell the gunner to prepare to traverse right or left. The gunner traverses the mortar all the way in the direction commanded and then back two turns (four turns on M120/M121) on the traversing handwheel. With the aid of the assistant gunner, the gunner moves the bipod legs until he is approximately re-laid on his direction stake. Using the traversing mechanism, the gunner then completes realigning the mortar and announces, "Up."

(4) When the mortar is laid, the FO issues his subsequent fire command, announcing the number of rounds to be fired and the amount of manipulation between each round.

8-31. In the example in Figure 8-2, the FO is located with the 81-mm mortar squad and measures the width of his target to be 75 mils. Using a map, he estimates the range to be 2,200 meters. Using the mil-relation formula, he determines the width of the target to be 165 meters and decides to attack the target with seven rounds. There will be six intervals between the seven rounds. Since the target is 75 mils wide, he determines the number of turns to be 7 1/2. To determine the number of turns between rounds, he divides the number of turns by the number of intervals (7 1/2 divided by 6 = 1.07). This is rounded off to the nearest one-half turn (one turn).

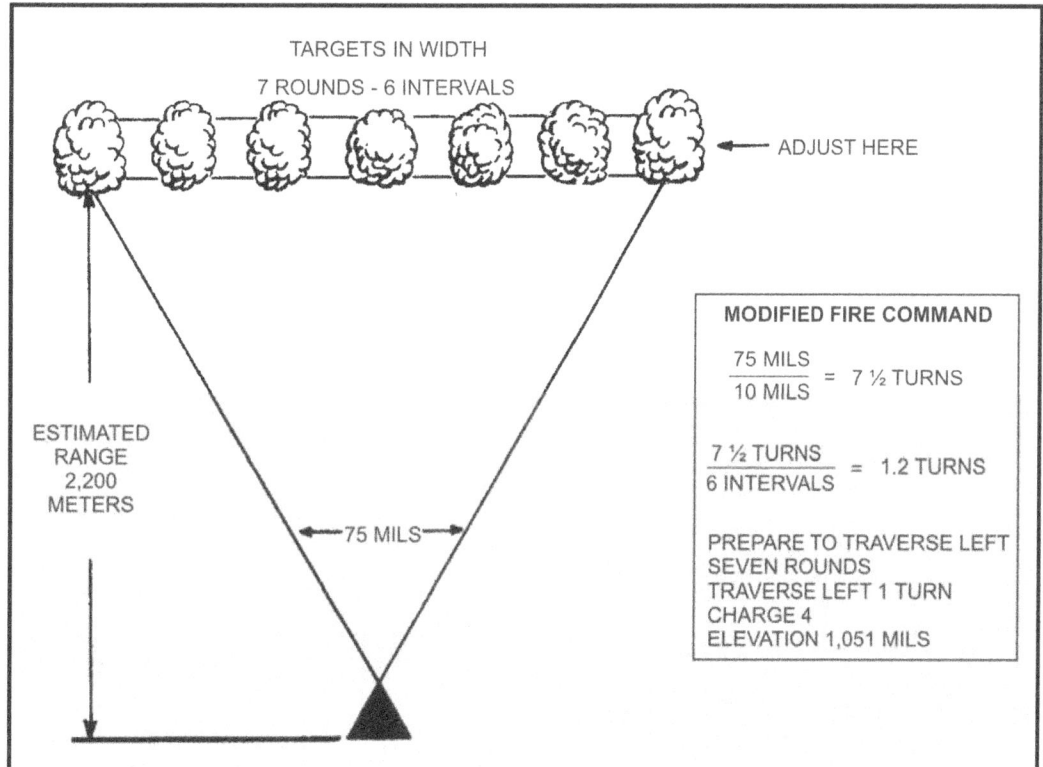

Figure 8-2. Traversing fire.

> NOTE: When determining the rounds for FFE, the FO applies the following rules:
>
> 1. For the 60- and 81-mm mortars, one round for each 30 meters and four rounds for each 100 meters.
>
> 2. For the 120-mm mortar, one round for each 60 meters and two rounds for each 100 meters.

ATTACK OF DEEP TARGETS

8-32. The FO uses searching fire to place effective fire on deep targets. To engage a deep target, the FO must adjust on one end of the target, normally the far end. When the FO anticipates using searching fire, he announces the range as a charge and elevation.

   (1) After adjustment is completed on one end of the target, the FO estimates the range to the other end of the target. He can use the firing table to determine the number of turns of the elevating crank needed to change the range 100 meters. If the FO does not have this information, he can decide the number of turns by determining the difference in elevation in mils that exists between the two ranges. He divides this difference by 5 or 10 to determine the number of turns needed on the elevating crank to cover the target.

   (2) The FO must then determine the number of rounds to be fired. Usually five rounds cover an area 100 meters deep, except at long ranges where dispersion is greater. Once he has determined the number of rounds to be fired, he determines the number of intervals between rounds. There will *always* be one less interval than the number of rounds fired.

   (3) The FO then divides the total number of turns required by the total number of intervals to determine the number of turns the gunner must make between each round (computed to the nearest one-half turn).

EXAMPLE (Adjustment of an 81-mm mortar, Figure 8-3)

If the FO has adjusted to the far end of the target and found it to be 1,000 meters, he estimates the near edge of the target to be 950 meters. By using the firing table, he determines that he must make five turns of the elevating crank to change the range 100 meters. Since the FO only wants to make a 50-meter change, he makes only half of the turns, which is 2 1/2 turns. If he did not have this information, he would determine that there is a 23-mil difference in elevation for the two ranges (elevation 1231 mils for range 1,000 meters, and elevation 1254 mils for range 950) and, by dividing by 10, that it would require two turns of the elevating crank.

The FO has determined that he will use two rounds to attack the target. There will be one interval between the two rounds fired. Then the FO divides the total number of turns required by the number of intervals, and rounds off the answer to the nearest one-half turn.

2 1/2 ÷ 1 (interval) = 2 1/2 turns between rounds

The FO is now ready to send the subsequent fire command to the gunner. The commands are—

TWO ROUNDS
SEARCH UP TWO AND ONE-HALF TURNS
ELEVATION ONE TWO THREE ONE (1231)

The gunner is told to search in the direction that the barrel moves. For this example, the barrel moves from 1231 mils to 1254 mils of elevation; therefore, the command is SEARCH UP.

Fire Without a Fire Direction Center

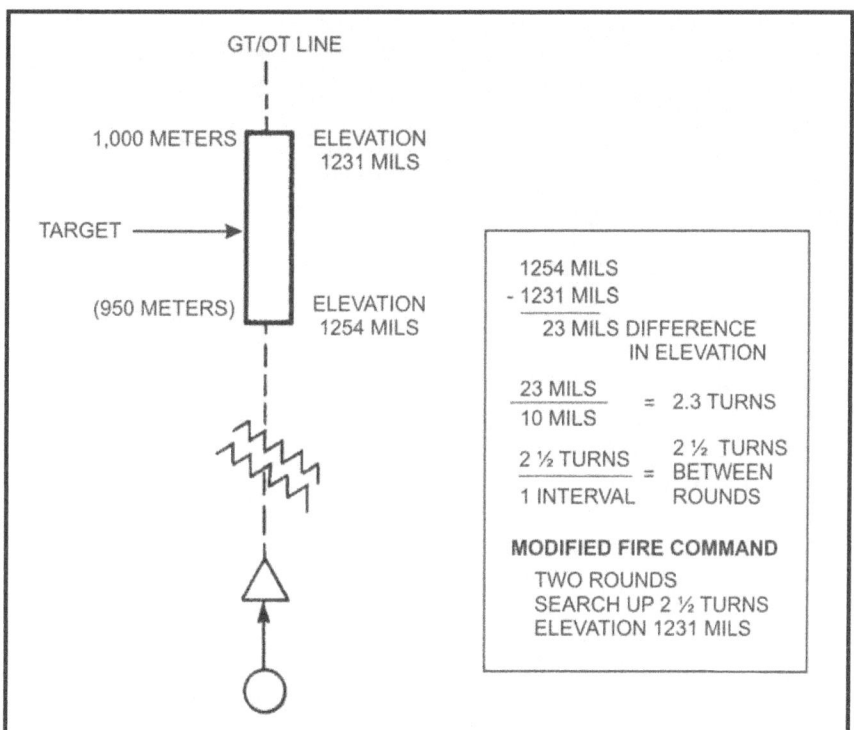

Figure 8-3. Searching fire.

SECTION II. DIRECT-LAY METHOD

In the direct-lay method of emplacing a mortar, the gunner sees the target through the mortar sight. No directional or aiming posts, FO, or FDC are used. The firing table should be used to try to obtain a first-round hit. If the first-round hit is not achieved, the firing table should be used to obtain a bracket. Depending on the location of friendly troops to the target, the bracket method, modified ladder method, or creeping method of adjustment apply.

STEP 1: INITIAL FIRING DATA

8-33. The elevation setting and charge selected should be obtained from a firing table. In the absence of a firing table, they can be determined through unit SOP or by other expedient techniques such as memorizing charge and elevation for 1,000, 2,000, and 3,000 meters; and firing with the charge and elevation setting closest to the estimated target range.

(1) Determine initial range by—
- Estimating.
- Using maps, photographic maps, and so on.
- Intersection.

(2) Place a 3200-mil deflection on the sight, and lay on the center of the target. With the appropriate elevation setting on the sight, center all bubbles by adjusting the lay of the barrel. Take appropriate actions to preclude damage to the sight and, with established charges, fire the first round. Replace the sight, if needed, and observe the burst of the round.

STEP 2: REFERRING THE SIGHT

8-34. Referring the sight centers the vertical line of the sight reticle on the burst.

(1) If the burst is over the target, turn the elevating crank up 4, 8, or 16 turns, depending on the gunner's sensing of the round, range to the target, and other possible factors. If the burst is short, turn the elevating crank down 4, 8, or 16 turns.

(2) Turn the sight elevation micrometer knob to center the elevation bubble. (If the deflection change requires a bipod displacement, the desired range change is maintained.)

(3) Re-lay the barrel on center of target (centering both bubbles by adjusting the barrel. Fire the second round and observe the burst.

NOTE: An alternate method of correction for range is to use the firing table.

STEP 3: BRACKETING THE TARGET

8-35. If the second round is a line shot and brackets the target, changing the elevation of the barrel half the number of turns used in STEP 2 splits the bracket. For example, if in STEP 2 the barrel was cranked up eight turns, now crank down four and fire the third round.

- If the second round is not a line shot but does bracket the target, refer the sight to center of burst and split the bracket by changing the elevation of the barrel half the number of turns used in STEP 2. Change the sight to the center elevation bubble, and then re-lay the barrel on center of target (centering both bubbles by changing the lay of the barrel).
- If the second round is not a line shot and does not bracket the target, repeat STEP 2 until a bracket is obtained.

STEP 4: FIRE FOR EFFECT

8-36. The appropriate actions of STEP 3 are repeated until an effect on the target is seen, then mortars fire for effect.

8-37. After obtaining hits, change the sight to center the elevation bubble and vertical line of the sight reticle on the target, and then record these data. Number the target and retain the number along with appropriate firing data. The mortar can be taken out of action, moved a short distance, and placed back into action with the mortar able to quickly and accurately attack the recorded target or other close targets.

8-38. If the mortar crew is fired upon during any of the above steps, the mortar can be displaced 75 to 100 meters with minimal effect on the fires as long as the elevation setting for the last round fired has been recorded or memorized. Once in the new position, use the recorded/memorized data as a starting point and then complete the interrupted step.

SECTION III. DIRECT-ALIGNMENT METHOD

When the FO/squad leader prepares the initial firing data, he uses the quickest and simplest method available. Initial data consist of a direction of fire and mortar-target range.

MORTAR DISMOUNTED

8-39. If the mortar is dismounted, the squad leader moves to a vantage point where he can see the targets and places out an aiming post. He directs a member of the mortar squad to place out a second aiming post (to be used as a baseplate assembly stake), aligning it with the first aiming post and target. The mortar crew mounts the mortar at the baseplate assembly stake, places 3200 mils deflection on the sight, and traverses the mortar, aligning the sight on the aiming post placed out by the squad leader. The gunner uses this aiming post as an aiming point or he may place out other aiming posts to be used as aiming points.

MORTAR MOUNTED

8-40. If the mortar is mounted, the squad leader moves to a vantage point on a line between the mortar and target. He places out an aiming post (direction stake) on which the gunner lays the mortar with the deflection set at 3200. The gunner uses this aiming post as an aiming point or he may place out other aiming posts to be used as aiming points.

Fire Without a Fire Direction Center

NOTE: The squad leader can also move to a vantage point behind the gun.

NATURAL OBJECT METHOD

8-41. When the tactical situation does not permit alignment of aiming posts and the mortar is rapidly placed into action, the squad leader can establish the mounting azimuth as follows: He selects an object with a clearly defined vertical edge that is situated in the general direction of fire. He directs the gunner to mount and lay the mortar on the edge of this object with the deflection scale set at 3200 mils. By using the aiming point as a reference point, he can place fire on a target to the right and left by determining the angle in mils between the aiming point and target. He directs the gunner to place the corresponding deflection on the mortar using the LARS rule.

SECTION IV. ADJUSTMENT OF RANGE

When firing without an FDC, the normal procedure for the adjustment of range is the establishment of a bracket along the GT line. A bracket is established when one group of rounds falls over and one group of rounds falls short of the adjusting point. The FO must establish the bracket early in the adjustment and then successively decrease the size of the bracket until it is appropriate to enter FFE.

SPOTTINGS

8-42. A spotting is the FO's determination of the location of the burst with respect to the target. Spottings are made for the range and the deviation from the GT line. The FO must be able to visualize the GT line so that he can orient the bursts to it.

RANGE SPOTTINGS

8-43. Definite range spottings are required to make a proper range adjustment. Any range spotting other than DOUBTFUL, LOST, or UNOBSERVED is definite.

Definite Range Spottings

8-44. Definite range spottings may include—
- A burst or group of bursts on or near the GT line.
- A burst(s) not on the GT line but located by the FO using his knowledge of the terrain or drifting smoke, shadows and wind patterns. Use of these spottings requires caution and good judgment.
- The location of the burst fragmentation pattern on the ground.
- A burst that appears beyond the adjusting point (over).
- A burst that appears between the FO and adjusting point (short).
- A round that bursts within the target area (target).
- A burst or center of a group of bursts that is at the proper range (range correct).

Doubtful, Lost, or Unobserved Range Spottings

8-45. Range spottings that are doubtful, lost, or unobserved usually cannot be used for adjustments and a bold shift in deviation or range should be made for the next round. These range spottings include—
- *Doubtful*. A burst that can be observed but cannot be determined as over, short, target, or range correct.
- *Lost*. A burst whose location cannot be determined by sight or sound.
- *Unobserved*. A burst not observed but known to have impacted (usually heard).
- *Unobserved Over or Short*. A burst not observed but known to be over or short. This spotting provides some information to assist the FO in the subsequent adjustment.

DEVIATION SPOTTINGS

8-46. A deviation spotting is the angular measurement from the target to the burst. There are three possible deviation spottings:
- *Line*. A round that impacts on the GT line
- *Left*. A round that impacts to the left of the GT line.
- *Right*. A round that impacts to the right of the GT line.

ERRATIC SPOTTINGS

8-47. A round that varies greatly from normal behavior is classified as an erratic round.

BRACKETING METHOD

8-48. When the first definite range spotting is obtained, the FO should make a range correction that is expected to result in a range spotting in the opposite direction—for example, if the first definite range spotting is SHORT, the FO should add enough to get an OVER with the next round. The inexperienced FO should use the guide in Table 8-1 to determine the initial range change needed to establish a bracket (Figure 8-4).

Table 8-1. Initial range change.

OT DISTANCE	MINIMUM RANGE CHANGE (ADD or DROP)
Up to 999 meters	100 meters
Over 1,000 to 1,999 meters	200 meters
2,000 meters and over	400 meters

8-49. Once a bracket has been established, it is successively decreased by splitting it in half until it is appropriate to enter FFE. FFE is usually requested in area fire when a 100-meter bracket is split.

8-50. The FO must use his knowledge and experience in determining the size of the initial and subsequent range changes. For example, if the FO adds 800 after an initial range spotting of SHORT and the second range spotting is OVER but the bursts are much closer to the target than the initial rounds, a range change of DROP 200 would be appropriate.

Fire Without a Fire Direction Center

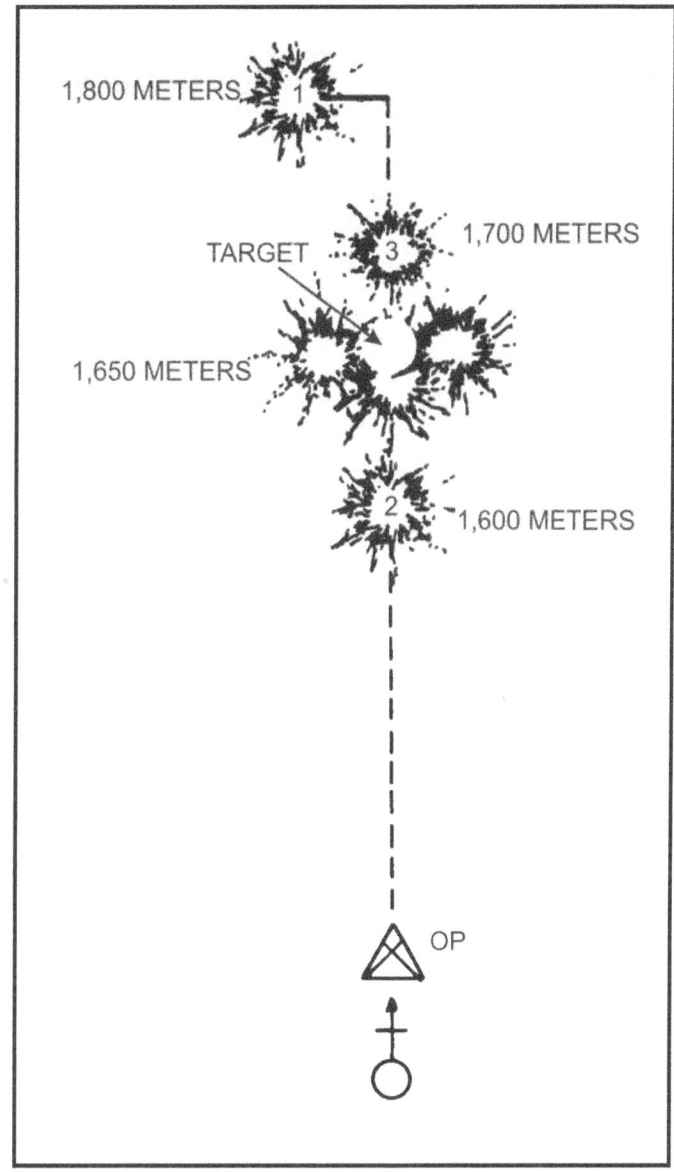

Figure 8-4. Bracketing method.

CREEPING METHOD OF ADJUSTMENT

8-51. When a danger close mission is requested, the creeping method of adjustment is used. When the FO requests an adjustment on a target that is within 400 meters of friendly troops, he adds a 200-meter safety factor to ensure that the first round does not fall short. When the initial round is spotted, he estimates the overage in meters. He then makes the correction for range by dropping half of the estimated overage. Once he has given a correction in turns equivalent to a drop in range of 50 meters, OR in the nearest number of half turns equivalent to 50 meters, he continues this same adjustment until he has a RANGE CORRECT, a TARGET, or a SHORT spotting. If, during the adjustment, a round falls short of the target, the FO continues the adjustment using the bracket method of adjustment.

LADDER METHOD OF ADJUSTMENT

8-52. Since surprise is an important factor in placing effective fire on a target, any form of adjustment that reduces the time interval between the burst of the first round for adjustment and FFE is useful. The

Chapter 8

ladder method of adjustment, a modification of the bracketing method, reduces the time interval and permits FFE to be delivered more rapidly. The ladder method may also be used by the FO when firing with an FDC (Figure 8-5). In the following example, the FO is located within 100 meters of the mortar position.

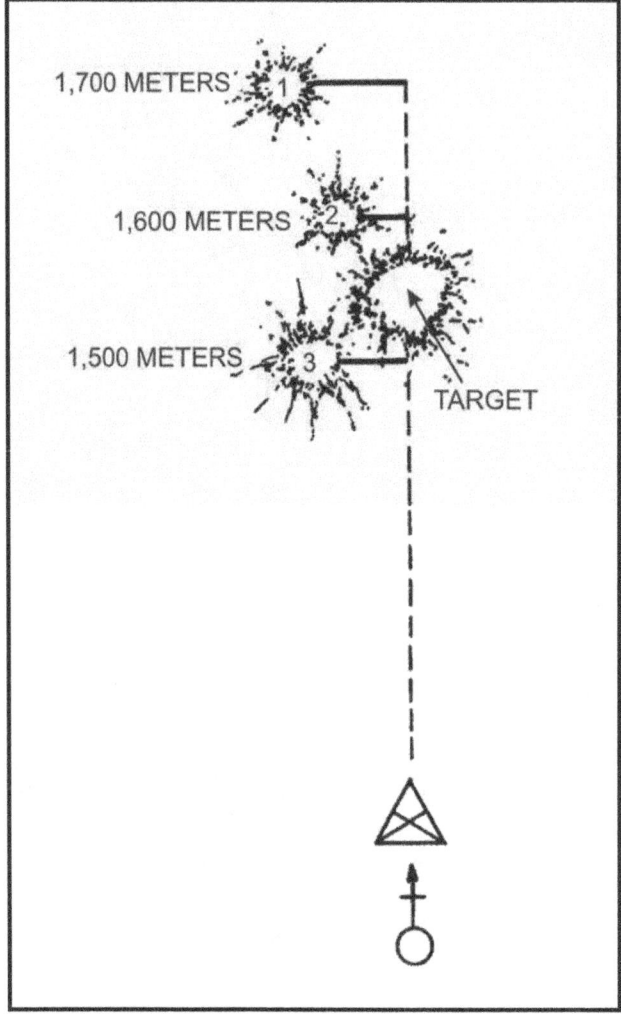

Figure 8-5. Ladder method of fire adjustment.

Fire Without a Fire Direction Center

EXAMPLE (Ladder method of adjustment, Figure 8-5)

The FO measures the deviation of a target from the registration point as right 30 mils. He estimates the GT range to be 1,600 meters. The size of the ladder is based on the minimum range change guide. To obtain a 200-meter ladder, the FO adds 100 meters to this estimated range to establish one range limit for the ladder and subtracts 100 meters from the estimated range to establish the other limit. This should result in a ladder that straddles the target. Three rounds are fired in this sequence: far, middle, and near at 10-second intervals. This helps the FO make a spotting, since no burst is obscured by the dust and smoke from a preceding burst. The FO checks his firing table to obtain the lowest charge and elevation for the far range, 1,700 meters (1098); the middle range, 1,600 meters (1140); and the near range, 1,500 meters (1178) and issues the following initial fire command:

    NUMBER ONE
    HE QUICK
    CHARGE 1
    THREE ROUNDS IN EFFECT
    100-METER LADDER
    FROM REGISTRATION POINT, RIGHT THREE ZERO (30)
    ONE ROUND, ELEVATION, ONE ZERO NINER EIGHT (1098)
    ONE ROUND, ELEVATION, ONE ONE FOUR ZERO (1140)
    ONE ROUND, ELEVATION, ONE ONE SEVEN EIGHT (1178)

The method-of-fire element in the normal initial fire command for the ladder contains the word "ladder." The word "ladder" tells the gunner that three rounds will be fired as follows: the first at the elevation announced (far range), the second at the announced middle elevation (target range), and the third at the last announced elevation (near range). The gunner indexes his sight for deflection at the base deflection minus 30, while the ammunition bearer prepares the rounds with the designated charge.

The average deviation of all three rounds is left 30 mils. The target is bracketed between the second and third bursts. The squad leader now has a 100-meter bracket of the target between 1,500 and 1,600 meters, and is ready to FFE. He issues the following subsequent fire command:

    THREE ROUNDS
    RIGHT THREE TURNS
    ELEVATION, ONE ONE SIX ZERO (1160)

ESTABLISHMENT OF A REFERENCE LINE AND SHIFTING FROM THAT LINE

8-53. The normal method of establishing initial direction when operating without an FDC is the direct-alignment method. After initial direction has been established, the FO should conduct a registration using only the direction stake. After he completes this registration, he establishes the GT line as a reference line. This is accomplished by referring the sight to the desired deflection (usually 3200 or 2800 mils) and then realigning the direction stake on this deflection. The following is an example of the FO located within 100 meters of the mortar position.

Chapter 8

EXAMPLE (Establishment of a Reference Line, Figure 8-6)

A FO has adjusted the fire of his squad on a target. A reference line has been established at 3200 mils on the azimuth scale. The FO observes another target and decides to adjust onto it. He estimates the GT range to be 1,200 meters, determines with his binoculars that the target is 60 mils to the right of the first target or registration point, and issues an initial fire command:

    NUMBER ONE
    HE QUICK
    ONE ROUND
    THREE ROUNDS IN EFFECT
    SHIFT FROM RP1, RIGHT FIVE ZERO (50)
    RANGE ONE TWO ZERO ZERO (1200)

The gunner sets his sight with a deflection of 3200 mils and applies the LARS rule (50 mils subtracted from 3200 mils equals a deflection of 3150). Looking at the firing table for HE M821 ammunition, he finds the charge to be 1 at elevation 1278. He announces the charge to the ammunition bearer and, at the same time, places the elevation on the sight. He lays the mortar and commands, FIRE.

The FO spots the first round burst as over the target and 25 mils to the left. The squad leader issues a subsequent fire command correcting the deflection and, using the minimum range change guide (Table 8-1), decreasing the range 200 meters between himself and the target. In doing so, the FO establishes a bracket. He then commands, RIGHT TWO ZERO (20), RANGE ONE ZERO ZERO ZERO (1000).

The second round bursts between the FO and target and on the GT line. The deflection is now correct, and a 200-meter bracket has been established. The squad leader's subsequent fire command is RANGE, ONE ONE ZERO ZERO (1100). The gunner selects the charge and elevation from his firing table, announces the charge to the ammunition bearer, and when all is ready he commands, FIRE.

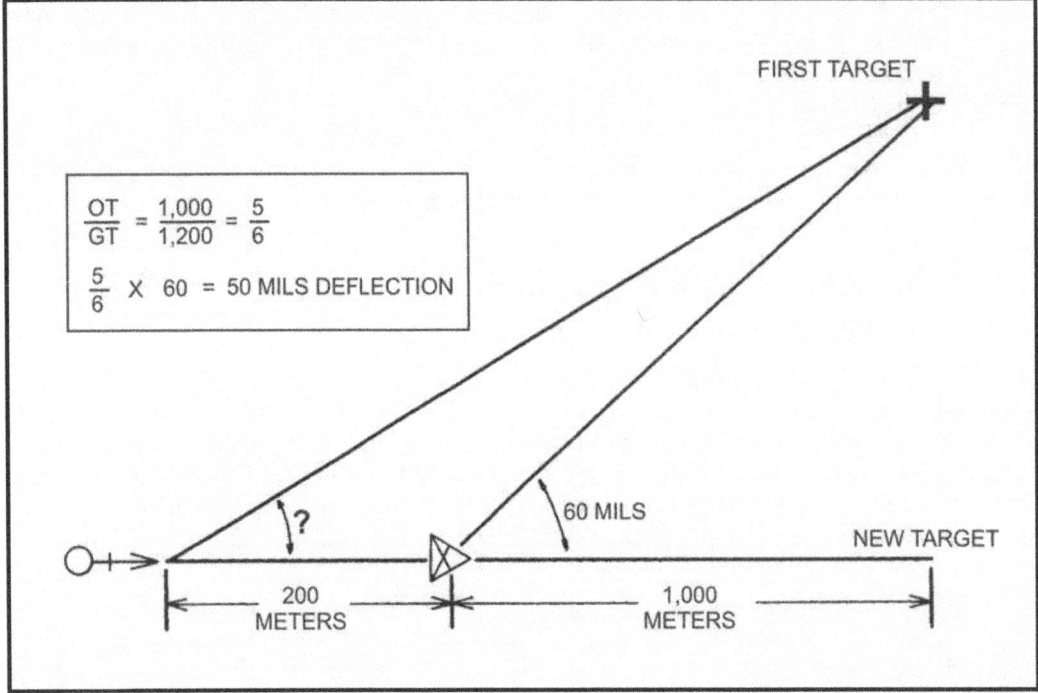

Figure 8-6. Adjusting fire onto a new target with the observer within 100 meters of the GT line.

This third round bursts beyond the target and on the GT line. A 100-meter range bracket has now been established. In his next fire command, the squad leader combines the adjustment with FFE:

    THREE ROUNDS
    RANGE, ONE ZERO FIVE ZERO (1050)

The bursting area of these rounds and their normal dispersion cover the target area with casualty-producing fragments. If the FFE fails to cover the target adequately, the squad leader makes any necessary changes in deflection or range and again orders FFE. This adjustment can also be fired using modified fire commands. The initial fire command for this mission would then be:

    NUMBER ONE
    HE QUICK
    ONE ROUND IN ADJUST
    THREE ROUNDS IN EFFECT
    FROM RP
    RIGHT FIVE ZERO (50)
    CHARGE 1
    ELEVATION ONE TWO SEVEN EIGHT (1278)

This page intentionally left blank.

Chapter 9
Gunner's Examination

The gunner's examination tests the proficiency of the gunner in five areas: mounting the mortar, making small deflection change, referring the sight and realigning aiming posts, making large deflection and elevation changes, and reciprocally laying the mortar. It is also a test of the three qualified assistants the candidate is allowed to choose. The candidate's success in the examination depends mainly on his ability to work harmoniously with these assistants. The examining board must consider this factor and ensure uniformity during the test. Units should administer the gunner's examination at least semiannually to certify crew proficiency.

SECTION I. PREPARATION

Preparation for the gunner's examination teaches the Soldier how to properly and accurately perform the gunner's duties. The squad leader is responsible for this preparation. In table of organization and equipment (TOE) units, squad members should be rotated within the squad so that each member can become proficient in all squad positions. Individual test scores should be maintained and consolidated to determine each squad's score. These squad scores can then be compared to build esprit de corps.

METHODS OF INSTRUCTION

9-1. The conditions and requirements of each step of the qualification course are explained and demonstrated. Then each candidate is given practical work and is constantly supervised by his squad leader to ensure accuracy and speed—accuracy is stressed from the start; speed is attained through repetition. The platoon leader/platoon sergeant monitors the instruction given by the squad leaders within the platoon. Demonstrations are usually given to the entire group. Squads also perform practical work under the supervision of the squad leader.

PRIOR TRAINING

9-2. A Soldier must be proficient in mechanical training, crew drill, and fire commands and their execution before he qualifies to take the examination.

PREPARATORY EXERCISES

9-3. The preparatory exercises for the gunner's examination consist of training in those steps found in the qualification course. After sufficient preparatory exercises, candidates are given the gunner's examination. Those failing the examination should be retrained for testing at a later date.

EXAMINING BOARD

9-4. The examining board consists of one officer and two senior NCOs who are proficient with the weapon. No more than one member is selected from the candidate's organization. (The commander who has authority to issue special orders appoints the board.) Scores are recorded on DA Form 5964-R, (Figure 9-1). A blank copy of this form is located in the back of this publication for local reproduction on 8 1/2- by 11-inch paper.

Chapter 9

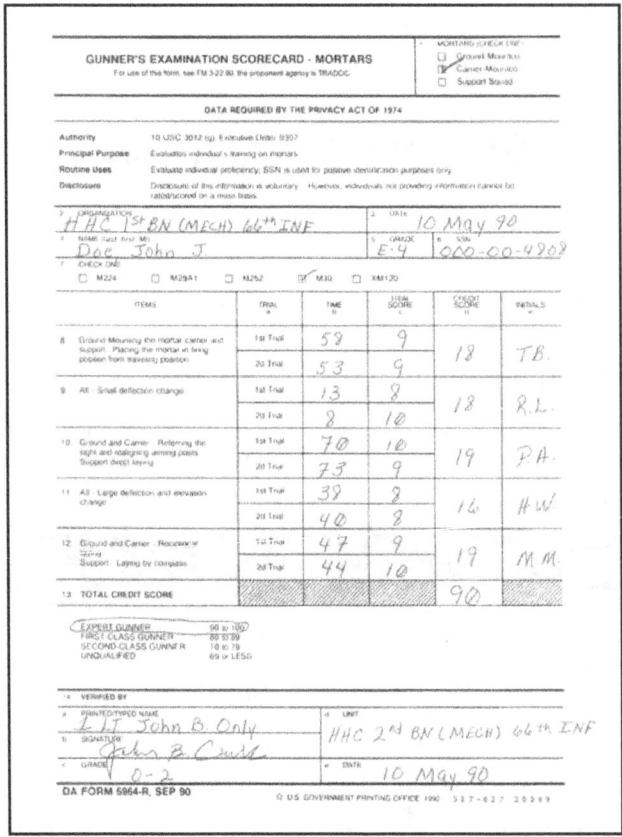

Figure 9-1. Example of completed DA Form 5964-R.

LOCATION AND DATE

9-5. Each unit armed with a mortar weapon system gives examinations semiannually. Other units may conduct examinations or allow their eligible members to take the qualification tests at nearby stations. (The commander authorized to issue special orders determines the date of the examination.) The area selected should be on flat terrain consisting of soil that allows for aiming posts to be easily positioned at 50 and 100 meters from the station position.

ELIGIBLE PERSONNEL

9-6. The following personnel are eligible to take the examination:
- Commissioned officers, NCOs, and enlisted men assigned to a mortar unit.
- Commissioned officers, NCOs, and enlisted men whose duties require them to maintain proficiency in the use of mortars, as determined by battalion and higher commanders.

QUALIFICATION SCORES

9-7. A candidate's earlier qualification ends when he is administered a record course with the mortar. He is classified according to his latest examination score as follows:
- Expert gunner   90 - 100
- First-class gunner   80 - 89
- Second-class gunner   70 - 79
- Unqualified   69 or less

Gunner's Examination

GENERAL RULES

9-8. Conditions should be the same for all candidates during the test. The examining board ensures that information obtained by a candidate during testing is not passed to another candidate, and that candidates do not receive sight settings or laying of mortars left by a previous candidate.

9-9. Unit equipment should be used in the examination; however, it should be the best available. Sight settings are considered correct when any part of the index coincides with any part of the line of graduation of the required setting.

9-10. The left side of the aiming post is used for alignment. The elevation and cross-level bubbles are considered centered when the bubbles are resting entirely within the outer etched lines on the vials.

9-11. The candidate is permitted to traverse the mortar to the middle point of traverse before each trial at laying the mortar, except at Station No. 5.

9-12. In any test that calls for mounting or emplacing the mortar, either by the candidate or the board, the surface emplacement is used. Digging is not allowed, and the rear of the baseplate assembly is not staked.

9-13. In time trials, the candidate does not receive credit for the trial if he performs any part of it after announcing, "Up."

9-14. The candidate selects his assistants from within his squad to participate in the test. When squad members are unavailable for testing, the candidate may select his assistants from outside the squad but from within his organization. The board makes sure that no unauthorized assistance is given the candidate during the examination.

9-15. A candidate is given three trials—one for practice and two for record. If he takes the first trial for record, then he must take the second trial for record even if he fails it. His credit score is the total of the two record trials. When he fails in any trial through the fault of an examiner, defective sight, mortar, mount, or other instrument used, that trial is void and the candidate is given another trial as soon as possible. If his actions cause the mortar to function unsatisfactorily during testing, he receives no credit for that portion of the test.

9-16. When a mechanical failure occurs and a mortar fails to maintain the lay after the candidate announces, "Up," a board member twists or pushes the mortar (taking up the play without manipulation) until the cross-level bubble is within the two outer etched lines. He then looks through the sight and, if the vertical line is within two mils of the correct sight picture, the candidate is given credit for that trial, as long as other conditions are met.

9-17. The candidate must repeat all commands. Commands should be varied between trials, using even and odd numbers, and right and left deflections.

Chapter 9

SECTION II. GUNNER'S EXAMINATION WITH THE GROUND-MOUNTED MORTAR

This examination tests the gunner's ability to perform basic mortar gunnery tasks with the ground-mounted mortar system.

SUBJECTS AND CREDITS

9-18. The examination consists of the following tests with maximum credit scores as shown.

Mounting the mortar	20 points
Small deflection change	20 points
Referring the sight and realigning aiming posts	20 points
Large deflection and elevation changes	20 points
Reciprocal laying	20 points

EQUIPMENT

9-19. The recommended equipment needed for the five stations includes five mortars, five sights, one aiming circle, eight aiming posts, and five stopwatches.

ORGANIZATION

9-20. The organization prescribed in Table 9-1 is recommended for the conduct of the gunner's examination. Variations are authorized, depending on local conditions and the number of Soldiers being tested.

PROCEDURE

9-21. The candidate carries his scorecard (Figure 9-1) from station to station. The evaluator at each station fills in the time, trial scores, and credit score, and initials the appropriate spaces.

Table 9-1. Organization for conducting gunner's examination (ground-mounted).

STATION	PHASE	EQUIPMENT	
		FOR CANDIDATE	FOR EXAMINING OFFICER
1	Mounting the mortar.	1 mortar 1 sight 1 baseplate stake	1 stopwatch
2	Small deflection change.	1 mortar 1 sight 2 aiming posts	1 stopwatch
3	Referring the sight.	1 mortar 1 sight 2 aiming posts	1 stopwatch
4	Large deflection and elevation change.	1 mortar 1 sight 2 aiming posts	1 stopwatch
5	Reciprocal laying.	1 mortar 1 sight 2 aiming posts	1 stopwatch 1 aiming circle

MOUNTING OF THE MORTAR

9-22. The candidate is tested at Station No. 1 on his ability to perform the gunner's duties in mounting the mortar.

Equipment

9-23. Prescribed in Table 9-1.

Conditions

9-24. The candidate is directed to mount the mortar with his authorized assistants. The conditions of the test are as follows:

All Mortars

9-25. The candidate arranges his equipment as outlined in Figures 9-2 through 9-4. The emplacement is marked before the examination.

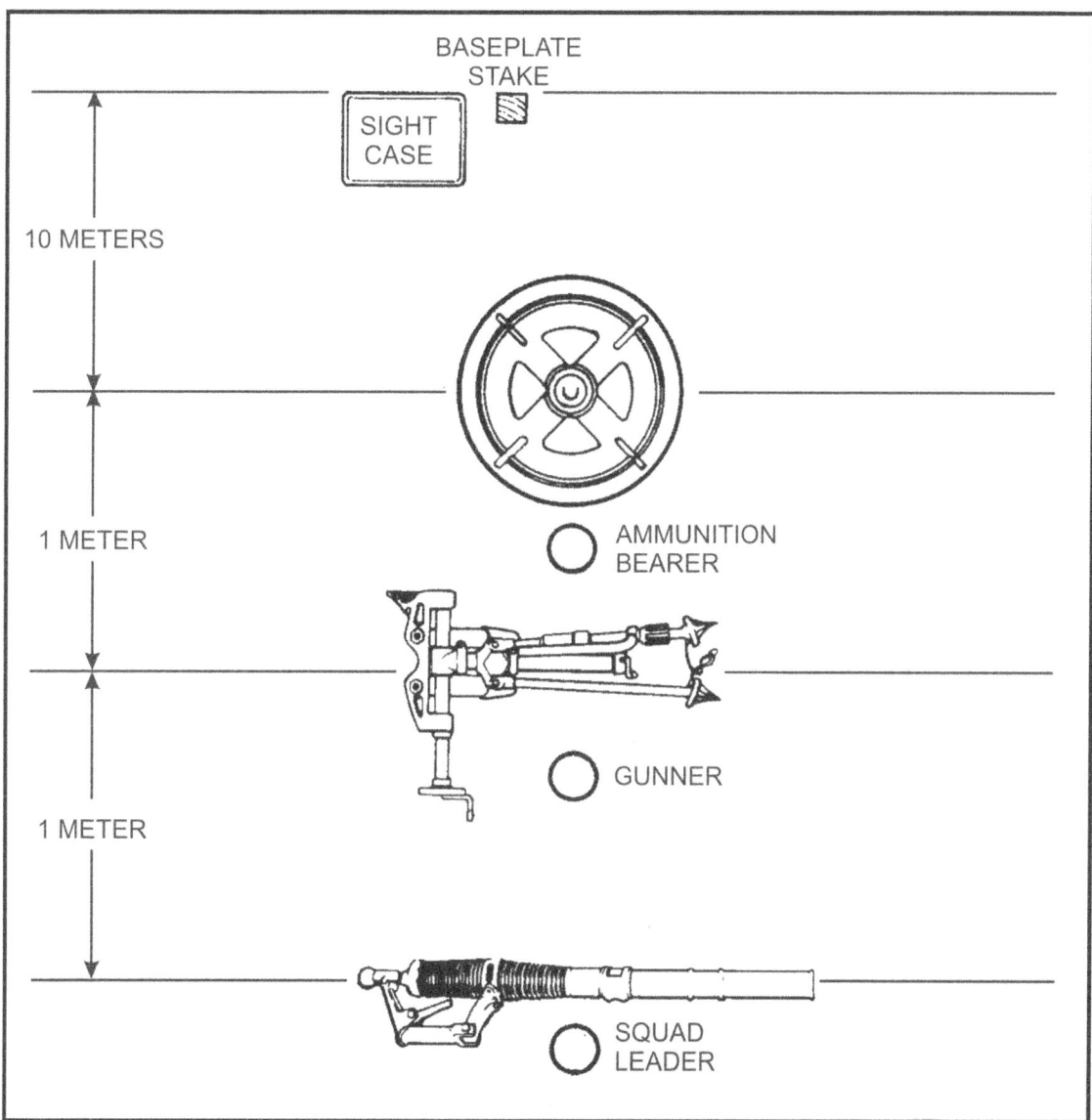

Figure 9-2. Diagram of equipment layout and position of personnel for the gunner's examination (60-mm mortar).

Chapter 9

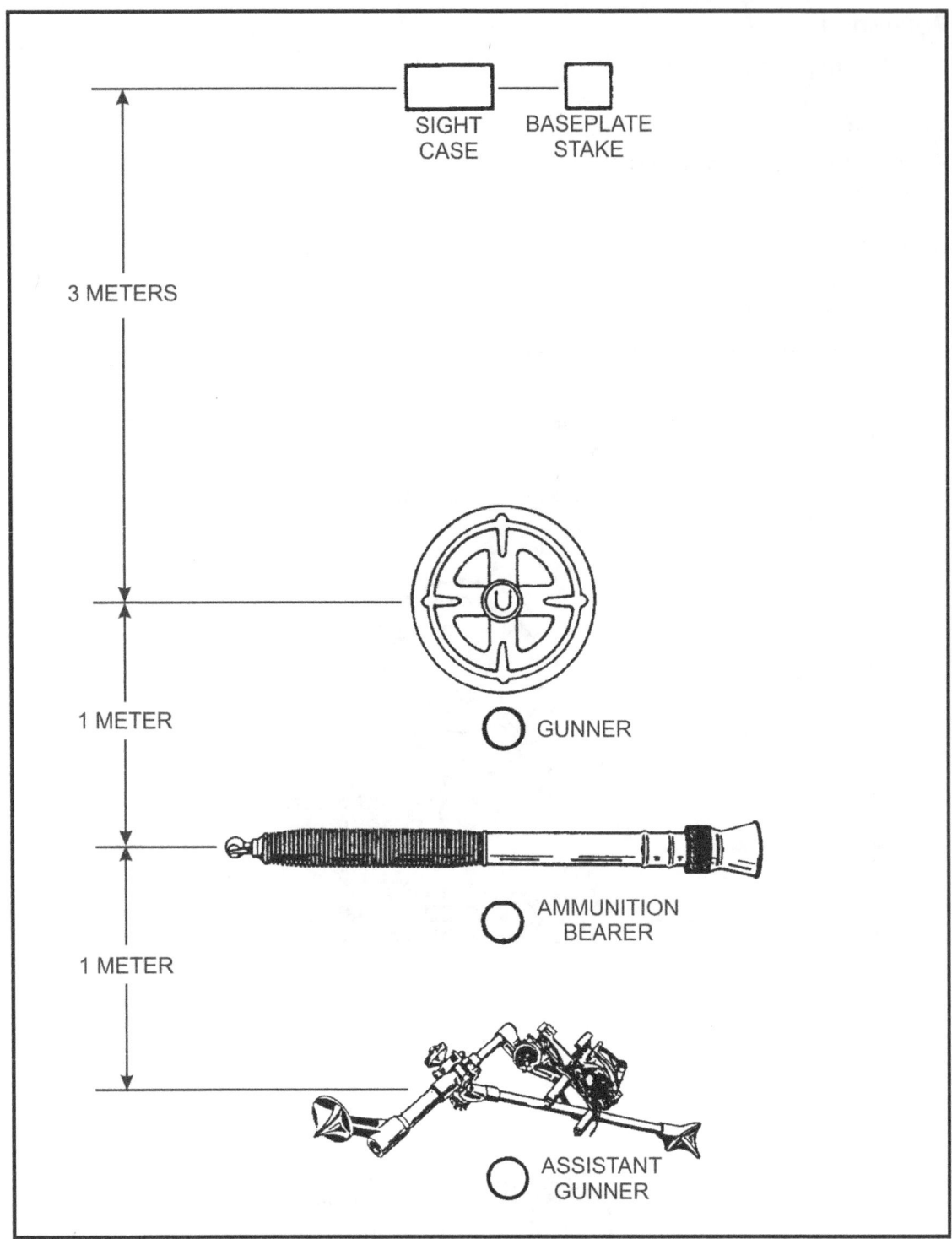

Figure 9-3. Diagram of equipment layout and position of personnel for the gunner's examination (81-mm mortar, M252).

Gunner's Examination

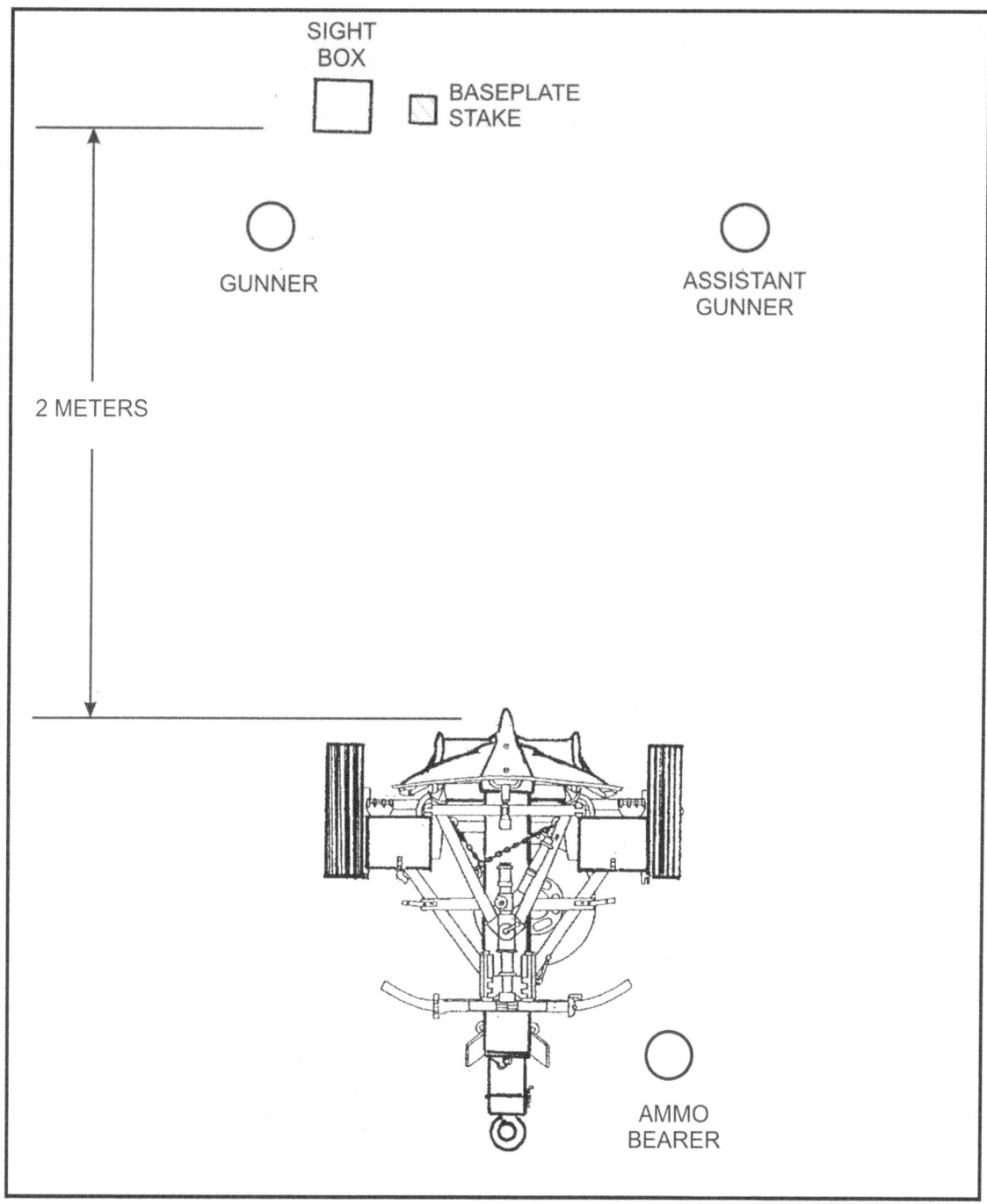

Figure 9-4. Diagram of equipment layout and position of personnel for the gunner's examination (120-mm mortar).

9-26. The mortar sight is seated in its case with 3800 mils set on the deflection scale and 1100 mils set on the elevation scale, and the sight box is closed and latched.

120-mm Only

9-27. Traverse extension is locked and centered.

Chapter 9

PROCEDURE

9-28. The candidate is given three trials. If he chooses to use the first (practice) as record, he must use the second as record. If he chooses to use the first trial as practice, he must use the second and third trials as record. His credit score for the test is the total of the two record trials.

(1) The candidate and his assistants take their positions. The candidate is instructed to mount the mortar at 3200 mils deflection and 1100 mils elevation.

(2) The evaluator points to the exact spot where the mortar is to be mounted. He indicates the initial direction of fire by pointing in that direction and gives the command ACTION, at which time the candidate begins mounting the mortar. After mounting the mortar, he should have 3200 mils deflection 1100 mils elevation.

(3) The assistants may manipulate the sight mount knob/cross-level handwheel and elevation hand crank. They may center the connection for the mortar locking pin assembly, but they MUST NOT manipulate the sight for deflection or elevation settings. After the sight is mounted, only the gunner may manipulate the elevation cross-level handwheel/ hand crank for deflection or elevation settings.

(4) When the test is completed, the candidate announces, "Up." Time is charged against him from the command ACTION to the announcement "Up."

SCORING

9-29. Scoring procedures are as follows:

(1) The candidate receives no credit when the—
- Time exceeds 1 minute and 15 seconds for the 120-mm or 90 seconds for all other mortars.
- Sight is not set correctly for deflection and elevation.
- Cross-level and elevation bubbles are not centered.
- Mortar locking pin, or the clevis lock pin is not fully locked.
- Connection for the mortar locking pin assembly (buffer carrier, 60-mm mortar) or the traversing slide assembly is off center more than two turns.
- Assistant manipulates the sight for a deflection or elevation setting.
- Baseplate is not positioned correctly in relation to the baseplate stake.
- Selector switch on the barrel is not on D for drop-fire (60-mm mortar only).
- Collar assembly is not positioned on the lower saddle (60-mm mortar only).
- Firing pin recess is not facing upwards on the barrel (81-mm mortar, M252 only).
- Traverse is more than four turns (120-mm only).
- Barrel clamp is not locked (120-mm only).
- Cross-level lock is not tight (120-mm only).
- Leg-locking handwheel is not wrist-tight (81-mm mortar, M252 only).
- Coarse cross-level nut is not wrist-tight (60-mm mortar only).
- Collar locking knob is not secured (hand-tight) to the barrel (60-mm and 81-mm mortars only).
- Bipod legs are not fully extended and the spread cable or chain is not taut (60-mm and 120-mm mortars only).

(2) When the mortar is correctly mounted within the prescribed limits, credit is given as follows:

Time (Seconds)		Point Credit for Each Trial
120-mm	60-/81-mm	
51 or less	65 or less	10
52 to 57	66 to 70	9
58 to 63	71 to 75	8
64 to 69	76 to 80	7
70 to 75	81 to 85	6
	86 to 90	5
76 or over	91 or over	0

SMALL DEFLECTION CHANGE

9-30. The candidate is tested at Station No. 2 on his ability to perform the gunner's duties when he is given commands that require a change in deflection.

EQUIPMENT

9-31. Prescribed in Table 9-1.

CONDITIONS

9-32. A mortar is mounted with the sight installed. The sight is laid on two aiming posts (placed out 50 to 100 meters from the mortar) on a referred deflection of 2800 mils and 1100 mils elevation. The mortar is center of traverse, and the vertical line of sight is on the left edge of both aiming posts.

   (1) The candidate is allowed to check the deflection set on the sight before each trial.

   (2) He is allowed to start each trial with his hand on the deflection knob

   (3) The change in deflection does not involve movement of the bipod assembly but causes the candidate to traverse the mortar at least 20 mils and not more than 60 mils

   (4) Traversing extension is locked and centered (120-mm).

PROCEDURE

9-33. The candidate is given three trials. If he chooses to use the first (practice) as record, he must use the second as record. If he chooses to use the first trial as practice, he must use the second and third trials as record. His credit score for the test is the total of the two record trials.

   (1) The candidate is given one assistant. A different command is given for each trial. The evaluator records the time and checks the candidate's work after each command has been executed.

   (2) The evaluator announces an initial command requiring a change in deflection of 20 to 60 mils and an elevation change of 35 to 90 mils. The candidate may proceed with the exercise as soon as the deflection element is announced. The evaluator announces the command in normal sequence and cadence.

   (3) No manipulation by the assistant is allowed.

   (4) Time is charged against the candidate from the announcement of the last digit of the elevation element until the candidate announces, "Up."

Chapter 9

SCORING

9-34. Scoring procedures are as follows:

(1) The candidate receives no credit when the—
- Time is 36 seconds or greater.
- Sight is not set correctly for deflection or elevation.
- Elevation bubble is not centered.
- Cross-level bubble is not centered.
- Assistant manipulates the mortar or sight for elevation or deflection.
- Vertical cross line of the sight is more than 2 mils off the correct sight picture.

(2) When the mortar is laid correctly within the prescribed limits, credit is given as follows:

Time (Seconds) All Mortars	Point Credit for Each Trial
20 or less	10
21 to 23	9
24 to 26	8
27 to 31	7
32 to 35	6
36 or over	0

REFERRING OF THE SIGHT AND REALIGNMENT OF AIMING POSTS

9-35. The candidate is tested at Station No. 3 on his ability to perform the gunner's duties in referring the sight and realigning the aiming posts.

EQUIPMENT

9-36. Prescribed in Table 9-1.

CONDITIONS

9-37. The mortar is mounted with the correct sight installed. The sight is laid on two aiming posts (placed out 50 and 100 meters from the mortar) on a referred deflection of 2800 mils and 1100 mils elevation.

(1) The mortar is within two turns of center of traverse (four turns for the 120-mm). The candidate receives an administrative command with a deflection of 2860 or 2740 mils. The mortar is then re-laid on the aiming posts using the traversing crank.

(2) The candidate checks the conditions before each trial and is allowed to start the test with his hand on the deflection knob of the sight.

(3) The change in deflection in the command must be less than 25 mils but greater than 5 mils. The elevation remains constant at 1100 mils.

(4) The candidate is allowed two assistants—one to place out aiming posts and one to move the bipod (mount). The assistants do not manipulate the sight or mortar for elevation or deflection.

(5) Traverse extension will not be used. It will remain locked in the center position.

PROCEDURE

9-38. The candidate is given three trials. If he chooses to use the first (practice) as record, he must use the second as record. If he chooses to use the first trial as practice, he must use the second and third trials as record. His credit score for the test is the total of the two record trials.

Gunner's Examination

(1) A different command is given for each trial. The evaluator records the time and checks the candidate's work after each command has been executed.

(2) When the candidate is ready, he is given a command—for example, REFER, DEFLECTION TWO EIGHT EIGHT ZERO (2880), REALIGN AIMING POSTS.

(3) The candidate repeats each element of the command, sets the sight with the data given in the command, and directs one assistant in realigning the aiming posts. Then he centers his traversing assembly and, with the help of the assistant gunner, moves the bridge or bipod (mount) assembly and re-lays on his aiming posts. After he lays the mortar on the realigned posts, he announces, "Up."

NOTE: This procedure ensures that, after a registration mission (using a parallel sheaf), the mortars have matching deflections.

(4) Time is taken from the announcement of refer and align aiming post to the candidate's announcement of "Up."

(5) The candidate's assistant may not leave the mortar position until he hears the word POSTS in the command REALIGN AIMING POSTS.

SCORING

9-39. Scoring procedures are as follows:

(1) No credit is given when the—
- Time exceeds 1 minute 15 seconds.
- Traversing crank is turned before the aiming posts are realigned.
- Sight is not set correctly for deflection or elevation.
- Mortar is not cross-leveled or correctly laid for elevation.
- Vertical line of the sight is more than 2 mils off the correct sight picture.
- Traversing assembly slide is more than two turns (four turns for the 120-mm) to the left or right of the center position.
- Assistant manipulates the sight or mortar for elevation or deflection.

(2) When the mortar is found to be correctly laid within the prescribed limits, credit is given as follows:

Time (Seconds) All Mortars	Point Credit for Each Trial
60 or less	10
61 to 65	9
66 to 70	8
71 to 75	7
76 or over	0

LARGE DEFLECTION AND ELEVATION CHANGES

9-40. The candidate is tested at Station No. 4 on his ability to perform the gunner's duties when he is given commands requiring a large change in deflection and elevation.

EQUIPMENT

9-41. Prescribed in Table 9-1.

Chapter 9

CONDITIONS

9-42. A mortar is mounted with the sight installed. The sight is laid on two aiming posts placed out 50 to 100 meters from the mortar on a referred deflection of 2800 mils and elevation of 1100 mils. The elevation change will be greater than 100 mils but less than 200 mils. The mortar is within two turns of center of traverse (four turns for the 120-mm).

(1) The candidate is allowed to check the deflection and elevation setting before each trial. He is allowed to start each trial with his hand on the deflection knob.

(2) The change in deflection involves movement of the bipod assembly and causes the candidate to shift the barrel not less than 200 mils and not more than 300 mils. The change in elevation causes him to elevate or depress the barrel from low range to high range, or vice versa.

(3) Traverse extension is locked and centered (for the 120-mm).

PROCEDURE

9-43. The candidate is given three trials. If he chooses to use the first (practice) as record, he must use the second as record. If he chooses to use the first trial as practice, he must use the second and third trials as record. His credit score for the test is the total of the two record trials.

(1) The candidate is given two assistants—one assistant may visually align the mortar, while the other shifts the bipod. The assistants neither manipulate the sight nor lay the mortar for deflection. A different command is given for each trial. The evaluator records the time and checks the candidate's mortar after each command has been executed.

(2) The evaluator announces a command for a change in deflection and elevation, that requires the movement of the bipod assembly and a change in the elevation range—for example: NUMBER ONE, HE QUICK, ONE ROUND, DEFLECTION THREE ZERO FOUR FIVE (3045), CHARGE, FOUR (4), ELEVATION, ONE ONE FOUR ZERO (1140).

(3) The candidate repeats each element of the command. As soon as the deflection element is given, he places the data on the sight and re-lays on the aiming point with a compensated sight picture. As soon as the mortar is laid, he announces, "Up." The assistants must remain in their normal positions until the deflection element is given.

(4) Time is taken from the announcement of the last digit of the elevation element of the fire command until the candidate announces, "Up."

SCORING

9-44. Scoring procedures are as follows:

(1) The candidate receives no credit when the—

- Time exceeds time exceeds 55 seconds for the 120-mm or 60 seconds for all other mortars.
- Sight is not set correctly for deflection or elevation.
- Mortar is not correctly laid for elevation.
- Mortar is not cross-leveled.
- Vertical line is more than 2 mils off the compensated or aligned sight picture.
- Traversing assembly slide is more than two turns (four turns for the 120-mm) to the left or right of the center position.
- Assistants make unauthorized movements or manipulations.
- Collar assembly is not positioned on the correct saddle for the announced elevation (60-mm).
- Traverse extension is not locked and centered (for the 120-mm).

(2) When the mortar is laid correctly within the prescribed limits, credit is given as follows:

Time (Seconds)		Point Credit for Each Trial
120-mm	60-/81-mm	
35 or less	35 or less	10
36 to 40	36 to 40	9
41 to 45	41 to 45	8
46 to 50	46 to 50	7
51 to 55	51 to 55	6
-	56 to 60	5
56 or over	61 or over	0

RECIPROCAL LAYING

9-45. The candidate is tested at Station No. 5 on his ability to perform the gunner's duties in laying a mortar for direction.

EQUIPMENT

9-46. Prescribed in Table 9-1.

STATION SETUP

9-47. The evaluator sets up the aiming circle about 25 meters to the left front of the station. He levels the instrument and orients the aiming circle so that the 0-3200 line is in the general direction the mortar is mounted. A direction stake is placed out about 25 meters in front of the mortar position.

CONDITIONS

9-48. The candidate is given one assistant to shift the bipod assembly. The assistant does not manipulate the sight or mortar in laying for elevation or deflection. The conditions of the test are as follows:

(1) All mortars are mounted at 3200 mils deflection and 1100 mils elevation. The mortar is laid on a direction stake on the initial mounting azimuth with the traversing mechanism centered.

(2) The mounting azimuth on which the candidate is ordered to lay the mortar is not less than 150 mils or more than 200 mils away from the initial mounting azimuth.

(3) The evaluator sets up the aiming circle about 25 meters to the left front of the mortar, with the instrument leveled and the 0-3200 line already on the mounting azimuth on which the mortar is to be laid.

(4) The candidate is allowed to start the test with his hand on the deflection knob. The assistants must remain in their normal positions until the evaluator gives the first deflection element.

(5) Traverse extension is locked and centered (120-mm).

PROCEDURE

9-49. The candidate is given three trials. If he chooses to use the first (practice) as record, he must use the second as record. If he chooses to use the first trial as practice, he must use the second and third trials as record. His credit score for the test is the total of the two record trials.

(1) The evaluator operates the aiming circle during this test. He lays the vertical line on the mortar sight and commands AIMING POINT THIS INSTRUMENT.

(2) The candidate refers his sight to the aiming point and replies AIMING POINT IDENTIFIED.

Chapter 9

(3) The evaluator then announces the deflection—for example, "Number one, deflection two three one five (2315)."

(4) The candidate repeats the announced deflection, sets it on his sight, and lays the mortar on the center of the aiming circle lens. He then announces, "Number one ready for recheck." The evaluator announces the new deflection immediately so that there is no delay.

(5) The operation is completed when the candidate announces, "Number one, zero (or one) mil(s), mortar laid."

(6) Time is taken from the last digit of elevation first announced by the evaluator until the candidate announces, "Number one, zero (or one) mil(s), mortar laid."

SCORING

9-50. Scoring procedures are as follows:

(1) The candidate receives no credit when the—
- Time exceeds 1 minute, 55 seconds.
- Sight is not set correctly for deflection or elevation.
- Elevation bubble is not centered.
- Cross-level bubble is not centered.
- Vertical line of the sight is not centered on the aiming circle lens.
- The mortar sight and the aiming circle deflection difference exceed 1 mil.
- Assistant performs unauthorized manipulations or movements.
- Traversing mechanism is more than two turns (four turns for the 120-mm) from center of traverse. Traverse extension is not locked in the center position.

(2) When the mortar is laid correctly within the prescribed limits, credit is given as follows:

Time (Seconds) All Mortars	Point Credit for Each Trial
55 or less	10
56 to 67	9
68 to 79	8
80 to 91	7
92 to 103	6
104 to 115	5
116 or over	0

SECTION III. GUNNER'S EXAMINATION WITH THE TRACK-MOUNTED MORTAR, M121

This examination tests the gunner's ability to perform basic mortar gunnery tasks with the track-mounted 120-mm mortar system.

SUBJECTS AND CREDITS

9-51. The examination consists of the following tests with maximum credit scores as shown.

Placing the mortar into a firing position from the traveling position	20 points
Small deflection change	20 points
Referring the sight and realigning the aiming posts	20 points
Large deflection and elevation changes	20 points
Reciprocally laying	20 points

Gunner's Examination

EQUIPMENT

9-52. The minimum equipment needed for the five stations includes five mortars, five M1064A3-series carriers, five sights, one aiming circle, eight aiming posts, and five stopwatches.

ORGANIZATION

9-53. The organization prescribed in Table 9-2 is recommended for the conduct of the gunner's examination. Variations are authorized, depending on local conditions and the number of men being tested.

Table 9-2. Organization for conducting gunner's examination (carrier-mounted).

STATION	PHASE	EQUIPMENT FOR CANDIDATE	EQUIPMENT FOR EXAMINING OFFICER
1	Placement of the mortar into firing position from the traveling position.	1 mortar carrier 1 mortar 1 sight	1 stopwatch
2	Small deflection change.	1 mortar carrier 1 mortar 1 sight 2 aiming posts	1 stopwatch
3	Referring of the sight and realignment of the aiming posts.	1 mortar carrier 1 mortar 1 sight 2 aiming posts	1 stopwatch
4	Large deflection and elevation changes.	1 mortar carrier 1 mortar 1 sight 2 aiming posts	1 stopwatch
5	Reciprocal laying.	1 mortar carrier 1 mortar 1 sight 2 aiming posts	1 stopwatch 1 aiming circle

PROCEDURE

9-54. The candidate carries his scorecard from station to station. The evaluator at each station fills in the time, trial scores, and credit score, and initials the appropriate spaces.

PLACEMENT OF MORTAR INTO A FIRING POSITION FROM TRAVELING POSITION, 120-mm MORTAR

9-55. The candidate is tested at Station No. 1 on his ability to perform quickly and accurately the gunner's duties in placing the mortar into the firing position from the traveling position.

EQUIPMENT

9-56. Prescribed in Table 9-2.

CONDITIONS

9-57. The mortar is secured in the traveling position by the mortar tie-down strap.
   (1) The sight is in its case, and the case is in its stowage position.
   (2) The candidate selects an assistant gunner.
   (3) The BAD is removed and stored properly for the 120-mm mortar system.

Chapter 9

(4) The mortar hatch covers are closed and locked (the ramp may be in the up or down position).

(5) The gunner and assistant gunner are seated in their traveling positions.

(6) The evaluator ensures that the candidate understands the requirement of the test and instructs him to report I AM READY before each trial.

PROCEDURE

9-58. The candidate is given three trials. If he chooses to use the first (practice) as record, he must use the second as record. If he chooses to use the first trial as practice, he must use the second and third trials as record. His credit score for the test is the total of the two record trials.

> **NOTE:** The traverse extension is not used during the gunner's examination. It remains locked in the center position.

(1) The evaluator positions himself inside or outside the carrier where he can best observe the action of the candidate. The evaluator's position should not interfere with the action of the candidate.

(2) The trial is complete when the candidate announces, "Up."

SCORING

9-59. Scoring procedures are as follows:

(1) The candidate receives no credit when the—
- Time exceeds 1 minute 15 seconds.
- Sight is not set at 3200 mils deflection and 1100 mils elevation
- Elevation and cross-level bubbles are not centered (within outer red marks).
- Turntable and traversing assembly slide are not centered. The traverse extension must also be centered and locked.
- The traversing lock handle is not locked.
- The white line on the barrel is not aligned with the white line on the buffer housing assembly.
- The mortar carrier rear hatch covers are not securely latched.
- The safety mechanism is not set on FIRE (F showing)
- The cross-level locking knob is not hand-tight.
- The buffer housing assembly is not positioned against the lower collar stop.
- The BAD knob is not hand-tight.
- The assistant manipulates the sight and/or mortar for elevation and/or deflection.

(2) When the mortar is found to be in the correct firing position within the prescribed limits, credit is given as follows:

Time (Seconds)	Point Credit for Each Trial
50 or less	10
51 to 57	9
58 to 63	8
64 to 69	7
70 to 75	6
76 or over	0

SMALL DEFLECTION CHANGE

9-60. The candidate is tested at Station No. 2 on his ability to perform the gunner's duties when he is given commands that require a change in deflection.

EQUIPMENT

9-61. Prescribed in Table 9-2.

CONDITIONS

9-62. The mortar is prepared for action with sight installed.

(1) The sight is laid on two aiming posts placed out 50 and 100 meters from the mortar on a referred deflection of 2800 mils and 1100 mils elevation. The turntable is centered, and the traversing mechanism is within four turns of center of traverse, and the traverse extension is centered and locked. The vertical line of the sight is on the left edge of both aiming posts.

(2) The change in deflection causes the candidate to traverse the mortar 20 to 60 mils for deflection and 30 to 90 mils for elevation.

(3) The candidate is allowed to begin the test with his hand on the deflection knob.

PROCEDURE

9-63. The candidate is given three trials. If he chooses to use the first (practice) as record, he must use the second as record. If he chooses to use the first trial as practice, he must use the second and third trials as record. His credit score for the test is the total of the two record trials.

(1) The evaluator announces an initial command requiring a change in deflection.

(2) The candidate repeats each element of the command, sets the sight with the data given, and traverses and cross-levels the mortar until he obtains the correct sight picture.

(3) Time is charged against the candidate from the announcement of the last digit of the elevation element until the candidate's announcement of "Up."

SCORING

9-64. Scoring procedures are as follows:

(1) The candidate receives no credit when—
- The time exceeds 35 seconds.
- The deflection is not indexed correctly.
- The elevation and cross-level bubbles are not centered within the outer lines.
- The vertical cross line of the sight is not within 2 mils of the left edge of the aiming post.
- The traverse extension is centered and locked in position.

(2) When the mortar is laid correctly within the prescribed limits, credit is given as follows:

Time (Seconds)	Point Credit for Each Trial
20 or less	10
21 to 23	9
24 to 26	8
27 to 31	7
32 to 35	6
36 or over	0

Chapter 9

REFERRING OF THE SIGHT AND REALIGNMENT OF AIMING POSTS

9-65. The candidate is tested at Station No. 3 on his ability to perform the gunner's duties in referring the sight and realigning the aiming posts.

EQUIPMENT

9-66. Prescribed in Table 9-2.

CONDITIONS

9-67. The sight is laid on two aiming posts (placed out 50 and 100 meters from the mortar) on a referred deflection of 2800 mils and 1100 mils elevation. The ramp is down with ammunition bearer in or outside the vehicle.

  (1) The mortar is within four turns of center of traverse. The candidate receives an administrative command with a deflection of 2860 or 2740 mils. The mortar is then re-laid on the aiming posts using the traversing crank.
  (2) The candidate checks the conditions before each trial. He is allowed to start the test with his hand on the deflection knob of the sight.
  (3) The change in deflection in the command must be less than 25 mils but greater than 5 mils. The elevation remains constant at 1100 mils.
  (4) The candidate selects two assistants—one assistant realigns the aiming posts and the other assists in moving the turntable and cross-leveling. The assistants do not manipulate the sight or mortar for elevation or deflection.
  (5) The traversing extension will not be used. It will remain locked in the center position.

PROCEDURE

9-68. The candidate is given three trials. If he chooses to use the first (practice) as record, he must use the second as record. If he chooses to use the first trial as practice, he must use the second and third trials as record. His credit score for the test is the total of the two record trials.

  (1) A different command is given for each trial. The evaluator records the time and checks the candidate's work after each command has been executed.
  (2) When the candidate is ready, he is given a command—for example, REFER, DEFLECTION TWO EIGHT EIGHT ZERO (2880), REALIGN AIMING POSTS.
  (3) The candidate repeats each element of the command, sets the sight with the data given in the command, and directs one assistant in realigning the aiming posts. Upon completion of these actions, the candidate centers the traversing assembly and, with the help of the other assistant, moves the turntable and re-lays on the aiming posts. After he lays the mortar on the realigned aiming posts, he announces, "Up."
  (4) Time is taken from the announcement of REFER, DEFLECTION TWO EIGHT EIGHT ZERO (2880), REALIGN AIMING POSTS until the candidate announces, "Up."
  (5) The candidate's assistants are not permitted to leave the carrier until the command REALIGN AIMING POSTS is given.

SCORING

9-69. Scoring procedures are as follows:

  (1) The candidate receives no credit when the—
      - Time exceeds 1 minute, 20 seconds.
      - Traversing assembly slide is turned before the aiming posts are realigned.
      - Traverse extension and turntable are not locked in the center position.
      - Sight is set incorrectly for deflection or elevation.

- Elevation and deflection bubbles are not centered.
- Sight picture is not correct.
- Traversing assembly slide is more than four turns to the left or right of the center position.
- Assistant manipulates the sight or mortar for elevation or deflection.

(2) When the mortar is laid correctly within the prescribed limits, credit is given as follows:

Time (Seconds)	Point Credit for Each Trial
60 or less	10
61 to 65	9
66 to 70	8
71 to 75	7
76 to 80	6
81 or over	0

NOTE: If for any reason either of the aiming posts fall before the candidate announces, "Up," the trial will be terminated and re-administered.

LARGE DEFLECTION AND ELEVATION CHANGES

9-70. The candidate is tested at Station No. 4 on his ability to perform the gunner's duties when he is given commands requiring a large change in deflection and elevation.

EQUIPMENT

9-71. Prescribed in Table 9-2.

CONDITIONS

9-72. The evaluator selects a deflection change that is at least 200 but not more than 300 mils off the referred deflection of 2800 mils and 1100 mils elevation.

(1) The change in deflection involves movement of the turntable. The change in elevation causes the candidate to elevate or depress the barrel from low range to high range, or vice versa. The change in elevation is not less than 100 mils and not more than 200 mils.
(2) The candidate selects two assistants.
(3) Traversing extension and turntable are locked at the center position.
(4) The candidate is allowed to check the deflection and elevation settings before each trial.
(5) The candidate is allowed to begin the test with his hand on the deflection knob.

PROCEDURE

9-73. The candidate is given three trials. If he chooses to use the first (practice) as record, he must use the second as record. If he chooses to use the first trial as practice, he must use the second and third trials as record. His credit score for the test is the total of the two record trials. He selects two assistants—one assistant may visually align the mortar, while the other elevates or depresses the standard assembly and assists in moving the turntable. The assistant does not manipulate the sight or lay the mortar for deflection.

(1) The evaluator announces a command that requires a change in deflection involving movement of the turntable and an elevation change involving movement of the elevating mechanism cam.
(2) The candidate is allowed to start the test with his hand on the deflection knob. He repeats each element of the fire command and sets the sight with the data given in the command.

Chapter 9

(3) As soon as the deflection element is announced, he can immediately place the data on the sight. The assistants must remain in their normal positions until the elevation element is given.

(4) The evaluator times the candidate from the announcement of the last digit of the elevation command to the candidate's announcement of "Up."

(5) A different deflection and elevation are given in the second trial.

SCORING

9-74. Scoring procedures are as follows:

(1) The candidate receives no credit when the—
- Time exceeds 65 seconds.
- Sight is not indexed correctly for deflection or elevation.
- Elevation and cross-level bubbles are not centered.
- Vertical line of the sight is more than 2 mils off the correct compensated sight picture.
- Traversing mechanism is more than four turns off center of traverse.
- Turntable is not in the locked position.
- Assistants make any unauthorized manipulation of the mortar or sightunit for elevation or deflection.
- Traversing extension is not locked in the center position.

(2) When the mortar is laid correctly within the prescribed limits, credit is given as follows:

Time (Seconds)	Point Credit for Each Trial
45 or less	10
46 to 50	9
51 to 55	8
56 to 60	7
61 to 65	6
66 or over	0

RECIPROCAL LAYING

9-75. The candidate is tested at Station No. 5 on his ability to quickly and accurately perform the gunner's duties in reciprocally laying the mortar.

EQUIPMENT

9-76. Prescribed in Table 9-2.

CONDITIONS

9-77. The mortar is prepared for action and laid on an initial azimuth by the evaluator and his assistants.

(1) The sight is set at 3200 mils deflection and 1100 mils elevation.

(2) The evaluator sets up the aiming circle about 75 meters from the carrier where it is visible to the gunner.

(3) The evaluator orients the aiming circle on an azimuth of not less than 150 mils or not more than 200 mils away from the initial azimuth.

(4) The candidate is allowed to begin the test with his hand on the deflection knob with the carrier engine running.

(5) A relay man is positioned halfway between the aiming circle and carrier to relay commands.

(6) The traversing mechanism is centered and the traversing extension is locked in the center position.

PROCEDURE

9-78. The candidate is given three trials. If he chooses to use the first (practice) as record, then he must use the second as record. If he chooses to use the first trial as practice, then he must use the second and third trials as record.
   (1) The evaluator operates the aiming circle during the test.
   (2) Once the candidate identifies the aiming point, the evaluator announces the deflection.
   (3) Time is started from the last digit of the first deflection announced by the evaluator.
   (4) When the candidate announces, "Ready for recheck," the evaluator immediately announces the new deflection.
   (5) The trial is complete when the gunner announces, "Zero mils (or one mil), mortar laid."

SCORING

9-79. Scoring procedures are as follows:
   (1) The candidate receives no credit when the—
   - Time exceeds 1 minute, 35 seconds.
   - Difference between the deflection setting on the sight and the last deflection reading from the aiming circle is more than 1 mil.
   - Elevation and cross-level bubbles are not centered.
   - Vertical reticle line of the sight is not centered on the lens of the aiming circle.
   - Traversing extension is not locked in the center position.
   - The mortar sight and the aiming circle are not sighted on each other with a difference of more than 1 mil between deflection readings.
   - Turntable is not centered and locked.
   (2) When the mortar is laid correctly, credit is given as follows:

Time (Seconds)	Point Credit for Each Trial
55 or less	10
56 to 67	9
68 to 79	8
80 to 90	7
91 to 95	6
96 or over	0

SUPPORT SQUAD

9-80. Support squads are located in cavalry units, task units, and infantry units. The gunner's examination for the support squad is the same as that used by the mortar section, except for the reciprocal laying, and refer and realign stations. The tests below are substituted respectively for the reciprocal laying and for the refer and realign stations. The entire refer and realign station is eliminated, and the procedures for direct lay are used.

RECIPROCAL LAYING

9-81. In this test (120-mm mortar only) the compass is substituted for the aiming circle.

Chapter 9

Conditions

9-82. The mortar is prepared for action and laid on an initial azimuth by the evaluator and his assistants.
- The turntable is centered with the sight set at 3200 mils deflection and 1100 mils elevation.
- The evaluator places the M2 compass on a stake about 75 meters from the mortar carrier and measures the azimuth to the mortar sight. He then selects a mounting azimuth from the azimuth measured to the mortar sight.
- The candidate selects an assistant gunner and driver.
- The evaluator ensures that the candidate understands the requirements of the test and instructs him to report I AM READY before each trial.

Procedure

9-83. The candidate is given three trials. If he chooses to use the first (practice) as record, he must use the second as record. If he chooses to use the first trial as practice, he must use the second and third trials as record. His credit score for the test is the total of the two record trials.
- The evaluator operates the compass during the test.
- When the candidate identifies the aiming point, the evaluator announces the deflection.
- When the gunner is laid back on the aiming point, he announces, "Up," and the evaluator commands REFER, DEFLECTION TWO EIGHT ZERO ZERO (2800), PLACE OUT AIMING POSTS.
- The ammunition bearer moves out as soon as the initial deflection has been announced by the evaluator and places out the aiming posts as directed by the gunner.
- The trial is complete when the gunner announces, "Up," after the aiming posts are in position.

Scoring

9-84. The scoring procedures are as follows:
- The candidate receives no credit when the—
  - Time exceeds 2 minutes, 5 seconds.
  - Deflection placed on the sight is incorrect.
  - Elevation and cross-level bubbles of the sight are not centered.
  - Turntable is not centered.
  - Aiming posts are not properly aligned.
  - Mortar is not within four turns of center of traverse.
- When the mortar is laid correctly, credit is given as follows:

Time (Seconds)	Point Credit for Each Trial
70 or less	10
71 to 81	9
82 to 92	8
93 to 103	7
104 to 114	6
115 to 125	5
126 or over	0

Gunner's Examination

RECIPROCAL LAYING (INFANTRY BRIGADE COMBAT TEAM MORTARS)

9-85. In this test, the sight-to-sight method is used to reciprocally lay the mortar.

Conditions

9-86. The mortar is prepared for action on an azimuth by the evaluator and his assistants.
- The sight is set at 3200 mils deflection and 1100 mils elevation.
- The evaluator sets up the base mortar about 35 meters from the test mortars where it is visible to the gunner.
- The evaluator orients the base mortar on an azimuth of not less than 150 mils or more than 200 mils away from the initial azimuth.
- The candidate selects an assistant gunner (optional for the 60-mm mortar).
- The candidate is allowed to begin the test with his hand on the deflection micrometer knob.
- The evaluator ensures that the candidate understands the requirements of the test, and he instructs him to report I AM READY before each trial.

Procedure

9-87. The candidate is given three trials. If he chooses to use the first (practice) as record, he must use the second as record. If he chooses to use the first trial as practice, he must use the second and third trials as record. His credit score for the test is the total of the two record trials.
- The evaluator positions himself at the base mortar and commands AIMING POINT THIS INSTRUMENT.
- The gunner refers his sight to the aiming point and replies, "Aiming point identified."
- The evaluator reads the deflection from the sight of the base mortar. He determines the back azimuth of that deflection by adding/subtracting 3200 mils and announces the deflection—for example, the deflection on the base mortar is 1200 mils. The evaluator adds 3200 mils to this deflection (1200 + 3200 = 4400 mils) and announces, "NUMBER ONE, DEFLECTION FOUR FOUR ZERO ZERO (4400)."
- The candidate repeats the announced deflection, sets it on the sight, and, with the help of his assistant gunner, lays the mortar on the center of the base mortar sight lens. He then announces, "Number one ready for recheck." The evaluator announces the new deflection as soon as possible so that there is no delay.
- The operation is completed when the candidate announces, "Number one, zero (or one) mil(s), mortar laid."

Scoring

9-88. The scoring procedures are as follows:
- The candidate receives no credit when the—
    - Time exceeds 1 minute, 55 seconds.
    - Deflection placed on the sight is incorrect.
    - Elevation and cross-level bubbles of the sight are not centered.
    - Mortar is not within two turns of center of traverse.
    - The sight and the base mortar sight are not sighted on each other with a difference of not more than 1 mil between deflection readings.

- When the mortar is laid correctly, credit is given as follows:

Time (Seconds) IBCT	Point Credit for Each Trial
55 or less	10
56 to 67	9
68 to 79	8
80 to 91	7
92 to 103	6
104 to 115	5
116 or over	0

Appendix A
Mortar Training Strategy

This appendix provides a comprehensive unit training strategy for training mortarmen. Leaders have the means to develop a program for training their mortar units to full mission proficiency. This training strategy applies to all mortars in all organizations of the U.S. Army. It must be adapted to a unit's mission, commander's guidance, local training resources, and unit training status.

TRAINING PHILOSOPHY

A-1. This training strategy synchronizes institutional and unit components to produce units that are trained to win on the battlefield. It includes the training documents, institutional training, unit training, and training resources needed to achieve and sustain the required outcome. It covers the skills required for individual, crew, leader, and collective proficiency and ensures that the strategy is linked horizontally within the career management field (CMF) and vertically between officer and enlisted. This strategy integrates information from several publications, including this manual, into a single-source document (see References).

UNIT MORTAR TRAINING

A-2. Technical and tactical proficiency is based on sound training. The importance of training the complete indirect fire team (skilled and proficient mortarmen and observers) must not be overlooked within the context of the battalion's overall training strategy. Unless leaders have a mortar background, they may not understand the distinct training requirements and tactical role of mortars. This ensures that they allocate priorities and resources required for effective training to mortars. Furthermore, leaders may also require training. This can be achieved by officer professional development (OPD) and noncommissioned officer professional development (NCOPD) instruction on mortars, which include both technical and tactical mortar subjects, and the call for and adjustment of indirect fires.

A-3. Once mortarmen and observers have mastered their own tasks, they must be fully integrated into the training exercises of the company, battalion, or both. However, mortars suffer from not having a training device (such as MILES) to simulate the terminal effects of mortar rounds. As a result, maneuver units tend to under-employ their supporting mortars. Despite the current absence of such devices, there are other techniques to assess the effects of indirect fire. (These are outlined FM 25-4.) Fire missions not specifically using enemy targets, such as registration and adjusting FPF, should also be routinely conducted in maneuver exercises.

A-4. A training plan that employs mortarmen or FIST personnel only as OPFOR riflemen is not effective for many reasons. Firstly, the indirect fire team is not being trained in the technical and tactical tasks pertinent to their mission. Secondly, riflemen are deprived of a valid training experience as OPFOR. Thirdly, maneuver units are not trained to employ their mortar for indirect fire support.

MORTAR TRAINING AT TRAINING BASE

A-5. The mortar unit training strategy begins with the training base. Leaders must know what skills mortarmen bring with them when they report to their unit. This forms the base to build mortar training in the unit. The career pattern for NCOs and officers is depicted in individual training. It entails alternating between the training base and units with progressively advanced levels of training and responsibility. Mortar training in the institution focuses on preparing the Soldier for these positions. Depending on the course, the training focus includes technical training in mortar skills, mortar familiarization, and mortar issues update (Table A-1).

Appendix A

ONE-STATION UNIT TRAINING (11C)

A-6. One-station unit training (OSUT) trains new Soldiers for their initial assignment in IBCT or HBCT units. Training is divided into two phases. Phase I (seven weeks) teaches common entry-level infantry tasks. Phase II continues to foster the self-discipline, motivation, physical readiness, and proficiency in combat survivability started in Phase I. The 11C Soldiers receive instruction in mortar systems to prepare them for their specific unit assignments. Soldiers receive familiarization on FDC and FO procedures and are required to pass a mortar gunner's examination to be awarded their MOS.

BASIC NONCOMMISSIONED OFFICER COURSE (11C)

A-7. Basic Noncommissioned Officer Course (BNCOC) teaches junior NCOs to lead, train, and direct subordinates in the maintenance, operation, and employment of weapons and equipment. The instruction includes FDC procedures, fire planning, tactical employment of mortars, and maintenance.

MANEUVER ADVANCED NONCOMMISSIONED OFFICER COURSE

A-8. Maneuver Advanced Noncommissioned Officer Course (MANCOC) prepares NCOs to lead a mortar platoon in combat as part of the battalion team. This includes fostering an understanding of the battalion task force concept and how it fights. Training that applies to mortars includes fire planning, FDC, and FO procedures.

INFANTRY MORTAR LEADER COURSE

A-9. Infantry Mortar Leader Course (IMLC) provides lieutenants and NCOs (sergeant through master sergeant) with the knowledge to supervise and direct the fire of a mortar platoon. Instructions include tactical employment of the mortar platoon, graphics, fire planning, mechanical training, FO procedures, and fire direction control procedures. Officers are awarded the additional skill identifier of 3Z. Commanders must ensure that IMLC graduates fill mortar leadership positions. The skills personnel have learned are complex and perishable and must be sustained in the unit.

INFANTRY BASIC OFFICER LEADER'S COURSE

A-10. Infantry Basic Officer Leader's Course (IBOLC) trains lieutenants in weapons, equipment, leadership, and tactics. It also teaches them how to instruct their subordinates in the maintenance, operation, and employment of weapons and equipment for combat. Students receive instruction in mechanical operation of the mortar as well as detailed instruction on FO procedures.

INFANTRY CAREER CAPTAIN'S COURSE

A-11. Infantry Career Captain's Course (ICCC) trains first lieutenants and captains in leadership, war fighting, and sustainment skills required to serve as company commanders and staff officers at battalion and brigade levels. Mortar training focuses on supervisory tasks.

PRE-COMMAND COURSE

A-12. Pre-command Course (PCC) is intended for field-grade officers (majors through colonel) designated for battalion and brigade command. Training consists of a review and update on mortar issues such as new equipment; tactics, techniques, and procedures (TTPs); battle drills; and safety.

Table A-1. Institution courses.

COURSE	SKILL LEVEL					COURSE FOCUS
	1	2	3	4	OTHER	
One-Station Unit Training	X	X				A
Basic NCO Course		X	X			C,D
Advanced NCO Course				X		C,D
Infantry Mortar Leader Course			X	X	X	B,C,D
Infantry Officer Basic Course					X	C
Infantry Career Captain Course					X	D
Pre-Command Course					X	C,D

A = MOS-PRODUCING
B = ADDITIONAL SKILL IDENTIFIER FOR OFFICERS
C = FAMILIARIZATION
D = REVIEW/UPDATE

TRAINING IN UNITS

A-13. A unit training program consists of initial and sustainment training. Both may include individual and collective skills. Resources, such as devices, simulators, simulations, ranges, and ammunition, further develop skills learned in the institution. The critical aspect of unit training is to integrate Soldiers into a collective, cohesive effort as a mortar squad or platoon member. Drills, situational training exercises (STXs), and live-fire drills develop these collective skills.

TRAINING PLAN DEVELOPMENT

A-14. Training plans are developed at higher headquarters and published in the form of command guidance so that subordinate units can develop their plans. The process begins with identifying the unit's mission-essential task list (METL). The METL contains all the collective tasks that a unit must perform to be successful in combat. (FM 7-0 contains specific information on the METL development process.)

Mission-Essential Task List Tasks

A-15. Commanders assess the unit's proficiency level in each METL task. Information for this assessment is obtained by reviewing past gunner and FDC examinations, ARTEP results, external evaluation, after-action reviews (AARs), and by observing the execution of current training.

A-16. Once the assessment is complete, the commander lists the tasks in priority. Tasks that are identified as untrained (U) and are critical to the mission have training priority, followed by tasks that need practice (P) and tasks that are trained (T) to standard. Resources (ranges, ammunition, equipment, and time) are requested to train those tasks that do not meet the standard (U and P), while sustaining the proficiency of the tasks that do meet the standard (T). The commander refines his plan in the form of a training guidance and training schedules. (FM 7-1 contains specific information on the training plan development process.)

Initial Training

A-17. Initial training trains Soldiers and units to a high degree of proficiency. Initial training ensures that each Soldier, squad, and platoon has the basic core skills proficiency for his skill level or the

Appendix A

collective team. Initial training must be trained correctly to a rigid standard so that proficiency will be retained longer. Decay in skill proficiency will occur due to a lack of available training time, skill difficulty, or personnel turnover.

Sustainment Training

A-18. Sustainment training reduces skill decay and maintains proficiency within the band of excellence described in FM 7-0. Retraining may be required if a long period elapses between initial and sustainment training. Once proficiency is demonstrated in a task or collective event, more difficult scenarios and exercises should be developed to train to a higher level of proficiency, while sustaining previously learned skills.

INTEGRATED TRAINING STRATEGY

A-19. Figure A-1 outlines a logical progression of events that a mortar platoon can adapt to their training strategy. Mortar squads and the FDC are dual-tracked to focus on their specific training needs. Both tracks must be integrated to develop a mortar platoon that fights as one unit. Individual and collective training must be evaluated against specific standards and discussed in AARs. Objective evaluations provide readiness indicators and determine future training requirements.

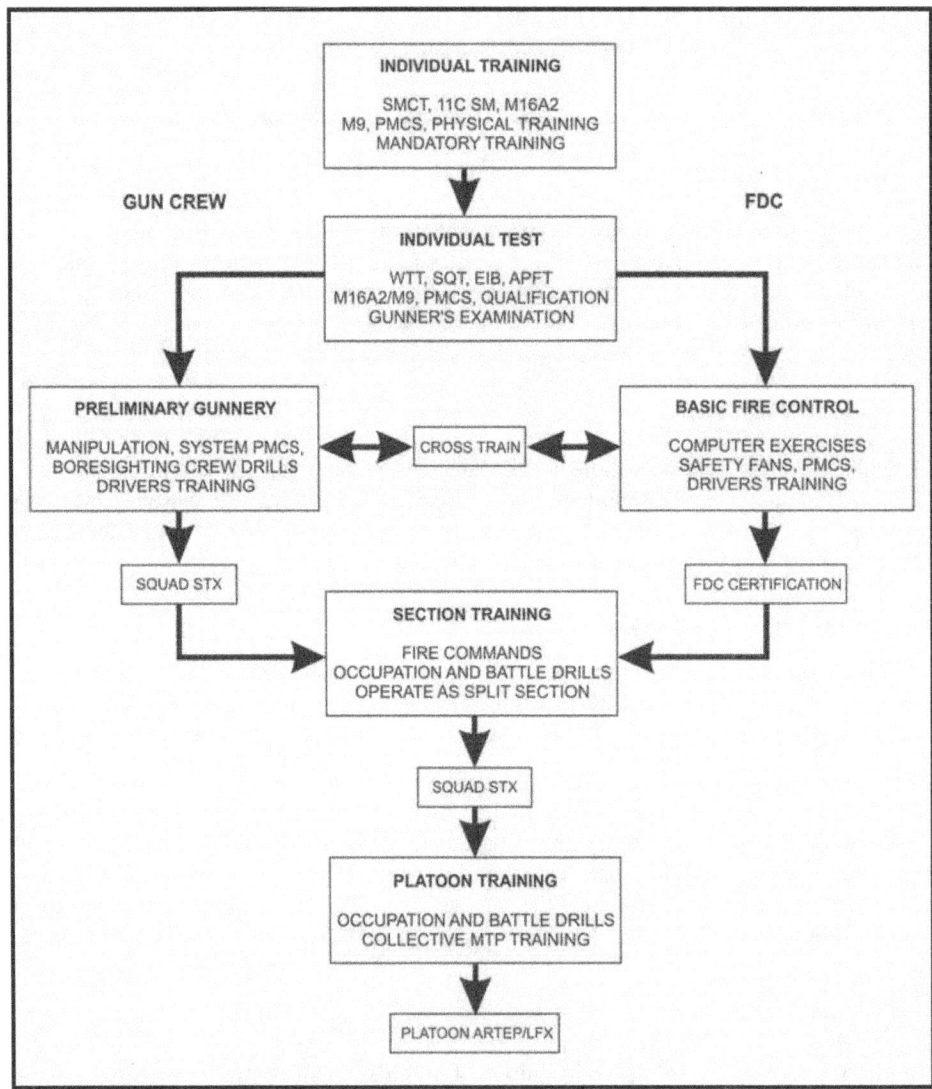

Figure A-1. Integrated training strategy.

Common Tasks

A-20. These tasks are found in STP 7-11B1-SM-TG, Skill Level 1 and in STP 7-11B24-SM-TG, Skill Levels 2, 3, and 4. These manuals contain the common tasks that all Soldiers must know, regardless of MOS or duty position, to help them fight, survive, and win in combat. Mastery of these common tasks is a prerequisite for individual training specific to mortars. The communications and land navigation common tasks are vital.

MOS 11C Tasks

A-21. These tasks are found in STP 7-11C14-SM-TG, Skill Levels 1-4 for MOS 11C. This manual contains the individual tasks specific to mortarmen. The trainer's guide provides leaders the information to develop the individual portions of a unit training plan. Each 11C task is listed in this guide, along with the following.

Training Extension Courses

A-22. Service schools use training extension courses (TECs) to provide their expertise directly to units in the field. This is accomplished through lessons in the form of booklets, video slides, audio tapes, videotapes, and Electronic Information Delivery System (EIDS) machines. These lessons focus on individual tasks and are stocked in unit learning centers and the local training support centers. TEC materials provide standardized instruction, which is helpful to Soldiers as they move from unit to unit. Preparation time is also saved. However, trainers must ensure the lessons directly and fully support the training objectives.

Army Correspondence Course Program

A-23. The Army Correspondence Course Program (ACCP) provides printed training courses through the mail to Soldiers. It is a valuable program for training the trainer, skill progression, and functional training for a specific specialty, MOS, or duty position. However, planning is needed. Leaders must identify courses that support the goals of the long-range training plan; Soldier's enrollment must then be arranged. (ACCP includes the infantry mortar platoon leaders' course.)

CROSS TRAINING

A-24. A Soldier's individual training tends to focus on his duty position. The assistant computer concentrates on FDC tasks while the squad leader concentrates on mortar mechanical tasks and leading the squad. However, this focus should not exclude other mortar training. Continuous training in duty-specific tasks can become boring to Soldiers and deprive them of gaining broad proficiency in their MOS. Casualties (whether in war or in training) can quickly render the mortars ineffective if key personnel are lost. These variables make cross training essential. For example, cross training ensures that a squad leader can assume the duties of a computer and that subordinates are ready to assume the roles of their supervisors.

COLLECTIVE TRAINING

A-25. Collective training includes squad, section, and platoon drills and exercises.

Squad Training

A-26. The core of squad training is crew drill. Squad-level mortar tasks are in the infantry MTP. This training can be performed in garrison (using devices or live ammunition). Once these tasks are mastered, performing them under different conditions (urbanized terrain, limited visibility, CBRN) increases the challenge. Cross training is accomplished at this level by rotating Soldiers among duty positions, such as squad leader and FDC positions, while providing coaching.

A-27. The foundation of squad training is sound individual training. If individual proficiency is effectively sustained, new Soldiers can be readily integrated into the unit during collective training.

Appendix A

They arrive at the unit proficient in specific tasks learned in the training base. These new Soldiers learn additional tasks while training with their experienced peers and their squad leaders.

Section and Platoon Training

A-28. The core section and platoon tasks are found in the ARTEP MTP. This training usually consists of an exercise in the field: live-fire exercise (LFX), STX (either alone or with a rifle company), or field training exercises (FTX) as part of the battalion. LFXs may involve subcaliber, full-range training cartridge (FRTC), short-range training round (SRTR), or service ammunition. STXs and FTXs may entail dry-fire, live-fire, or devices, either alone or in combination.

A-29. The FIST must be a part of this training. An LFX must never take place without the FISTs who typically deploy with the supported companies. The complete indirect fire team (see Chapter 1) must train together for maximum efficiency. This teaches the capabilities, limitations, and unique requirements of operating mortars.

A-30. Another important area is the mortar's role in overall task force operations. This collective training mainly involves leaders in an FTX. However, mortars must be considered along with other fire support assets when conducting a map exercise (MAPEX), command field exercise (CFX), tactical exercise without troops (TEWT), or command post exercise (CPX).

Collective Training Resources

Drill Books

A-31. Crew and battle drills are published in a pocket-sized ARTEP manual called a drill book for each unit organized under a different TOE. (See ARTEP 7-90-Drill.)

- Battle drills are a specific category of collective tasks performed at squad, section, or platoon level. They are vital to the mortar's success in combat. Battle drills are mostly independent of mission, enemy, terrain and weather, troops and support available, time available, civil considerations (METT-TC) and require minimal leader actions to execute. They are usually executed or initiated on a cue such as an enemy action or a simple leader order. Battle drills are standardized throughout the U.S. Army and may not be modified in training. The mortar unit is required to be proficient in all battle drills contained in the drill book. Less critical drills are published in other sources such as training circulars or field manuals.

- Mortar drills are divided into two general areas. The first are those battle drills that previously were termed "crew drills." These focus on the mechanical manipulation of the mortar such as small deflection changes or removing a misfire. Full proficiency in these tasks is a prerequisite for performing fire support missions. The second area encompasses those battle drills essential to combat survival. These include such tasks as React to Chemical Attack, React to Indirect Fire, or React to Nuclear Attack.

Mission Training Plan

A-32. The mission training plan (MTP) is a descriptive ARTEP document for training mortarmen to critical wartime mission proficiency. It gives the mortar platoon or section a clear description of "what" and "how" to train. This is achieved through comprehensive, detailed training and evaluation outlines (T&EOs); guidance on training exercises; and other related training management aids. While its focus is on collective training, the MTP also provides matrixes that identify individual tasks, common 11C SM tasks, and Military Qualification Standard (MQS) tasks. Like the drill book, the mortar MTP applies to platoons or sections organized under a specific TOE (see ARTEP 7-90-MTP).

Battalion-Level Training Model

A-33. The battalion-level training model (BLTM) is a means to quantify the cost of maintaining training readiness. This cost is expressed in terms of types of training events, their annual frequency, and the equipment miles/hours expended. This model is used to forecast and resource requirements to

Mortar Training Strategy

support the unit's specified training readiness level. It does not, however, prescribe what training a unit must conduct to maintain this level. Rather, the BLTM provides a basis for understanding the trade-off between a unit's training resources and its training strategy. This helps leaders program training alternatives to achieve and maintain combat readiness. The frequency of training events under the BLTM is reflected in the battalion's long-range training plan.

EXAMPLE ANNUAL MORTAR TRAINING PROGRAMS

A-34. Figure A-2 depicts an example IBCT battalion mortar training program, and Figure A-3 depicts an example HBCT battalion mortar training program.

JANUARY	FEBRUARY	MARCH
IND WPN QUAL/SUST	SQD/SEC/FDC DRILL (LITR)	POST SUPPORT/BN CPX
CREW WPN QUAL/SUST	PLT FTX	SQD/SEC/FDC DRILL
MAINTENANCE	CO FTX	(SRTR)
SQD/SEC/FDC DRILL	BN FTX	ITEP/CMT
(SRTR)	DEPEX	CO TEWT
PLT/SEC LFX	MAINTENANCE	
GUNNER'S EXAM	PLT STX (81-mm ONLY)	
FDC CERTIFICATION	SUPPORT CO STX	

APRIL	MAY	JUNE
IND WPN QUAL/SUST	SQD DRILL	POST SUPPORT
CREW WPN SUST	PLT STXs	BN CPX
MAINTENANCE	CO FTXs	CO MAPEX
SQD/SEC/FDC DRILL (LITR)	DEPEX	ITEP (EIB)
SEC/PLT LFX	BN FTX	SQD/SEC/FDC DRILL (LITR)
	MAINTENANCE	
	SUPPORT CO FTX	

JULY	AUGUST	SEPTEMBER
NG & ROTC SUPPORT	IND WPN QUAL/SUST	SQD DRILL
SQD/SEC/FDC DRILL	CREW WPN QUAL/SUST	MAINTENANCE
(SRTR)	BN FTX	CO FTXs
	CALFEX/LFX	BN CFX
	MAINTENANCE	BN FTX (EXT EVAL)
	FDC EXAM	SQD/SEC/FDC DRILL
	GUNNER'S EXAM	(SRTR/LITR)
	SQD/SEC/FDC DRILL (LITR)	PLT STX (81-mm)
		DEPEX

OCTOBER	NOVEMBER	DECEMBER
POST SUPPORT	IND WPN QUAL/SUST	POST SUPPORT
ITEP (WTT) CMT	CREW WPN/SUST	ITEP
BN TEWT	DEPEX	CMT
CO MAPEXs	MAINTENANCE	SQD/SEC/FDC DRILL (LITR)
CPX (81-mm)	SQD/SEC/FDC DRILL	
SQD/SEC/FDC DRILL (LITR)	(SRTR)	
	GUNNER'S EXAM	
	FDC CERTIFICATION	

Figure A-2. Example training program for IBCT battalion.

Appendix A

JANUARY CMT ITEP BN CPX MAINTENANCE SQD DRILLS PLT STX SQD/SEC FDC DRILL (SRTR) IND WPN QUAL/SUST FDC CERTIFICATION	**FEBRUARY** MAINTENANCE CO FTX BN FTX DEPEX SQD/SEC/FDC DRILL PLT FTX	**MARCH** MAINTENANCE POST SUPPORT/CO MAPEX BN TEWT ITEP CMT IND WPN QUAL/SUST SQD/SEC/FDC DRILL (SRTR)
APRIL GUNNERY QUAL (MTA) CALFEX (MTA) MAINTENANCE GUNNER'S EXAM LFX (M60, .50-CAL, 120-mm) CALFEX/LFX DRILL (60-mm SUBCAL)	**MAY** ITEP BN CPX PLT STX MAINTENANCE SQD/SEC BATTLE DRILLS (SRTR/LITR) GUNNER'S EXAM	**JUNE** CO FTX BN FCX BN DEPEX BN FTX (EXT EVAL [HTA]) MAINTENANCE SQD/SEC/FDC DRILL BN EXT EVAL/LFX
JULY POST SUPPORT ITEP (EIB) BN MAPEX IND WPN QUAL/SUST CMT SQD/SEC/FDC DRILL (SRTR) MAINTENANCE	**AUGUST** IG INSPECTION ITEP BN CPX MAINTENANCE SQD/SEC/FDC DRILL (60-mm SUBCAL) PLT STX FDC EXAM GUNNER'S EXAM	**SEPTEMPER** CO FTX MAINTENANCE BN DEPEX BN FTX-(REFORGER) SQD/SEC/FDC DRILL (SRTR) PLT STX
OCTOBER MAINTENANCE POST SUPPORT CO TEWT ITEP (WTT) CMT SQD/SEC/FDC DRILL	**NOVEMBER** IND WP QUAL/SUST GUNNERY QUAL (MTA) MAINTENANCE BN CFX SQD/SEC/FDC DRILL (SRTR) FDC EXAM GUNNERY QUAL (.50-CAL, M60) LFX	**DECEMBER** MAINTENANCE DEPEX POST SUPPORT ITEP IND WPN QUAL/SUST CMT SQD/SEC/FDC DRILL

Figure A-3. Example training program for HBCT battalion.

TRAINING EVALUATION

A-35. Evaluation cannot be separated from effective training. It occurs during the top-down analysis when planners develop the training plan. Planners use various sources of information to assess their unit's individual and collective training status. Evaluation is continuous during training. Soldiers receive feedback through coaching and AARs. Leaders also assess their own training plan and the instructional skills of their subordinate leaders. After training, leaders evaluate by sampling training or reviewing AARs. Much of this evaluation is conducted informally. Formal evaluations occur under the Individual Training and Evaluation Program (ITEP) and the Army Training and Evaluation Program (ARTEP) to assess individual and collective training respectively.

INDIVIDUAL TRAINING

Commander's Evaluation

A-36. The commander's evaluation is routinely conducted in units. Commanders select and evaluate individual tasks that support their unit mission and contribute to unit proficiency. This may be performed through local tests or assessments of Soldier proficiency on crucial MOS tasks or common tasks. The commander's evaluation is based on year-round, constant evaluation by the chain of command. It is supported by the MOS 11C Soldier's manuals, trainer's guides, and job books.

Warrior Task Test

A-37. The warrior task test (WTT) is a hands-on test that evaluates basic survival and combat tasks. It is taken directly from STP 7-11B1-SM-TG, STP 7-11B24-SM-TG, and STP 7-11C14-SM-TG. The WTT gives the unit commander regular, objective feedback on common task proficiency.

Gunner's Examination

A-38. The gunner's examination is a continuation of the mortar-based drills in which a mortarman's proficiency as a gunner is established. The examination is contained in Chapter 9 of this manual. It includes tests, equipment, conditions, testing procedures, scoring, and administrative procedures. It focuses on the individual qualification of the Soldier in the role of a gunner. However, the gunner's success also depends on the collective performance of his assistants. Within these limitations, evaluators should try to standardize the examination. The BLTM specifies that the squad leader, gunner, and assistant gunner should pass the gunner's exam semiannually. All gunners should have a current qualification before an LFX (whether using service or subcaliber ammunition).

FDC Certification

A-39. This provides commanders a means to verify that their FDC mortarmen have the knowledge and skills for their positions: squad leader, FDC computer, section sergeant, platoon sergeant, and platoon leader. Certification helps ensure that ammunition is wisely expended and that training is conducted safely and effectively. Mortarmen are certified when they receive a passing score on the two-part examination. (See FM 3-22.91 for FDC certification.)

COLLECTIVE TRAINING

Army Training and Evaluation Program

A-40. The aim of collective training is to provide units the skills required to perform unit-level tasks. The ARTEP is the overall program for this collective training. It prescribes the collective tasks that a unit must successfully perform to accomplish its mission and to survive in combat. These tasks include conditions and performance standards, and they are located in MTPs and drill books.

Appendix A

External Evaluation

A-41. The commander formally determines the status of his collective training through external evaluation. The external evaluation gives the commander an objective appraisal of this status by using mortar expertise found outside the normal chain of command. The external evaluation is not a test in which a unit passes or fails; it is a diagnostic tool for identifying training strengths and weaknesses. It must be emphasized that an external evaluation is not a specific training event but a means to evaluate a training event. Mortar units undergo external evaluations during an LFX, FTX, or a combination thereof. The unit may be evaluated alone, as part of its parent unit, or with other mortar units. The MTP provides guidance on planning, preparing, and conducting an external evaluation.

Evaluation of the Indirect Fire Team

A-42. The members of the indirect fire team must train and correctly execute their respective tasks to successfully complete any fire mission. However, only as a last resort should the fire mission be deleted from the evaluation. Evaluators should determine the reason why any fire mission fails to meet standards in order to determine where additional training is required. The indirect fire team should be given the opportunity to successfully complete the fire mission. This can be accomplished in the following ways:

- Allow the mission to continue if the detected error will still result in the rounds impacting within the safety limits. The team must train to accomplish the mission by finding and correcting any errors based on the round's impact. The appropriate evaluator should intervene only if the team prepares to fire incorrect data that is out of the safety area or when ammunition is constrained.
- Start the fire mission over. Although ammunition constraints during live-fire may not permit this, tasks can be repeated using devices or, less preferably, dry-fire.
- The evaluator corrects the error when the mission data would result in rounds fired out of the safety area. The FO evaluator at the observation post can change the CFF or correction to reflect proper procedures. The FDC evaluator may correct the improperly computed firing data while the mortar squad evaluator may correct improperly set data or a faulty sight picture.

Appendix B
Training Devices

The most efficient and direct method of teaching conduct of fire is by firing service ammunition under field conditions. However, ammunition training allowances and range limitations often restrict such training. When properly used, mortar training devices provide realistic training at reduced costs and at more accessible locations.

SECTION I. FULL-RANGE TRAINING CARTRIDGE, M931

The full-range training cartridge (FRTC) provides realistic training for all members of the indirect fire team at a reduced cost.

DESCRIPTION

B-1. The M931 FRTC is a non-dud producing 120-mm training cartridge used to simulate the M934 HE cartridge for M120 and M121 120-mm mortars.

B-2. The weight, charges, fuzing, and ballistic characteristics of the cartridge are similar to those of the HE cartridge. The M781 fuze for the M931 produces a flash, a bang, and a smoke signature that provide audio and visual feedback to the mortar crew and FO. The M931 has the following general characteristics:
- Maximum range: 7,200 meters.
- Minimum range: 200 meters.
- Maximum rate of fire: 16 rounds per minute (first minute).
- Sustained rate of fire: 4 rounds per minute.

B-3. The cartridge allows the entire mortar team—FO, FDC, and guns—to practice the actions required for a fire mission. These actions include—
- FO CFF and adjustment.
- FDC calculations and procedures.
- Gun crew preparation of the round, and the manipulation and firing of the mortar.

B-4. A major advantage of using the M931 FRTC is the elimination of the time required to retrieve and clean training devices such as the short-range training cartridge (SRTC). It also does not exceed 75 percent of the M934 cartridge's unit production cost.

B-5. A disadvantage of using the FRTC is the need to treat it as a regular service mortar cartridge when firing; it has the same range and firing restrictions as a service round.

B-6. The M931 consists of the following major components:
- Projectile body assembly.
- M233 propelling charges (with M47 propellant).
- M1005 ignition cartridge (with M44 propellant).
- M34 fin assembly.
- M781 point detonating practice fuze.

PROCEDURES

B-7. The procedures to fire the FRTC are the same as for the M934 service cartridge.

Appendix B

SECTION II. SHORT-RANGE TRAINING ROUND, M880

The M880 SRTR cartridge provides an economical means of realistically training personnel in all phases of mortar gunnery. The SRTR enables units to train in locations where mortar training facilities are limited or do not exist. Trainers should exercise imagination and ingenuity in developing training situations for employing the SRTR. This section provides the technical information required to employ the SRTR.

TRAINING WITH THE SHORT-RANGE TRAINING ROUND, M880

B-8. The M880 SRTR is a training round for the 81-mm mortar (M29 and M252) and the 120-mm mortar (when used with M303 and M313 81-mm subcaliber inserts). It can be fired, recovered, refurbished, and refired. When fired, the SRTR travels from 47 to 458 meters, according to its charge and elevation. Upon impact, the SRTR emits a flash, a bang, and smoke discharge, but no fragmentation.

TRAINING

B-9. The SRTR is effective for training all elements of the indirect fire team. It provides the FO with an audio and visual signal by which he can spot rounds and make adjustments. The FDC computes data received from the FO, who observes the impact of the SRTR on a 1:10 scaled range. The gunner places data obtained from the FDC on the mortar and the assistant gunner drops an SRTR down the barrel. The entire indirect fire team functions the same as if they were firing service ammunition with one exception—the FO observes the SRTR impact on a 1:10 scaled range instead of the service ammunition at a normal distance in an impact area.

CHARACTERISTICS

B-10. The characteristics of the SRTR are as follows:
- Overall length: 14.5 inches.
- Firing weight: 6.84 pounds.
- Maximum range: 458 meters.
- Minimum range: 47 meters.

NOTE: The 1:10 scale is based on 81-mm 300-series ammunition.

ADVANTAGES

B-11. The SRTR is used with all 81-mm mortars and 120-mm mortars (when used with M303 and M313 81-mm subcaliber inserts). It also accommodates prescribed crew drill procedures, and it can be used by both HBCT and IBCT elements. The SRTR has the following advantages:
- Can be fired in training areas where firing service ammunition is prohibited.
- Saves time by eliminating long-distance travel to suitable mortar ranges.
- Saves in the cost of firing service ammunition.
- Requires only a brief period of time for personnel to learn proper operation and maintenance.
- Allows the indirect fire team to use the same procedures required to fire service ammunition, excluding appropriate charge and fuze settings.
- Uses all of the equipment required to fire service ammunition.

RANGE

B-12. A scale of 1:10 is used to relate the range and deviation between the SRTR projectile and the standard 81-mm service projectile. The minimum range requirement for firing, through charge 3, is an area 700 to 900 meters in depth and 300 meters in width. This is necessary to accommodate the desired number of targets and provide a maneuver area to operate in. When using 81-mm mortars, the mortars can be no closer than 4 meters apart, but can be as far apart as the range permits (check with the local range control regulations). The range must be cleared of all unexploded ammunition, concrete, and steel objects (Figure B-1).

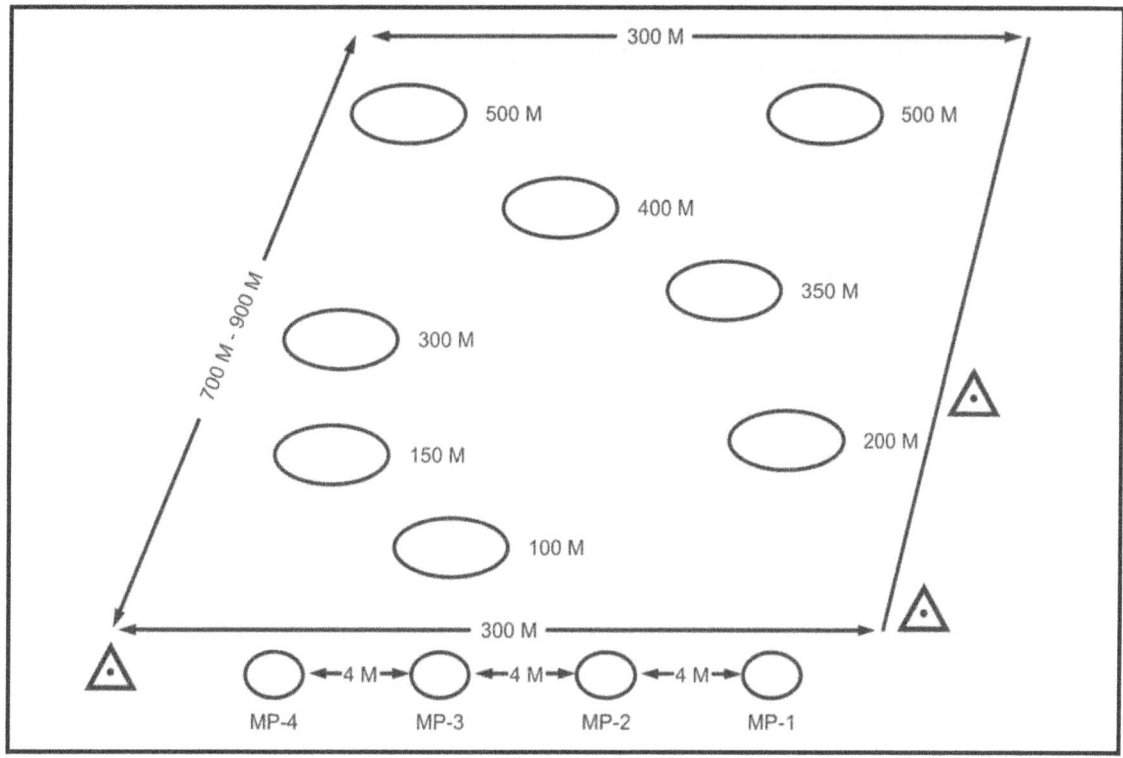

Figure B-1. Scaled range for short-range training round, M880.

COMPONENTS

B-13. The SRTR consists of eight major components (Figure B-2).
- Fuze, point detonating M775.
- Projectile body.
- Obturator band.
- Plastic charge plugs.
- Dud plugs.
- Fin assembly.
- Ignition cartridge, M987.
- Breech plug assembly.

Appendix B

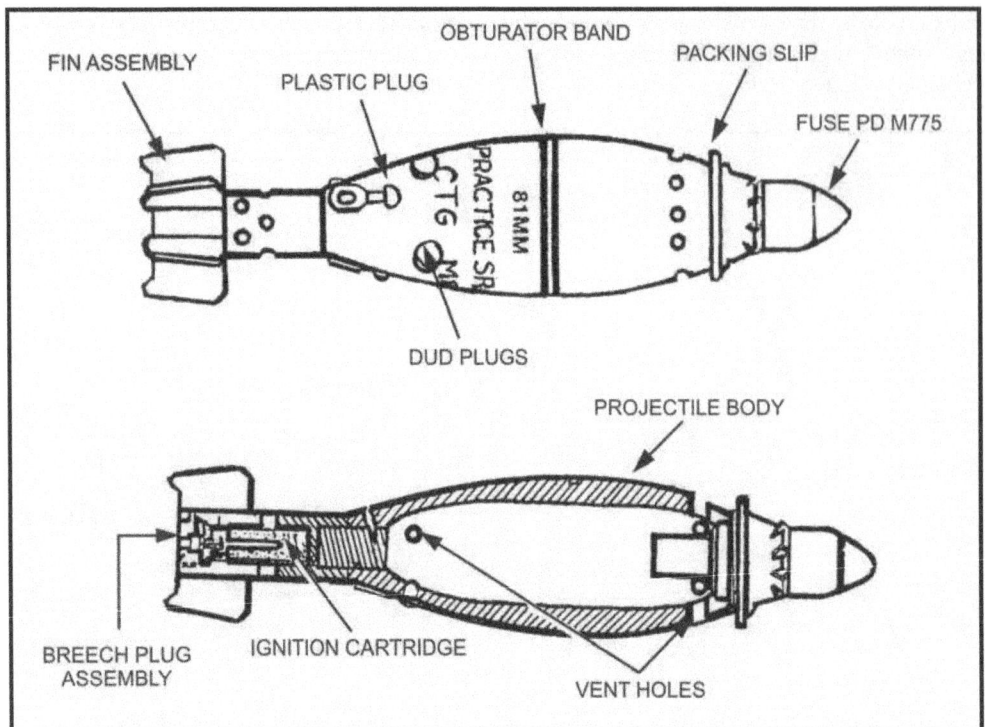

Figure B-2. Short-range training round, M880—practice round.

Projectile Body

B-14. The projectile body is made from cast iron and is machined to accept the M775 fuze, the tail fin, three charge plugs, three dud plugs, and the obturator band. It is painted blue with white lettering and one brown band to identify it as a practice round. The charge and dud plugs are fitted into holes drilled in the base of the projectile body. The obturator band is plastic and shaped like a split ring so that it will expand to fit over the projectile body and will seat into the band groove. The M775 fuze assembly includes a pyrotechnic spotting cartridge, which is retained in the fuze body by a threaded aluminum cup. This assembly is threaded into the nose of the projectile body.

Fuze, M775

B-15. The M775 fuze replicates the M734 multioption fuze used with service ammunition. Even though the M775 has four fuze settings for ammunition bearer training, it will only function upon impact. The tail fin assembly is threaded onto the stud at the base of the projectile body. Inside the tail fin is the propellant cartridge, which is a primed 12-gauge shotgun shell casing loaded with a double base propellant. The breech plug/firing pin is threaded into the rear of the tail fin and retains the propellant cartridge in the fin cavity.

Training Devices

PREPARATION

B-16. The M880 is issued ready to fire. When the round is unpacked, ensure the following:
- No red is showing at the top of the fuze.
- The packing clip is in place.
- The obturator band is present and flush with the body.
- The three plastic charge plugs are seated in the body.
- The three dud plugs are seated in the body.
- The ignition cartridge is inside the fin assembly.
- The breech plug is seated in the fin assembly.

> **WARNINGS**
>
> If any red can be seen above the fuze windshield, place the fuze in the dud pit and dispose of the fuze in accordance with the unit SOP.
>
> If the packing clip is loose, attempt to place the packing clip back into the fuze. If it will not go back into the fuze, place the fuze into the dud pit and dispose of the fuze in accordance with the unit SOP.
>
> Inspect the breech plug firing pin for freedom of movement. If it does not move freely, a misfire could occur or premature ignition of the ignition cartridge could take place when assembling the round.

B-17. To prepare a newly issued round for firing, set the desired charge by pulling one of the plastic charge plugs from the body of the round. Set the desired fuze setting by rotating the multioption fuze clockwise or counterclockwise until the desired fuze setting is lined up with the notch on the body of the round. Present the round to the squad leader for his inspection. When the assistant gunner is ready to fire the round, pull the packing clip from the round and hand it to him.

> **WARNING**
>
> Do not pull the packing clip from the round until the assistant gunner is ready to fire the round.

NOTE: The fuze setting is for training only and will not affect the functioning of the fuze.

B-18. Charge settings for the round are as follows:
- Charge 0—pull all the plastic charge plugs from the body.
- Charge 1—pull two of the plastic charge plugs from the body.
- Charge 2—pull one of the plastic charge plugs from the body.
- Charge 3—do not pull any of the plastic charge plugs from the body.

Appendix B

REFURBISH THE ROUND

B-19. To refurbish a previously fired round, ensure that the round is clean. All portholes must be free of dirt to allow the gas to escape, and all of the threads must be clean.

    (1) Inspect the refurbishing kit to ensure that all equipment is present.
- Three plastic charge plugs.
- One obturator band.
- One M987 ignition cartridge.
- One breech plug.
- One M775 point detonating fuze.
- Three dud plugs.

    (2) Assemble the round as follows:
- Install the three plastic plugs with the tangs pointed to the rear of the projectile body.
- Install the three dud plugs.
- Secure the fin assembly onto the projectile body hand-tight.
- Install the obturator band onto the projectile body.

CAUTION

If the obturator band does not seat flush or below the surface of the projectile body, send the body back to the maintenance table for further cleaning. If the obturator band still does not seat flush, the body must be disposed of.

- Install the fuze assembly hand-tight and flush with the top of the round.
- With the round laying horizontally on the table, install the M987 ignition cartridge into the tail fin assembly.

WARNING

Do not force the ignition cartridge into the tail fin assembly.

- Install the breech plug by hand and use the breech plug wrench to secure the breech plug hand-tight.

    (3) Place the round back into its canister until it is ready to be fired.

NOTE: Do not refurbish the M880 round if it will not be fired that same day.

    (4) To fire the round, follow the same procedures as if the round was newly issued.

CLEAN AND SERVICE THE ROUND

B-20. Use the following procedures to clean and service the SRTR.

    (1) Fill two trash cans with water and leave another trash can empty.

    (2) Dip the spent M880 into the first water bucket and scrub off the heavy dirt with a wire brush.

    (3) Place the M880 round on the work table and secure the projectile body with the strap wrench.

    (4) Remove the fuze (if necessary, use the long handle pliers), check to see that it was detonated, and discard.

> **WARNING**
>
> When the fuze is removed from the body, it must be inspected to ensure that it has detonated by inspecting the coin to see if it is missing.

(5) Using the breech plug wrench, unscrew the breech plug and discard.

(6) Using the fin wrench, unscrew the fin assembly from the projectile body.

(7) Place the tail fin assembly upright on the cartridge block, place the punch into the tail fin assembly, strike the punch with the ball peen hammer, and drive out the spent ignition cartridge. Discard the spent ignition cartridge.

(8) Place the tail fin assembly into the second wash bucket and clean it with a wire brush for the final cleaning. Ensure that all dirt and residue are removed from the round.

(9) Remove and discard all remaining plastic plugs and the obturator band from the projectile body.

(10) Place the projectile body into the second wash bucket and remove all dirt and residue. Take the M16 bore brush and thoroughly clean the vent holes and the thread.

(11) With a clean, dry rag, dry both components of the M880. Ensure no dirt or residue is present on the projectile body and tail fin assembly.

(12) Examine the M880 for cracks and any type of damage that may prevent the projectile from being refired. If there is evidence of cracks or some type of damage, discard the entire projectile body and tail fin assembly.

(13) If the M880 is not going to be fired the same day, lightly oil the projectile body and tail fin assembly. Screw the tail fin assembly back on the body and store it in its canister.

NOTE: The M880 must be recovered, cleaned, and serviced within 24 hours after it is fired.

RECOVERY PROCEDURES

B-21. Perform the following to recover a previously fired M880 round from downrange.

> **WARNING**
>
> Failure to follow the proper recovery procedures may result in personal injury.

(1) Ensure the range is closed and all weapon systems have ceased firing.

(2) Issue a shovel, a grappling hook, heavy duty leather work gloves, and an ammunition box to the assistants.

(3) Walk down range with the assistants and locate the M880 round.

(4) Pick up the round and verify that the round has functioned by observing that all of the dud plugs on the body are missing. Have a second Soldier verify that all of the dud plugs are missing.

Appendix B

> **WARNING**
>
> When picking up the M880, ensure that your hand and fingers are not covering the vent holes.

(5) If all of the dud plugs are missing, place the round in the ammunition box.

> **WARNING**
>
> If any of the dud plugs are present and the fuze has not functioned, place the round in a marked pit and follow the defuzing procedures outlined in paragraph B-22.

NOTE: Digging the round out with the shovel may be necessary. The grappling hook may also be used to assist in recovering the round.

(6) Ensure that each hole left by the M880 round is filled.
(7) Once the M880 rounds have been recovered, transport them to the maintenance table.

DEFUZE A DUD ROUND

B-22. A dud M880 round may be defuzed when first located or at the end of that day's firing as long as it is recovered within 24 hours after it has been fired. To defuse a dud M880 round—

(1) Set the defusing device (boom box) on the platform with the holes facing downrange and away from the troops. Pull out the four retaining pins that secure the top portion of the boom box, and then pull the top off the base of the boom box.
(2) Turn the round until the packing clip portion of the round is facing upward.
(3) With the chisel over the fuze, secure the top portion of the boom box to the base, and then attach the four retaining pins.
(4) Kneeling on the opposite side of the holes on the boom box, strike the chisel with the 4-pound hammer until the chisel moves freely from the base of the boom box to the top of the boom box.

> **WARNINGS**
>
> All personnel must be on the solid side of the boom box before the chisel is struck with the hammer.
>
> When defuzing an M880 round, ear plugs and goggles must be worn.

(5) Remove the cover of the boom box and inspect the fuze to verify that the spotting charge has functioned.

Training Devices

> **WARNING**
>
> If the spotting charge has not functioned or the fuze has pulled away from the thread well, transport it to the dud pit and dispose of it in accordance with the unit SOP.

(6) Remove the M880 round from the boom box and return the cartridge to the firing line for cleaning.

PARTS

B-23. The equipment shown in Table B-1 is available.

Table B-1. Supply data for short-range training round, M880.

DESCRIPTION	NATIONAL STOCK NO.	PART NO.	DODIC NO.
Cartridge, 81-mm Practice: M880	1315-01-216-7070		1315-C876
Refurbishment kit, M80	1315-01-219-3936	19200	1315-C045

TRAINING CONSIDERATIONS

B-24. SRTR training enables the unit commander to observe the actions of the indirect fire team and the tasks being performed by the indirect fire team. This enhances the ability of the commander to evaluate the proficiency of all elements of the training, to identify and isolate problems, and to begin corrective training. All members of the team can view the overall operation to understand each member's part.

B-25. The time used for preparation determines the benefits gained from training with the SRTR. Three steps must be accomplished before training begins:

(1) Build a scaled firing range (permanent, semi-permanent, or temporary).
(2) Construct a map of the firing range.
(3) Train the personnel with the SRTR. Personnel should be trained in the recovery, refurbishment, and maintenance of the M880 before range firing begins.

B-26. Gunners should not be allowed to observe the impact of the SRTR, unless firing a direct-lay mission. Normally, mortars are mounted in a defilade position so that mask exists between the impact area and the firing position. If mask does not exist, using a referred deflection of 0700 when firing is recommended.

B-27. When positioning 81-mm mortars, the lateral distance between the center of each baseplate should have a minimum distance of 4 meters (40 meters when firing service ammunition, 1:10 scale for the SRTR = 4 meters).

B-28. The SRTR can also be fired from the 120-mm mortar using the M303 81-mm subcaliber insert (see section III) and the M313 81-mm subcaliber insert (see section IV). When positioning 120-mm mortars, the lateral distance between the center of each baseplate should be a minimum of 6 meters (60 meters when firing service ammunition, 1:10 scale for the SRTR = 6 meters).

CONSTRUCTION OF A SCALED MAP

B-29. The range limitations associated with the M880 SRTR make a standard 1:50,000-scale military map difficult to use. Therefore, a new map 1/10 the size (1:5,000) must be constructed (Figure B-3) using the following procedures.

Appendix B

(1) Use a blank sheet of paper to draw grid squares the same size as the 1:50,000-scale map and renumber them based on 100 meters per square.

(2) Determine the 8-digit grid coordinates to the mortar position on a 1:50,000-scale military map and convert it to a 1:5,000 scale (Figure B-3).

NOTE: The first number of a grid coordinate is the 10,000-meter designator; the second number is the 1,000-meter designator; the third number is the 100-meter designator; and the fourth number is the 10-meter designator.

- To make a 1:5,000-scale map, drop the 10,000-meter designator and use the 1,000-meter designator as the first number for each grid. For example, a mortar location of 07368980 becomes 736980. The 7 and the 9 will precede each grid location since they identify the 1,000-meter increments. To make an 8-digit grid from this 6-digit grid, add a 0 at the end of the easting and northing grid location. For example, the mortar grid becomes 73609800. The last 0 is the 1-meter designator.

- Before constructing the 1:5,000-scale map, determine the direction of fire from the mortar position to the RP. (Surveying a point on the range should be done at least one day before training. If a point on the range has already been surveyed, it can be used to make the map.) For example, if the direction of fire is southeast, the mortar position would be in the northwest corner of the grid sheet. Using the example grid coordinates, the most westerly grid line would be 73 and the second most northerly grid line would be 98. Number the rest of the grid lines accordingly.

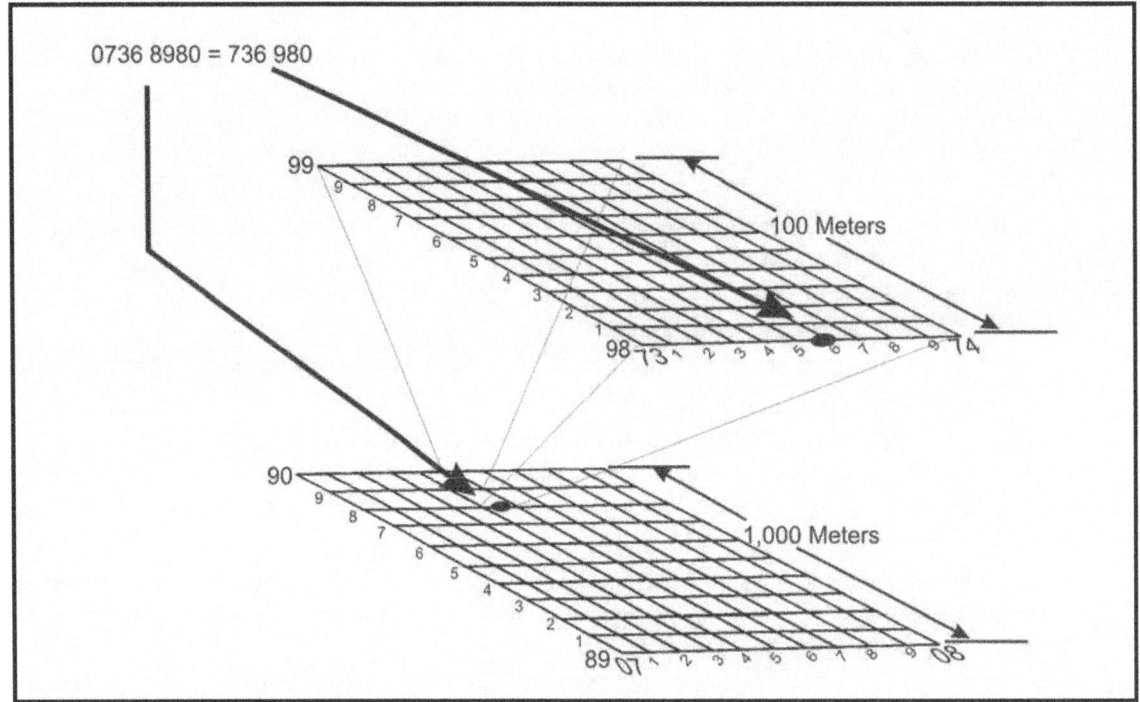

Figure B-3. Converting 1:50,000 grid to 1:5,000 grid.

(3) To plot the mortar position on the 1:5,000-scale map, use the same protractor that is used for the 1:50,000-scale map, but instead of the protractor measuring 1,000 meters, it will measure 100 meters.

(4) To determine the direction of fire, place a target downrange, set up the M2 aiming circle at the base gun position, and measure an azimuth to the target. To accurately plot targets on the map, measure the distance by pace count from the basepiece to the target and then plot it on the map. This procedure allows the creation of a surveyed firing chart. The FO's location

can be plotted in the same manner. After all plotting is done, give a copy of the map to the FO (Figure B-4). He can call for fire using this map the same as if firing service ammunition.

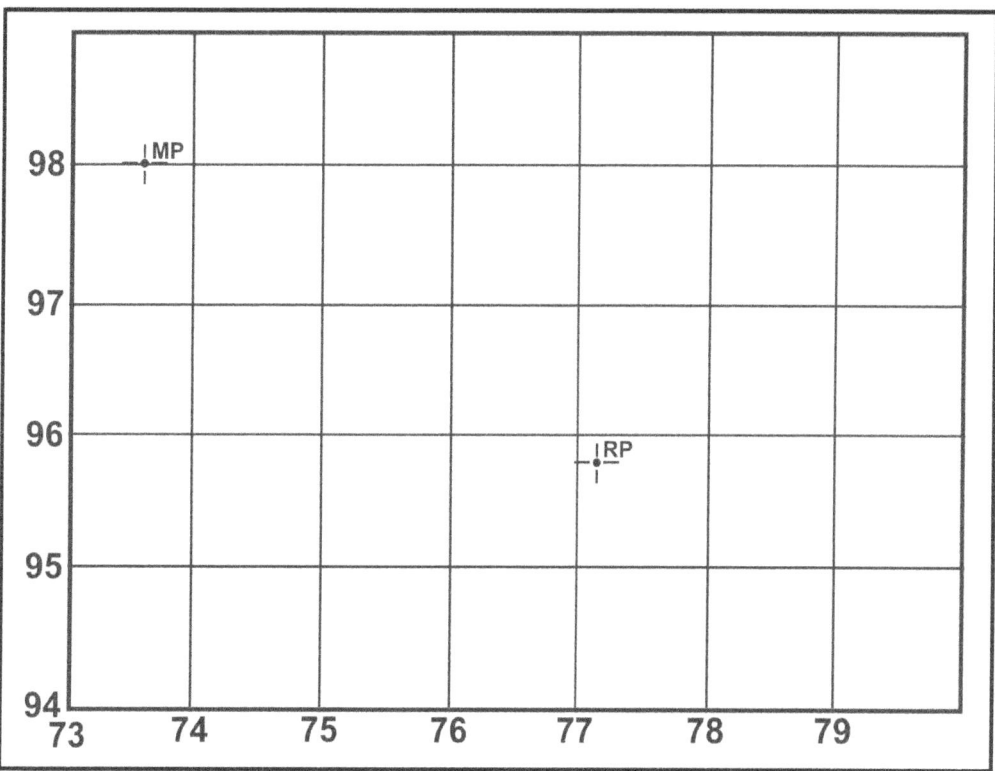

Figure B-4. Plotting targets on the 1:5,000-scale map.

(5) Before putting the map on the M16 plotting board, determine the grid intersection to represent the pivot point in the same manner as if firing service ammunition.

- Place the grid system on the plotting board the same as if firing service ammunition. Number every other black line. Every small green square equals 5 meters and every black square equals 50 meters.
- Place the mortar position and target on the plotting board the same as if firing service ammunition. The grids plotted are to the nearest 1 meter instead of 10 meters.

(6) To determine the mounting azimuth, line up the mortar position and the RP. Read the azimuth on the azimuth scale at the top of the plotting board. Round off the azimuth to the nearest 50 mils, and superimpose this azimuth over the referred deflection. Number the azimuth scale every hundred mils using the LARS rule (Figure B-5).

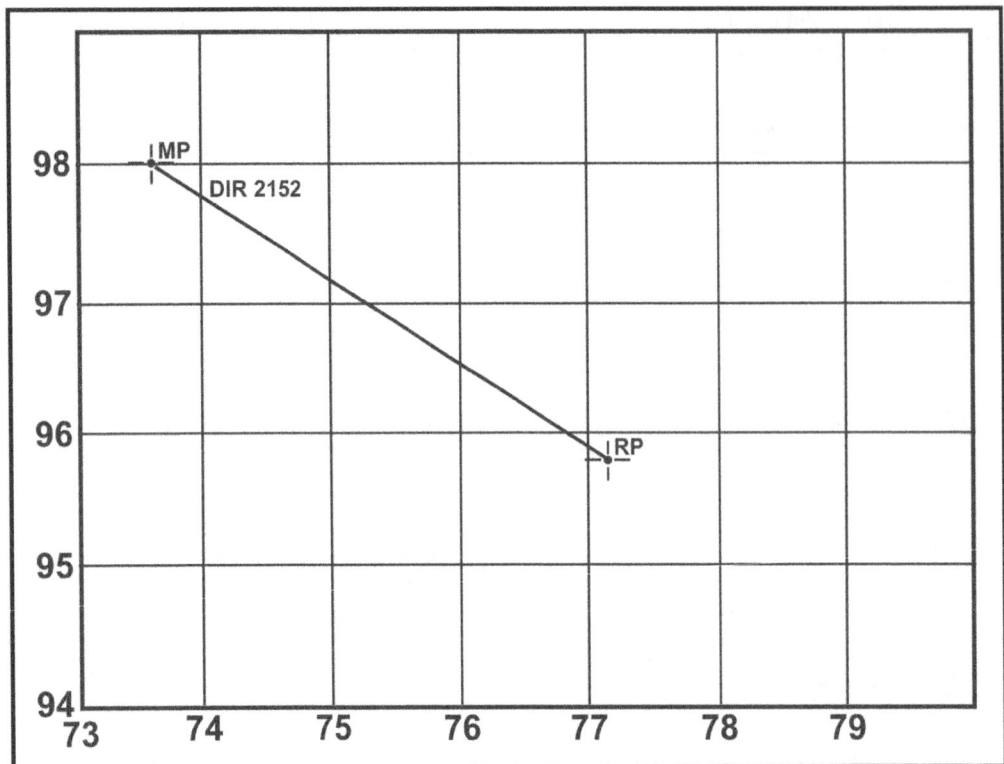

Figure B-5. Determining direction from mortar position to the registration point.

(7) When firing the M880 SRTR, range control must know exactly where the firing position is. The 1:50,000-scale grid must be annotated on DA Form 2188-R, labeled, and placed in the battle position (BP) box (Figure B-6).

DATA SHEET

For use of this form, see FM 23-91. The proponent agency is TRADOC.

SETUP	WEAPON DATA			FO DATA		
TIME OUT:	UNIT: E CO 2/29 IN	WPN:	WPN:	FO	ALT	GRID
TGT PRFX: AB		DIR:	DIR:			
TGT NO: 0100-0150	81 mm CAR: ☐ YES	DIS:	DIS:			
ALARM: ☐ ON ☐ OFF	(M880) ☒ NO					
MIN E:	BP: 1/50,000THS	WPN:	WPN:			
MIN N:	0736 8980	DIR:	DIR:			
GD: ☐ E ☐ W	E: 7360	DIS:	DIS:			
LAT: ☐ + ☐ –	N: 9800					
LISTEN: ☐ ON ☐ OFF	ALT:					
BIT RATE:	AZ: 2152/2150	WPN:	WPN:			
KEY TONE:	DEF: 2800	DIR:	DIR:			
BLK: ☐ SNG ☐ DBL	ELE:	DIS:	DIS:			
OWNER ID:						

AMMUNITION DATA

TEMPERATURE ______ TYPE: ☐ HE ☐ WP ☐ ILL ☐ CS ☒ TNG (M880)

LOT NUMBER									
WEIGHT									
ON HAND									
RECEIVED									

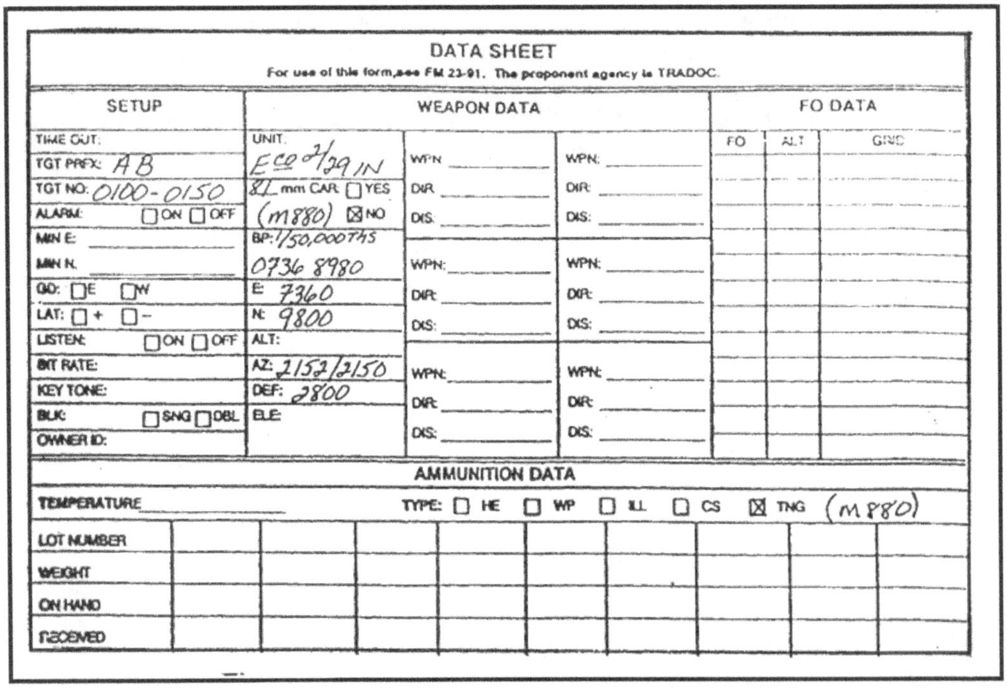

Figure B-6. Example completed DA Form 2188-R.

(8) DA Form 2399 is also altered. For safety, as well as range control's information, "M880" should be placed in the "SHELL AND FUZE" spaces in both the FDC ORDER and INITIAL FIRE COMMAND (Figure B-7).

FDC ORDER		INITIAL CHART DATA		INITIAL FIRE COMMAND		ROUNDS EXPENDED
MORTAR TO FFE	SEC	DEFLECTION	2807	MORTAR TO FOLLOW	SEC	
MORTAR TO ADJ	#2	DEFLECTION CORRECTION	□L □R	SHELL AND FUZE	HEQ (M880)	
METHOD OF ADJ	1 RD	RANGE	206	MORTAR TO FIRE	#2	
BASIS FOR CORRECTION		VWALT CORRECTION	□+ □-	METHOD OF FIRE	1 RD IN ADJ. 2 RDS IN FFE.	
SHEAF CORRECTION						
SHELL AND FUZE	HEQ (M880)	RANGE CORRECTION	□ □	DEFLECTION	2807	①
METHOD OF FFE	2 RDS			CHARGE	0	
RANGE LATERAL SPREAD		CHARGE/RANGE	0	TIME SETTING		
ZONE		AZIMUTH	2143	ELEVATION	1201	
TIME OF OPENING FIRE	W/R	ANGLE T	1630			

Figure B-7. Example completed DA Form 2399 showing SHELL AND FUZE entries in the FDC ORDER and INITIAL FIRE COMMAND columns.

SAFETY

B-30. Although the SRTR is safe to handle and fire, the following safety precautions must be enforced:

> **WARNING**
>
> The SRTR is inert and can be stored and handled as a weapon until it is loaded with the refurbishment kit at which time it will be handled as a live mortar round. When loaded and ready to fire, all safety precautions used during handling of service ammunition are observed.

- The SRTR refurbishment kit contains a propellant and smoke charge, and is always stored and handled as live ammunition.
- Practice ammunition is never fired over the heads of troops.
- Personnel never go forward of the firing line until a cease fire has been called by the range OIC or safety officer.

NOTE: Duds are disposed of as prescribed in paragraph B-22.

MALFUNCTIONS AND REMOVAL OF A MISFIRE

B-31. After failure to fire, misfire removal procedures are followed to remove the round from the weapon. Misfire procedures for live rounds also apply when firing the SRTR. (See Chapter 3, paragraphs 3-37 to 3-40 for a detailed discussion of malfunctions.)

Appendix B

B-32. If the primer cap of the SRTR is dented, the whole round is placed in the dud pit for disposal in accordance with the local SOP.

B-33. If the primer is not dented, the round can be refired upon determination of the cause of the misfire.

SECTION III. SUBCALIBER INSERT, M303

The M303 subcaliber insert (Figure B-8) provides realistic live-fire training at a relatively low cost to the 120-mm indirect fire team while firing 81-mm mortar ammunition. It uses 300-series HE, WP, and ILLUM ammunition. Currently, the FDC uses the M16 plotting board for plotting.

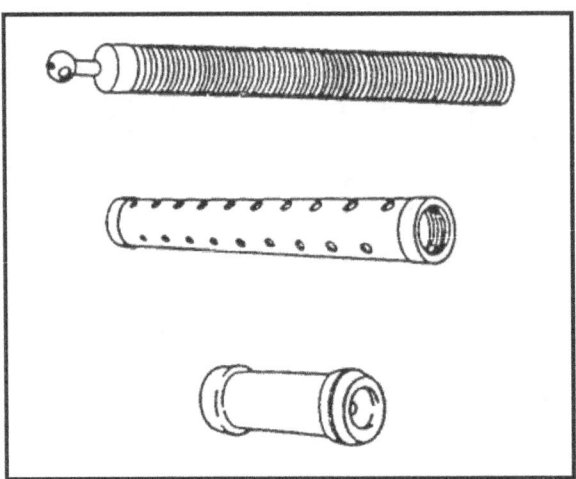

Figure B-8. Subcaliber insert, M303.

CHARACTERISTICS

B-34. The cannon is an M29A1 cannon assembly that has been refurbished. This is the only M29 cannon approved to use the M303 subcaliber insert. The cannon is 51 inches long and weighs 28 pounds. It has a smooth bore with one circular cooling ring and a removable firing pin.

B-35. The insert sleeve is 42 inches long and weighs 5.9 pounds. It is tubular shaped with cooling holes. It has a threaded end on top and a smooth end on the bottom. The sleeve slides over the cannon and is threaded onto the cannon until it is flush with the breech cap. It is held in place by three hex screws on the muzzle end. Approximately 2 inches of the M29 cannon will be showing at the muzzle end.

> NOTE: After every fire mission, the hex screws on the upper sleeve ring should be checked for tightness to ensure the interior of the 120-mm cannon is not damaged.

B-36. The filler block is 15.9 inches long and weighs 17 pounds. It is tubular shaped with the top and bottom the same diameter as the interior of the 120-mm cannon. The top and bottom have a concave slope with a threaded hole, which is used in the removal of the filler block.

B-37. The M303 subcaliber insert comes with a filler block adapter, a 0.0625-inch hex wrench, and a firing pin wrench.

B-38. The 300-series ammunition has a minimum range of 73 meters and a maximum range of 4,789 meters. It can be fired at a sustained rate of three (charge 8) or five (charge 6) rounds per minute. It can be fired indefinitely at a periodic rate of fire. The 300-series ammunition has a maximum rate of fire of 12 (charge 8) for two minutes and 12 (charge 6) for five minutes.

MAINTENANCE

B-39. Maintenance must be performed on the M303 subcaliber insert before, during, and after firing. Use the following procedures to perform this maintenance. (For more detail, see TM 9-1015-254-13&P.)

BEFORE

Cannon

B-40. Update the DA Form 2408-4 to reflect the day's firing. Ensure that all standards for bore scope and pullover gauging have not been exceeded (within the past 180 days).

B-41. Check the M120/M121 mortar for cracks, broken welds, rust, and missing or damaged parts.

B-42. Check the M29A1 cannon for foreign matter in the barrel, then wipe dry. Clean and lubricate the exterior surface. Check for bulges, dents, and visible cracks. Check for evidence of gas leakage (grayish color) around the firing pin located in the base of the spherical projectile.

B-43. Check the insert sleeve for cracks, broken welds, rust, and missing or damaged parts. Check that the insert sleeve assembles onto the cannon assembly, can be secured with set screws, and is free of any nicks and burrs.

Filler Block

B-44. Check the filler block for cracks, broken welds, rust, and nicks or burrs.

B-45. Check that the filler block fits with breech plug of the M29A1 cannon, seats in the base of the 120-mm bore, and can be employed or retrieved using the adapter tool.

DURING

B-46. Check the bore after approximately every 50 rounds for bulges, dents, and visible cracks. Check for evidence of gas leakage around the firing pin (grayish discoloration).

B-47. Dry swab the cannon bore after firing every 10 rounds or after every end of mission (EOD).

AFTER

B-48. Check DA Form 2408-4 to ensure it reflects the day's firing.

B-49. Ensure rifle bore cleaning compound (RBC) is used to thoroughly clean the cannon bore after firing and for two consecutive days thereafter. Be sure to wipe dry and lubricate with general purpose lubricant (GPL) after each cleaning.

> **NOTE:** For nonfiring periods, clean and lubricate the cannon on a weekly basis.

B-50. Check the insert sleeve for cracks, broken welds, rust, nicks, burrs, and damaged parts.

B-51. Check the filler block for cracks, broken welds, rust, nicks, burrs, and damaged threads.

B-52. Intervals are based on usual conditions. For unusual operating conditions, lubricate more often. When the weapon is not being used, the intervals may be extended if proper lubrication procedures have been followed. Clean the cannon bore weekly with RBC, and lubricate the cannon, firing pin, insert sleeve, and filler block weekly with GPL.

Appendix B

> **WARNING**
>
> Dry cleaning solvent is flammable. Do not clean parts near an open flame or in a smoking area. Dry cleaning solvent evaporates quickly and has a drying effect on the skin. When used without protective gloves, this chemical may cause irritation to, or cracking of, the skin.

NOTE: Before firing, dry swab the cannon bore of the M29A1 and the 120-mm cannon. Wipe dry the exterior of all components of the subcaliber insert.

MISFIRE PROCEDURES

B-53. See TM 9-1015-254-13&P for M303 misfire procedures.

SECTION IV. SUBCALIBER TRAINER, M313

The M313 subcaliber trainer (Figure B-9) allows the 120-mm fire team to conduct low cost, realistic live-fire training. This trainer fires 81-mm mortar ammunition and uses 300- and 800-series HE, smoke, and ILLUM ammunition and 300-series WP. When using this trainer, the FDC plots targets using an M16 plotting board.

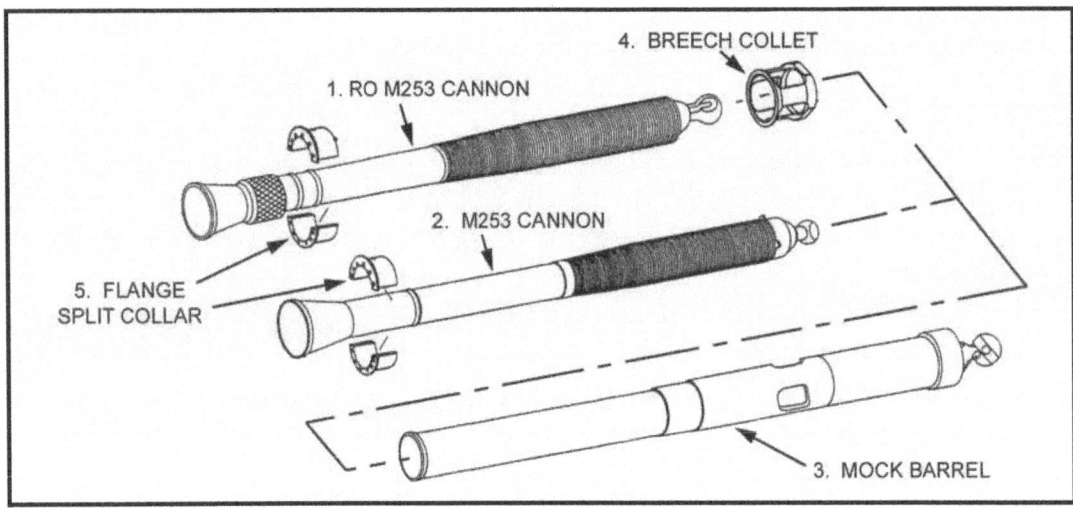

Figure B-9. Subcaliber trainer, M313.

CHARACTERISTICS

B-54. The M313 120-mm subcaliber trainer consists of an 81-mm mortar cannon, a 120-mm mock barrel, a breech collet, and a flange split collar. This paragraph describes the characteristics of the M313 subcaliber trainer and its components.

Cannon Tube

B-55. Two types of 81-mm cannon tubes are authorized for use with the M313: the RO M253 (Figure B-9, 1) and the M253 cannon (Figure B-9, 2) assemblies. The RO M253 cannon is 55 inches long and weighs 35 pounds. The M253 cannon is 55 inches long and weighs 29.5 pounds. Both types have a smooth bore with a circular cooling ring and a removable firing pin.

MOCK BARREL

B-56. The mock barrel (Figure B-9, 3) replaces the M298 cannon tube in the 120-mm mortar system. The barrel is 77.7 inches long and weighs 84 pounds.

BREECH COLLET

B-57. Exclusive to the RO M253 cannon, the breech collet (Figure B-9, 4) centers and stabilizes the tube in the mock barrel. The breech collet is 7 inches long and weighs 5 pounds.

FLANGE SPLIT COLLAR

B-58. The flange split collar (Figure B-9, 5) is used with both cannons to center the cannon in the mock barrel.

ASSEMBLY

B-59. The mock barrel is mounted on a 120-mm mortar baseplate and bipod assembly. The breech collet is installed in the mock barrel (RO M253 only). The flange split collar is placed around the muzzle of the 81-mm cannon, and the cannon is lowered into the mock barrel until it is seated on the mortar filler block. Ensure that the breech cap of the cannon tube seats correctly in the concave breech surface of the mortar filler block. The assistant gunner dry swabs the bore of the 81-mm mortar cannon; the gunner installs the sightunit and levels the mortar; and the crew awaits a fire mission.

OTHER COMPONENTS

B-60. The M313 subcaliber trainer comes with a tool bag, two cleaning brushes, a wire mesh cleaning pad, an M25 fuze setter, a pair of gloves, a head section cleaning tool, a firing pin, a copy of TM 9-1015-255-13&P, a firing pin wrench, an M18 fuze wrench, a 1-3/4 inch fuze setting wrench, and a pipe strap wrench.

MINIMUM AND MAXIMUM RANGE

B-61. The 300-series ammunition has a minimum range of 73 meters and a maximum range of 4,789 meters. It can be fired indefinitely at a sustained rate of fire of 3 (charge 8), 5 (charge 6), or 4 (charge 4 with the M374A3/M375A3) rounds per minute. The 300-series ammunition has a maximum rate of fire of 12 (charge 8) rounds per minute for two minutes and 12 (charge 6) rounds per minute for five minutes.

B-62. The 800-series ammunition has a maximum range of 5,608 meters. It can be fired at a sustained rate of fire of 15 rounds per minute indefinitely and has a maximum rate of fire of 30 rounds per minute for two minutes.

MAINTENANCE

B-63. Preventive maintenance must be performed on the M313 subcaliber trainer before, during, and after firing to ensure that the trainer is in good operating condition and ready for its primary mission. Use the following procedures to perform this maintenance. (For more detail, see TM 9-1015-255-13&P.)

BEFORE

Cannon

B-64. Before firing, dry swab the cannon bore of the mock barrel, and wipe dry the exterior of all components of the 81-mm subcaliber insert.

B-65. Review DA Form 2408-4 or DA Form 5988-E (Equipment Inspection Maintenance Worksheet [EGA]) to ensure that it reflects previous firings. Before firing the cannon, ensure that it has been bore scoped and pullover gauged within the past 180 days, and be sure that these procedures are performed every 180 days during the firing period. Notify field-level maintenance if bore scoping and pullover gauging are needed.

B-66. Check for cracks, broken welds, rust, and missing, loose, or damaged parts.

B-67. Check the cannon/BAD for foreign matter in the barrel, and then wipe dry. Clean and lubricate the exterior surface. Check for bulges, broken welds, dents, rust, and visible cracks. Check for evidence of gas leakage (grayish discoloration) around the firing pin or breech cap.

Mock Barrel

B-68. Check for cracks, broken welds, rust, nicks, burrs, and missing or damaged parts.

Flange Split Collar

B-69. Check for cracks, rust, nicks, and burrs.

Breech Collet

B-70. Check for cracks, rust, nicks, and burrs.

NOTE: When firing subcaliber training rounds, bore scope the M313 every 100 rounds. Record your inspections using DA Form 2408-4 or DA Form 5988-E.

DURING

B-71. Remove the 81-mm mortar cannon from the mock barrel after every 50 rounds (approximately), and check for bulges, dents, and visible cracks. Check for evidence of gas leakage (grayish discoloration) around the firing pin or breech cap.

B-72. Dry swab the cannon bore after firing every 10 rounds (approximately) or at the end of every fire mission.

AFTER

Cannon

B-73. Update DA Form 2408-4 or DA Form 5988-E to reflect the day's firing. Have the weapon bore scoped once 5,000 rounds have been fired and every 500 rounds thereafter.

B-74. Check the cannon for bulges, dents, and visible cracks. Check for evidence of gas leakage (grayish discoloration) around the firing pin or breech cap.

B-75. Use RBC to thoroughly clean the cannon bore after firing and for two consecutive days thereafter. Once clean, wipe it dry and lightly coat it with GPL.

NOTE: For nonfiring periods, clean and lubricate the cannon weekly.

Mock Barrel

B-76. Check for cracks, broken welds, rust, nicks, burrs, and missing or damaged parts.

Flange Split Collar

B-77. Check for cracks, rust, nicks, and burrs.

Breech Collet

B-78. Check for cracks, rust, nicks, and burrs.

> **WARNING**
>
> 1. Dry cleaning solvent is flammable. Do not clean parts near an open flame or in a smoking area.
>
> 2. Dry cleaning solvent evaporates quickly and has a drying effect on the skin. When used without protective gloves, this chemical may cause irritation to, or cracking of, the skin.
>
> 3. Cleaner, lubricant, and preservative (CLP) is not authorized for use on the M313 trainer.

MISFIRE PROCEDURES

B-79. See TM 9-1015-255-13&P for M313 misfire procedures.

This page intentionally left blank.

Glossary

AAR	after-action review
AC	alternating current
ACA	airspace coordination area
ACCP	Army Correspondence Course Program
ACU	Army combat uniform
ADR	address
AFATDS	Advanced Field Artillery Tactical Data System
AR	Army regulation
ARTEP	Army Training and Evaluation Program
ASP	ammunition supply point
AZ	azimuth
BAD	blast attenuator device
BIT	built-in test
BLTM	battalion-level training model
BNCOC	Basic Noncommissioned Officer Course
BP	battle position
CBRN	chemical, biological, radiological, nuclear
CFF	call for fire
CFL	coordinating fire line
CFX	command field exercise
CH	charge
CHN	channel
CI	commander's interface
CLP	cleaner, lubricant, and preservative
CMF	career management field
COMSEC	communications security
CONUS	continental United States
CPX	command post exercise
DA	Department of the Army
DAP	distant aiming point
DC	direct current
DD	driver's display
DEF	deflection
DLY	delay
DOF	direction of fire
DMD	digital messaging device
EDC	error detection and correction
EIDSA	Electronic Information Delivery System
EL	elevation
EOD	explosive ordnance disposal
FC	fires cell
FDC	fire direction center
FFE	fire for effect
FH	frequency hopping
FIST	fire support team
FLOT	forward line of own troops
FM	field manual
FO	forward observer

Glossary

FOS	Forward Observer System
FP	firing point
FPF	final protective fire
FRTC	full-range training cartridge
FSCM	fire support coordination measure
FT	firing table
FTX	field training exercise
GD	gunner's display
G-M	grid-magnetic
GPL	general purpose lubricant
GPS	Global Positioning System
GT	gun-target
GUI	graphic user interface
HBCT	heavy brigade combat team
HE	high-explosive
HMMWV	high-mobility multipurpose wheeled vehicle
IBCT	infantry brigade combat team
ICCC	Infantry Career Captain's Course
ILLUM	illumination
IMLC	Infantry Mortar Leader Course
IMP	impact
IMU	inertial measurement unit
IBOLC	Infantry Basic Officer Leader's Course
IP	Internet protocol
IR	infrared
ITEP	Individual Training and Evaluation Program
LARS	left, add; right, subtract
LCD	liquid crystal display
LED	light-emitting diode
LFX	live-fire exercise
LWCMS	Lightweight Company Mortar System
MANCOC	Manuever Advanced Noncommissioned Officer Course
MAPEX	map exercise
MBC	mortar ballistics computer
MET	meteorological data
METL	mission-essential task list
METT-TC	mission, enemy, terrain and weather, troops and support available, time available, civil considerations
MFCS	Mortar Fire Control System
MILES	Multiple Integrated Laser Engagement System
mm	millimeter
MOC	method of control
MOF	method of fire
MOS	military occupational specialty
MQS	Military Qualification Standard
MT	mechanical time
MTO	message to observer
MTP	mission training plan
MTSQ	mechanical time superquick
NA	not applicable

NAD	net access delay
NATO	North Atlantic Treaty Organization
NCO	noncommissioned officer
NCOPD	noncommissioned officer professional development
NFA	no fire area
NiMh	nickel metal hydride
NSB	near-surface burst
OBSNUM	observer number
OIC	officer in charge
OpACK	operationally acknowledge
OPD	officer professional development
OPFOR	opposing force
OpRDY	operationally ready
OSUT	one-station unit training
OT	observer-target
PCC	Pre-Command Course
PD	point detonating; pointing device
PDA	power distribution assembly
PE_D	probable errors deflection
PE_R	probable errors range
PLGR	precision lightweight GPS receiver
PROX	proximity
PRN	printer
PRX	proximity (used on fuze settings)
PTM	plain text message
QWERTY	the most common keyboard layout on English language-typing computers and keyboards; takes its name from the first six letters in the keyboard's top row of letters
RTO	radio-telephone operator
RBC	rifle bore cleaning compound
RDX	cyclonite
RP	red phosphorus; reference point
RPO	range patrol officer
S3	battalion/brigade level plans, training, and operations office/officer
SDZ	surface danger zone
SEL	select
SINCGARS	Single Channel Ground and Airborne Radio System
SM	Soldier's manual
SOP	standing operating procedure
SQ	superquick
SRTC	short-range training cartridge
SRTR	short-range training round
STANAG	Standardized NATO Agreement
STP	Soldier's training publication
STX	situational training exercise

Glossary

T&EO	training and evaluation outline
TCIM	tactical communication interface modem
TEC	Training Extension Course
TEWT	tactical exercise without troops
TG	trainer's guide
TM	technical manual
TNT	trinitrotoluene
TOE	table of organization and equipment
TP	training practice
TTP	tactics, techniques, and procedures
URN	unit reference number
U.S.	United States
USAREUR	U.S. Army Europe
V, v	volt
VMS	vehicle motion sensor
VT	variable time
WP	white phosphorus
WTT	warrior task test

References

DOCUMENTS NEEDED
These documents must be available to the intended users of this publication.

ARMY PUBLICATIONS
AR 385-63, *Range Safety {MCO 3570.1B}*. 19 May 2003.
ARTEP 7-90-Drill, *Battle Drills for the Infantry Mortar Platoon, Section, and Squad*. 26 July 2002.
ARTEP 7-90-MTP, *Mission Training Plan for the Infantry Mortar Platoon, Section, and Squad*. 1 April 2005.
DA Pamphlet 385-63, *Range Safety*. 10 April 2003.
FT 60-P-1, *Mortar, 60-mm: M224, Firing Cartridge, HE, M720; Cartridge, HE, M49A4; Cartridge, HE, XM888; Cartridge, TP, M50A3; Cartridge, WP, M302A1 and Cartridge, Illuminating M83A3*. 22 March 1980.
STP 7-11B1-SM-TG, *Soldier's Manual and Trainer's Guide, MOS 11B, Infantry, Skill Level 1*. 6 Aug 2004.
STP 7-11B24-SM-TG, *Soldier's Manual and Trainer's Guide, MOS 11B, Infantry, Skill Levels 2, 3, and 4*. 6 Aug 2004.
STP 7-11C14-SM-TG, *Soldier's Manual and Trainer's Guide, MOS 11C, Indirect Fire Infantryman, Skill Levels 1, 2, 3, and 4*. 6 Aug 2004.
TM 9-1010-223-10, *Operator's Manual for Lightweight Company Mortar, 60-mm: M224*. 15 September 1998.
TM 9-1015-249-10, *Operator's Manual for: Mortar, 81-mm, M252*. 30 October 1987.
TM 9-1015-250-10, *Operator's Manual for Mortar, 120mm: Towed M120 and Mortar, 120mm: Carrier-Mounted M121*. 19 August 1996.
TM 9-1015-254-13&P, *Operator's, Unit, and Direct Support Maintenance Manual with Repair Parts and Special Tools List for Mortar, Subcaliber Insert: M303*. 2 March 1994.
TM 9-1220-248-10, *Operator's Manual for Mortar Fire Control System, M95*. 31 October 2005.
TM 9-1290-262-10, *Operator's Manual for Aiming Circle, M2 W/E and M2A2 W/E*. 15 April 1981.
TM 9-1290-333-15, *Operator's, Organizational, Direct Support, General Support and Depot Maintenance Manual (Including Repair Parts and Special Tools List): Compass, Magnetic, Unmounted: M2*. 7 November 1963.
TM 9-1315-249-12&P, *Operator's and Unit Maintenance Manual (Including Repair Parts and Special Tools List) for 81-mm Mortar Training Device, 81-mm Sabot (Inert) M1 and 22-mm Sub-Caliber Practice Cartridges M744, M745, M746, and M747*. 1 September 1990.
TM 9-2350-277-10, *Operator's Manual for Carrier, Personnel, Full-Tracked, Armored, M113A3; Carrier, Command Post, Light Tracked, M577A3; Carrier, Smoke Generator, Full Tracked, M1059A3; Carrier, Mortar, 120-mm M121, Self Propelled, M1064A3; Carrier, Standardized Integrated Command Post System (SICPS) M1068A3; Carrier, Mechanized Smoke Obscurant M58*. 2 January 2001.

ARMY FORMS
DA Form 2188-R, *Data Sheet (LRA)*.
DA Form 2399-R, *Computer's Record (LRA)*.
DA Form 2408-4, *Weapon Record Data*.
DA Form 5964-R, *Gunner's Examination Scorecard–Mortars*.
DA Form 5988-E, *Equipment Inspection Maintenance Wroksheet (EGA)*.

READINGS RECOMMENDED
These sources contain relevant supplemental information.
FM 3-22.91, *Mortar Gunnery*. 18 January 2005.
FM 4-25.11, *First Aid {NTRP 4-02.1.1; AFMAN 44-163(I); MCRP 3-02G}*. 23 December 2002.

References

FM 6-30, *Tactics, Techniques, and Procedures for Observed Fire*. 16 July 1991 (to be republished as FM 3-09.30).
FM 7-0, *Training the Force*. 22 October 2002.
FM 7-1, *Battle Focused Training*. 15 September 2003.
FM 7-90, *Tactical Employment of Mortars*. 09 October 1992.
FM 25-4, *How to Conduct Training Exercises*. 10 September 1984.
FT 81-AR-2, *Firing Tables for Mortar, 81mm, M252, Firing Cartridge, HE, M821 Cartridge, HE, M889 Cartridge, RP, M819 Cartridge, Illum, M853A1 Cartridge, TP, M879 Cartridge, HE, M821A1 and M821A2 Cartridge, HE, M889A1 and Mortar, 81mm, M29A1 Mortar, 81mm, M252 Mortar, 120mm, M120 and M121 with 81mm, Insert Training Device Firing Cartridge, TP (SR), M880*. 1 June 1997.
TB 43-0250, *Ammunition Handling, Storage and Safety*. 1 March 2003.

INTERNET WEBSITES

U.S. Army Publishing Agency, http://www.army.mil/usapa
Reimer Doctrine and Training Digital Library, http://www.adtdl.army.mil

Index

NUMBERS

60-mm mortar, M224, 3-2 (illus)
  ammunition (*see also* ammunition), 3-16 to 3-22
    care and handling, 3-22 to 3-23
    cartridge preparation, 3-21 to 3-22
    classification and types of ammunition, 3-16 to 3-18
    fuzes, 3-19 to 3-21
  components, 3-2 to 3-5
    cannon assembly, M225 (*see also* cannon assembly), 3-3, 3-3 (illus)
    baseplate (*see also* baseplate)
      M7, 3-5, 3-5 (illus)
      M8, 3-5, 3-5 (illus)
    bipod assembly, M170 (*see also* bipod assembly), 3-3 to 3-4, 3-4 (illus)
  operation, 3-6 to 3-16
    carrying, 3-16
    deflection and elevation changes (*see also* deflection and elevation changes), 3-8 to 3-9, 3-9 (illus)
    dismounting (*see also* dismounting the mortar), 3-15 to 3-16
    malfunctions (*see also* malfunctions), 3-10 to 3-11
    misfire, removal (*see also* misfire, removal), 3-11 to 3-15
    mounting of the mortar (*see also* mounting of the mortar), 3-6 to 3-7
    premount checks (*see also* premount checks), 3-6
    referring the sight and realigning the aiming posts (*see also* referring the sight and realigning the aiming posts), 3-9 to 3-10, 3-10 (illus)
    safety checks before firing (*see also* safety checks before firing), 3-7 to 3-8
  squad and section organization and duties (*see also* squad and section duties and organization), 3-1
  tabulated data, 3-2, 3-2 (table)

81-mm mortar, M252, 4-3 (illus)
  ammunition (*see also* ammunition), 4-16 to 4-27
    care and handling, 4-25 to 4-27, 4-26 (illus), 4-26 (illus)
    cartridge preparation, 4-24
    classification and types of ammunition, 4-16 to 4-21
    fuzes, 4-21 to 4-24
  components, 4-3 to 4-6
    baseplate, M3A1 (*see also* baseplate), 4-6, 4-6 (illus)
    cannon assembly, M253 (*see also* cannon assembly), 4-5, 4-5 (illus)
    mount, M177 (*see also* mount, M177), 4-5 to 4-6, 4-5 (illus)
  operation, 4-7 to 4-16
    deflection and elevation changes (*see also* deflection and elevation changes), 4-10 to 4-11
    dismounting (*see also* dismounting the mortar), 4-16
    malfunctions (*see also* malfunctions), 4-12
    misfire, removal (*see also* misfire, removal), 4-12 to 4-16, 4-13 (illus), 4-14 (illus), 4-15 (illus)
    mounting of the mortar (*see also* mounting of the mortar), 4-8 to 4-9, 4-9 (illus)
    premount checks (*see also* premount checks), 4-7 to 4-8, 4-8 (illus)
    referring the sight and realigning the aiming posts (*see also* referring the sight and realigning the aiming posts), 4-12
    safety checks before firing (*see also* safety checks before firing), 4-10
  squad and section organization and duties (*see also* squad and section duties and organization), 4-1 to 4-2, 4-2 (illus)
  tabulated data, 4-4, 4-4 (table)

120-mm mortars, M120 and M121, 5-3 (illus)
  ammunition (*see also* ammunition), 5-38 to 5-45
    care and handling, 5-44 to 5-45
    cartridge preparation, 5-43 to 5-44
    classification and types of ammunition, 5-38 to 5-40
    fuzes, 5-40 to 5-43
  components, 5-3 to 5-8
    baseplate, M9 (*see also* baseplate), 5-8, 5-8 (illus)
    bipod assembly, M190 (*see also* bipod assembly), 5-7, 5-7 (illus)
    bipod assembly, M191 (*see also* bipod assembly), 5-6, 5-6 (illus)
    cannon assembly, M298 (*see also* cannon assembly), 5-4 to 5-5, 5-5 (illus)
  operation, carrier-mounted, 5-26 to 5-38
    deflection and elevation changes (*see also* deflection and elevation changes), 5-29 to 5-30
    misfire, removal (*see also* misfire, removal), 5-30 to 5-37, 5-33 (illus)
    mortar carrier, M1064A3 (*see also* mortar carrier, M1064A3), 5-23 to 5-26, 5-23 (illus), 5-24 (illus)
    mounting mortar from a carrier- to a ground-mounted position, 5-27 to 5-28
    placing mortar into action, 5-26 to 5-27
    premount checks (*see also* premount checks), 5-26
    reciprocally laying the mortar carrier section, 5-38
    safety checks before firing (*see also* safety checks before firing), 5-28 to 5-29
    taking out of action, 5-37 to 5-38
  operation, ground-mounted, 5-8 to 5-23
    deflection and elevation changes (*see also* deflection and elevation changes), 5-12 to 5-14
    loading and firing, 5-21 to 5-22
    malfunctions (*see also* malfunctions), 5-14
    misfire, removal (*see also* misfire, removal), 5-14 to 5-20, 5-16 (illus), 5-17 (illus), 5-18 (illus), 5-20 (illus)
    placing mortar into action, 5-10 to 5-11
    premount checks (*see also* premount checks), 5-8 to 5-10

Index

safety checks before firing (*see also* safety checks before firing), 5-11 to 5-12
  taking out of action, 5-22 to 5-23
squad organization and duties (*see also* squad and section duties and organization), 5-1 to 5-2, 5-2 (illus)
tabulated data, 5-4, 5-4 (table)

A

aiming circles, M2 and M2A2, 2-5 to 2-18, 2-6 (illus), 2-7 (illus)
  accessory equipment, 2-9
    aiming circle cover, 2-9
    backplate, 2-9
    cloth cover, 2-9
    instrument light, M54, 2-9
    lamp holder and remover, 2-9
    plumb bob, 2-9
    tripod, M24, 2-9
  care and maintenance, 2-18
  characteristics, 2-5, 2-5 (table)
  declination constant, 2-11 to 2-13
    declination station, 2-11
      declinating an aiming circle at a declination station, 2-11 to 2-12
      declinating an aiming circle when a declination station is not available, 2-12 to 2-13, 2-13 (illus)
      use of the grid-magnetic angle in a new area without a declination station, 2-12
    when to declinate the aiming circle, 2-13
  description, 2-5
  disassembly, 2-17
  measuring of the horizontal angle between two points, 2-14
  orienting
    by orienting angle, 2-17, 2-17 (illus)
    of the 0-3200 line on a given grid azimuth, 2-15, 2-15 (illus)
    of the 0-3200 line on a given magnetic azimuth, 2-15 to 2-16
    of the instrument on grid north to measure grid azimuth to objects, 2-14
  setup and leveling, 2-10 to 2-11
    leveling screws, 2-10 (illus)
    set-up distance from objects, 2-11 (table)

use, 2-6
  azimuth and elevation scales, 2-8
  levels, 2-8
  magnetic compass needle, 2-8
  notation strip, 2-8
  orienting and elevating mechanisms, 2-7
  telescope, 2-8
  verification of the lay of the platoon, 2-16, 2-16 (illus)

aiming posts, M14 and M1A2, 2-30, 2-30 (illus)

aiming post lights, M58 and M59, 2-31, 2-31 (illus)

ammunition (*see also* United States mortars), 1-3, 1-13 to 1-16
  60-mm mortar, M224 (*see also* 60-mm mortar, M224)
    care and handling, 3-22 to 3-23
    cartridge preparation, 3-21 to 3-22
      adjusting the propellant charge, 3-22
      inspecting fin assemblies, 3-21
      preparing to fire, 3-21
      unfired cartridges, 3-22
      unpacking, 3-21
    classification and types of ammunition, 3-16 to 3-18
      high-explosive, 3-17, 3-17 (table)
      illumination, 3-17, 3-17 (table)
      smoke, white phosphorus, 3-18, 3-18 (table)
      training practice, 3-18, 3-18 (table)
    fuzes, 3-18 to 3-21
      mechanical time superquick fuze, M776, 3-20
      multioption fuzes
        M734, 3-19, 3-19 (illus)
        M734A1, 3-20
      point-detonating fuzes
        M527-series, 3-20
        M745, 3-20
        M779, 3-20
        M935, 3-19
        M936, 3-20
      point-detonating/delay fuze, M783, 3-20
      practice fuze, M775, 3-20
      time fuze, M65-series, 3-20

81-mm mortar, M252 (*see also* 81-mm mortar, M252)
  care and handling, 4-25 to 4-27, 4-26 (illus)
  cartridge preparation, 4-24 to 4-25
  classification and types of ammunition, 4-16 to 4-21
    authorized cartridges, 4-17 to 4-20
      high-explosive, 4-18, 4-18 (table)
      illumination, 4-19, 4-19 (table)
      smoke, white phosphorus, 4-20, 4-20 (table)
      training practice, 4-20, 4-20 (table)
  functioning, 4-17
  identification, 4-16 to 4-17
    ammunition lot number, 4-17
    color code, 4-17
    markings on container, 4-17
    markings on rounds, 4-17
  fuzes, 4-21 to 4-24
    dummy fuzes, 4-24
    fuze wrench, 4-24
    mechanical time fuzes, 4-23
      M772A1, 4-23
      M84, 4-23
      M84A1, 4-23
    multioption fuzes, 4-23
      M734, 4-23 to 4-24
      M734A1, 4-24
    point-detonating fuzes, 4-21
      M524-series, 4-21
      M526-series, 4-21
      M567, 4-21
      M935, 4-21
    proximity fuzes, 4-22 to 4-23
      burst height, 4-22
      care, handling, and preservation, 4-22
      climatic effects, 4-22
      crest clearance, 4-22
      disposal precautions, 4-22
      installation, 4-23
120-mm mortars, M120 and M121 (*see also* 120-mm mortars, M120 and M121)
  care and handling, 5-44 to 5-45
  cartridge preparation, 5-43 to 5-44
    unpacking, 5-43
    inspection, 5-43
    preparation to fire, 5-43 to 5-44
    unfired cartridges, 5-44

Index

classification and types of
ammunition, 5-38 to 5-40
  high-explosive, 5-39, 5-39 (table)
  illumination, , 5-39, 5-39 (table)
  smoke, white phosphorus, 5-40, 5-40 (table)
  training practice, 5-40, 5-40 (table)
fuzes, 5-40 to 5-43
  mechanical time superquick fuze, M776, 5-41, 5-41 (illus)
    resetting, 5-41
    setting, 5-41
  multioption fuzes
    M734, 5-42, 5-42 (illus)
      resetting, 5-42
      setting, 5-42
    M734A1, 5-42, 5-42 (illus)
  point-detonating fuzes
    M745, 5-43, 5-43 (illus)
    M935, 5-41, 5-41 (illus)
      resetting, 5-41
      setting, 5-41
care and handling, 1-13 to 1-14
  burning of unused propellant charges, 1-13 to 1-14
  fuzes, 1-14
  projectiles/cartridges, 1-13
  segregation of ammunition lots, 1-14
cartridges, 1-4
  care and handling, 1-13
  characteristics, 1-4 (table)
  color codes, 1-15, 1-15 (table)
  field storage, 1-15 to 1-16, 1-16 (illus)
  fuzes, 1-4
    care and handling, 1-14
  propellant charges, 1-4
    burning unused propellant charges, 1-13 to 1-14

B

baseplate
  M3A1 (*see also* 81-mm mortar, M252), 4-6, 4-6 (illus)
  M7 (*see also* 60-mm mortar, M224), 3-5, 3-5 (illus)
  M8 (*see also* 60-mm mortar, M224), 3-5, 3-5 (illus)
  M9 (*see also* 120-mm mortars, M120 and M121), 5-8, 5-8 (illus)

bipod assembly
  M170 (*see also* 60-mm mortar, M224), 3-3 to 3-4, 3-4 (illus)
    collar assembly, 3-3
    elevating mechanism, 3-4
    left leg assembly, 3-4
      locking sleeve, 3-4
      fine cross-leveling nut, 3-4
    right leg assembly, 3-4
    shock absorbers, 3-3
    spread cable, 3-4
    traversing mechanism, 3-3 to 3-4
  M191 (*see also* 120-mm mortars, M120 and M121), 5-6, 5-6 (illus)
  M190 (*see also* 120-mm mortars, M120 and M121), 5-7, 5-7 (illus)

boresight, 2-24 to 2-30
  boresight method of calibration, 2-26 to 2-28
    elevation setting, 2-27
    deflection setting, 2-27 to 2-28, 2-28 (illus)
    removal, 2-28
  calibration for deflection M2 aiming circle, 2-28 to 2-30
    aiming point method, 2-29 to 2-30, 2-30 (illus)
    angle method, 2-29, 2-29 (illus)
  installation, 2-25
  M115, 2-25 to 2-26, 2-26 (illus)
    components, 2-25 to 2-26
      body, 2-25
      leveling bubbles, 2-25
      second cross-level bubble, 2-26
      telescope, 2-25
    tabulated data, 2-26, 2-26 (table)
  M45-series, 2-24 to 2-25, 2-25 (illus)
    components, 2-24
      body, 2-24
      clamp assembly, 2-24
      elbow telescope, 2-24
      strap assemblies, 2-24
      telescope clamp, 2-24
    tabulated data, 2-24, 2-24 (table)
  principles of operation, 2-25
  sight calibration, 2-26

C

cannon assembly
  M225 (*see also* 60-mm mortar, M224), 3-3, 3-3 (illus)
  M253 (*see also* 81-mm mortar, M252), 4-5, 4-5 (illus)

M298 (*see also* 120-mm mortars, M120 and M121), 5-4 to 5-5, 5-5 (illus)

commander's interface (*see also* mortar fire control system), 6-3 (illus)
  battery specifications, 6-9
  keys, controls, and indicators, 6-2 to 6-5
    battery 1 and battery 2 indicators, 6-4
    blackout key, 6-4
    control, alternate, and escape keys, 6-4
    enter key, 6-4
    function keys, 6-2, 6-3 (table)
    F1 and F2, 6-4
    keyboard backlighting control, 6-4
    number lock key and indicator, 6-4
    mouse, 6-4
    power indicator, 6-5
    right, left, up, and down arrow keys, 6-4
    screen brightness intensity buttons, 6-4
    Windows key, 6-4
  processing capabilities, 6-9

compass, M2, 2-1 to 2-5, 2-1 (illus), 2-3 (illus), 2-4 (illus)
  characteristics, 2-1
  description, 2-2
    angle-of-sight mechanism, 2-2
    azimuth scale and adjuster, 2-2
    compass body assembly, 2-2
    front and rear sight, 2-2
    magnetic needle and lifting mechanism, 2-2
  use, 2-2 to 2-5
    declinating from a surveyed declination station free from magnetic attractions, 2-4 to 2-5
    measuring an angle of sight or vertical angle from the horizontal, 2-4
    measuring a grid azimuth, 2-4
    measuring a magnetic azimuth, 2-3 to 2-4
    using a field-expedient method to declinate the M2 compass, 2-5
  reciprocal laying, 2-38 to 2-39

Index

crew (*see also* United States mortars), 1-3

D

deflection and elevation changes
  60-mm mortar, M224 (*see also* 60-mm mortar, M224)
    large changes, 3-8 to 3-9, 3-9 (illus)
    small changes, 3-8
  81-mm mortar, M252 (*see also* 81-mm mortar, M252)
    large changes, 4-11
    small changes, 4-10 to 4-11
  120-mm mortars, M120 and M121, carrier-mounted (*see also* 120-mm mortars, M120 and M121)
    large changes, 5-29 to 5-30
    small changes, 5-29
  120-mm mortars, M120 and M121, ground-mounted (*see also* 120-mm mortars, M120 and M121)
    large changes, 5-13 to 5-14
    small changes, 5-12 to 5-13
  ground-mounted mortar, in the gunner's examination (*see also* gunner's examination)
    large changes, 9-11 to 9-13
      conditions, 9-12
      equipment, 9-11
      procedure, 9-12
      scoring, 9-12 to 9-13
    small changes, 9-9 to 9-10
      conditions, 9-9
      equipment, 9-9
      procedure, 9-9
      scoring, 9-10
  track-mounted mortar, M121, in the gunner's examination (*see also* gunner's examination)
    small changes, 9-16 to 9-17
      conditions, 9-17
      equipment, 9-16
      procedure, 9-17
      scoring, 9-17

direct-alignment method (*see also* fire direction center, firing without), 8-10 to 8-11
  mortar dismounted, 8-10
  mortar mounted, 8-10 to 8-11
  natural object method, 8-11

direct-lay method (*see also* fire direction center, firing without), 8-9 to 8-10
  step 1: initial firing data, 8-9
  step 2: referring the sight, 8-9 to 8-10
  step 3: bracketing the target, 8-10
  step 4: fire for effect, 8-10

dismounting the mortar
  60-mm mortar, M224 (*see also* 60-mm mortar, M224), 3-15 to 3-16
  81-mm mortar, M252 (*see also* 81-mm mortar, M252), 4-16

driver's display (*see also* mortar fire control system), 6-7 to 6-8, 6-7 (illus)
  dimmer knob, 6-8
  liquid crystal display, 6-8
  toggle switch, 6-8
  vehicle motion sensor, 6-8, 6-8 (illus)

E

errors in firing, reporting (*see also* loading and firing), 2-47

equipment (*see also* United States mortars), 1-3

F

fire commands (*see also* loading and firing)
  execution, 2-43 to 2-45
    charge, 2-45
    deflection, 2-45
    elevation, 2-45
    method of fire, 2-43 to 2-44
      at my command, 2-44
      do not fire, 2-44
      searching fire, 2-44
      section right (left), 2-44
      traversing fire, 2-44
      volley fire, 2-43
    mortar(s) to fire, 2-43
    mortar(s) to follow, 2-43
    shell and fuze, 2-43
    time, 2-45
  repeating and correcting, 2-46 to 2-47
    initial fire command, 2-47
    subsequent fire command, 2-47
  sequence of transmission, 2-43 (table)

  subsequent, 2-46
    cease firing/check fire, 2-46
    deflection, 2-46
    elevation, 2-46
    end of mission, 2-46

fire direction center, firing without, 8-1 to 8-17
  adjustment of range (*see also* range, adjustment), 8-11 to 8-17
  direct-alignment method (*see also* direct-alignment method), 8-10 to 8-11
  direct-lay method (*see also* direct-lay method), 8-9 to 8-10
  fire procedures (*see also* fire procedures), 8-1 to 8-9

fire procedures (*see also* fire direction center, firing without), 8-1 to 8-9
  advantages and disadvantages, 8-1
  attack
    deep targets, 8-8, 8-9 (illus)
    example, 8-8 to 8-9
    wide targets, 8-6 to 8-8, 8-7 (illus)
  firing data, 8-1
  fire commands, 8-3 to 8-5
    normal fire commands, 8-3
      direction, 8-3
      elevation (range), 8-3
      gunner's correction, 8-3
    modified fire commands, 8-4
    example, 8-5
  fire control, 8-5
  initial fire commands, 8-3
  movement to alternate and supplementary positions, 8-5
  observer corrections, 8-2
    forward observer more than 100 meters from mortar position, 8-2, 8-2 (illus)
    example, 8-2
    forward observer within 100 meters of mortar position, 8-2
  reference line, 8-6
  squad conduct of fire, 8-5 to 8-6
  squad use of smoke and illumination, 8-6
    illumination, 8-6
    smoke, 8-6

full-range training cartridge, M931 (*see also* training devices), B-1
  description, B-1
  procedures, B-1

Index

G

general doctrine, 1-1 to 1-2
    effective mortar fire, 1-1
    mortar positions, 1-1 to 1-2

gunner's display (*see also* mortar fire
    control system), 6-7, 6-7 (illus)
    BRT and DIM keys, 6-7
    locking clamp, 6-7
    TEST key, 6-7

gunner's examination (*see also* training
    evaluation), 9-1 to 9-23, A-9
    preparation, 9-1 to 9-3
        DA Form 5964-R, 9-2 (illus)
        eligible personnel, 9-2
        examining board, 9-1
        general rules, 9-3
        location and date, 9-2
        methods of instruction, 9-1 to 9-3
        preparatory exercises, 9-1
        prior training, 9-1
        qualification scores, 9-2
    with the ground-mounted mortar, 9-4
        to 9-14
        deflection and elevation changes
            (*see also* deflection and
            elevation changes)
            large changes, 9-11 to 9-13
                conditions, 9-12
                equipment, 9-11
                procedure, 9-12
                scoring, 9-12 to 9-13
            small changes, 9-9 to 9-10
                conditions, 9-9
                equipment, 9-9
                procedure, 9-9
                scoring, 9-10
        equipment, 9-4
        mounting of the mortar, 9-4 to 9-9
            conditions, 9-5
                120-mm only, 9-7
                all mortars, 9-5 to 9-7
            equipment and layout, 9-5
                60-mm mortar, 9-5 (illus)
                81-mm mortar, 9-6 (illus)
                120-mm mortar, 9-7 (illus)
            procedure, 9-8
            scoring, 9-8 to 9-9
        organization, 9-4, 9-4 (table)
        procedure, 9-4
        reciprocal laying, 9-13 to 9-14
            conditions, 9-13
            equipment, 9-13
            procedure, 9-13 to 9-14
            scoring, 9-14
            station setup, 9-13
        referring of the sight and realigning
            the aiming posts, 9-10 to 9-11
            conditions, 9-10
            equipment, 9-10
            procedure, 9-10 to 9-11
            scoring, 9-11
        subjects and credits, 9-4
    with the track-mounted mortar,
        M121, 9-14 to 9-23
        deflection and elevation changes
            (*see also* deflection and
            elevation changes)
            large changes, 9-19 to 9-20
                conditions, 9-19
                equipment, 9-19
                procedure, 9-19
                scoring, 9-20
            small changes, 9-16 to 9-17
                conditions, 9-17
                equipment, 9-16
                procedure, 9-17
                scoring, 9-17
        equipment, 9-14
        infantry brigade combat team
            mortars
            reciprocal laying, 9-22 to 9-23
                conditions, 9-22
                procedure, 9-23
                scoring, 9-23
        organization, 9-15, 9-15 (table)
        placement of mortar into a firing
            position from traveling
            position, 120-mm mortar,
            9-15 to 9-16
            conditions, 9-15
            equipment, 9-15
            procedure, 9-16
            scoring, 9-16
        procedure, 9-15
        reciprocal laying, 9-20 to 9-21
            conditions, 9-20
            equipment, 9-20
            procedure, 9-20 to 9-21
            scoring, 9-21
        referring of the sight and
            realignment of aiming posts,
            9-17 to 9-19
            conditions, 9-18
            equipment, 9-17
            procedure, 9-18
            scoring, 9-18 to 9-19
        subjects and credits, 9-14
    support squad
        reciprocal laying, 9-21 to 9-22
            conditions, 9-21
            procedure, 9-22
            scoring, 9-22

I

indirect fire team, 1-2 to 1-4, 1-2 (illus)
    applications, 1-2
    team mission, 1-2 to 1-3
    United States mortars (*see also*
        United States mortars), 1-3 to 1-4,
        1-4 (table)

L

laying of the section, 2-32 to 2-41
    parallel sheaf, 2-32 (illus)
    placing out aiming posts, 2-39 to 2-41
        any direction fire capability, 2-39
        arm-and-hand signals, 2-40 to
            2-41 (illus)
        procedures when aiming posts
            cannot be laid on the prescribed
            referred deflection, 2-39 to 2-40
        referred deflection is 2800 mils,
            2-39
        referred deflection is not 2800
            mils, 2-39
    reciprocal laying, 2-33 to 2-39,
        2-33 (illus)
        on a grid azimuth, 2-34 to 2-36,
            2-35 (illus), 2-36 (illus)
        on a magnetic azimuth, 2-36
        using the M2 compass, 2-38 to
            2-39
        using the mortar sights, 2-37 to
            2-38, 2-37 (illus), 2-38 (illus)
        using the orienting angle, 2-37

loading and firing, 2-42 to 2-48
    arm-and-hand signals (*see also*
        signals, arm-and-hand), 2-45,
        2-46 (illus)
    execution of fire commands (*see
        also* fire commands), 2-43 to 2-45
    firing of the ground-mounted mortar,
        2-42
    night firing (*see also* night firing),
        2-47 to 2-48
    repeating and correcting of fire
        commands (*see also* fire
        commands), 2-46 to 2-47
    reporting of errors in firing (*see also*
        errors in firing, reporting), 2-47

Index

subsequent fire commands (*see also* fire commands), 2-46
target engagement, 2-42 to 2-43

M

malfunctions
  60-mm mortar, M224 (*see also* 60-mm mortar, M224), 3-10 to 3-11
    cookoff, 3-11
    hangfire, 3-11
    misfire, 3-11
  81-mm mortar, M252 (*see also* 81-mm mortar, M252), 4-12
  120-mm mortars, M120 and M121, ground-mounted (120-mm mortars, M120 and M121), 5-14
  short-range training round, M880 (*see also* training devices), B-13 to B-14

misfire, removal
  60-mm mortar, M224 (*see also* 60-mm mortar, M224)
    conventional mode, 3-11 to 3-14
    handheld mode, 3-14 to 3-15
  81-mm mortar, M252 (*see also* 81-mm mortar, M252), 4-12 to 4-16, 4-13 (illus), 4-14 (illus), 4-15 (illus)
  120-mm mortars, M120 and M121, ground-mounted (*see also* 120-mm mortars, M120 and M121)
    using a cartridge extractor, 5-14 to 5-18, 5-16 (illus), 5-17 (illus), 5-18 (illus)
    using the barrel tip method, 5-18 to 5-20, 5-20 (illus)
  120-mm mortars, M120 and M121, carrier-mounted (*see also* 120-mm mortars, M120 and M121)
    using a cartridge extractor, 5-30 to 5-34, 5-33 (illus)
    using the barrel tip method, 5-34 to 5-37

subcaliber insert, M303 (*see also* subcaliber insert), B-16
subcaliber insert, M313 (*see also* subcaliber insert), B-19

mortar carrier, M1064A3 (*see also* 120-mm mortars, M120 and M121), 5-23 to 5-26, 5-23 (illus), 5-24 (illus)
  description, 5-23 to 5-24
  maintenance, 5-26
  mortar and vehicular mount, 5-25
  tabulated data, 5-25, 5-25 (table)

mortar fire control system, 6-1 to 6-30, 6-2 (illus)
  additional functions, 6-21 to 6-30
    alerts function, 6-28, 6-28 (illus)
    ammo/status function, 6-21 to 6-24
      "Ammo Fire Unit" screen, 6-22, 6-22 (illus)
      "Ammo Roll-up" screen, 6-23, 6-23 (illus)
      "Status Fire Unit" screen, 6-24, 6-24 (illus)
    check fire function, 6-25
    plain text messages function, 6-26 to 6-27
      "Read" screen, 6-26, 6-26 (illus)
      "Send" screen, 6-27, 6-27 (illus)
    plot function, 6-29, 6-29 (illus), 6-30 (illus)
  commander's interface (*see also* commander's interface), 6-2 to 6-5, 6-3 (illus)
  common actions, 6-10 to 6-11
  conducting fire missions, 7-1 to 7-16
    pointing device (*see also* pointing device), 7-1 to 7-4,
      "Boresight" Screen, 7-4, 7-4 (illus)
      "Status" screen, 7-2 to 7-3, 7-2 (illus)
    fire commands, 7-10 to 7-13
      end of mission, 7-13
      initial fire command, 7-11, 7-11 (illus)
      subsequent fire command, 7-12, 7-12 (illus)
    final protective fires, 7-14 to 7-16
      fire a stored final protective fire mission, 7-16, 7-16 (illus)
      fire command for an assigned final protective fire, 7-14 to 7-15, 7-14 (illus), 7-15 (illus)
    navigation and emplacement, 7-5 to 7-10, 7-5 (illus)

navigation to fire area, 7-9 to 7-10, 7-9 (illus), 7-10 (illus)
navigation to waypoint, 7-5 to 7-8, 7-6 (illus), 7-7 (illus), 7-8 (illus)
data initialization and system configuration, 6-12 to 6-21
  "Channel A" and "Channel B" screens, 6-20 to 6-21, 6-20 (illus)
  "Configuration" screen, 6-15, 6-15 (illus)
  "Data" screen, 6-16, 6-16 (illus)
  "Geographic Reference" screen, 6-17, 6-17 (illus)
  "Mounting Azimuth and Reference" screen, 6-19, 6-19 (illus)
  "Position" screen, 6-18, 6-18 (illus)
  "Unit List" screen, 6-14, 6-14 (illus)
description, 6-1
driver's display (*see also* driver's display), 6-7 to 6-8, 6-7 (illus)
gunner's display (*see also* gunner's display), 6-7, 6-7 (illus)
log-in procedures, 6-12, 6-12 (illus)
pointing device (*see also* pointing device), 6-6, 6-6 (illus)
power distribution assembly (*see also* power distribution assembly), 6-5 to 6-6, 6-5 (illus)
Soldier's graphic user interface, 6-9 to 6-10, 6-11 (illus)
startup, 6-12

mortar range safety checklist (*see also* safety procedures), 1-5 to 1-7
  items to check
    after firing, 1-7
    before firing, 1-5 to 1-7
    during firing, 1-7

mount, M177 (*see also* 81-mm mortar, M252), 4-5 to 4-6, 4-5 (illus)
  bipod, 4-6
  elevating mechanism, 4-6
  traversing mechanism, 4-6

mounting of the mortar
  60-mm mortar, M224 (*see also* 60-mm mortar, M224), 3-6 to 3-7
  81-mm mortar, M252 (*see also* 81-mm mortar, M252), 4-8 to 4-9, 4-9 (illus)

Index

N

night firing (*see also* loading and firing), 2-47 to 2-48

P

pointing device (*see also* mortar fire control system), 6-6, 7-1 to 7-4, 6-6 (illus)
   "Boresight" Screen, 7-4, 7-4 (illus)
   "Status" screen, 7-2 to 7-3, 7-2 (illus)
   manually updating pointing device position fields, 7-2 to 7-3
   pointing device data precedence rules, 7-3
     after alignment, 7-3
     during alignment, 7-3
     restart, 7-3
     shut down, 7-3
   pointing device status, 7-2

power distribution assembly (*see also* mortar fire control system), 6-5 to 6-6, 6-5 (illus)
   FAULT LED indicator, 6-6
   power on switch, 6-5
   switch and LED indicator for commander's interface, 6-5
   switch and LED indicator for driver's display, 6-5
   switch and LED indicator for Global Positioning System, 6-5
   switch and LED indicator for gunner's display, 6-5
   switch and LED indicator for pointing device, 6-6
   switch and LED indicator for printer, 6-6

premount checks
   60-mm mortar, M224 (*see also* 60-mm mortar, M224), 3-6
     ammunition bearer, 3-6
     gunner, 3-6
     squad leader, 3-6
   81-mm mortar, M252 (*see also* 81-mm mortar, M252), 4-7 to 4-8
     ammunition bearer, 4-7
     assistant gunner, 4-7
     gunner, 4-7
     squad leader, 4-7
       layout of equipment, 4-8 (illus)
   120-mm mortars, M120 and M121, carrier-mounted (*see also* 120-mm mortars, M120 and M121), 5-26
   120-mm mortars, M120 and M121, ground-mounted (*see also* 120-mm mortars, M120 and M121), 5-8 to 5-10
     bipod assembly, 5-9
     M67 sight unit, 5-10
     M1100 mortar trailer, 5-10
     mortar baseplate, 5-9
     mortar cannon, 5-8 to 5-9

R

range, adjustment (*see also* fire direction center, firing without), 8-11 to 8-17
   bracketing method, 8-12 to 8-13, 8-12 (table), 8-13 (illus)
   creeping method of adjustment, 8-13
   establishment of a reference line and shifting from that line, 8-15
     example, 8-16 to 8-17, 8-16 (illus)
   ladder method of adjustment, 8-13 to 8-14, 8-14 (illus)
     example, 8-15
   spottings, 8-11 to 8-12
     deviation spottings, 8-12
     erratic spottings, 8-12
     range spottings, 8-11
       definite range spottings, 8-11
       doubtful, lost, or unobserved range spottings, 8-11

referring the sight and realigning the aiming posts
   60-mm mortar, M224 (*see also* 60-mm mortar, M224), 3-9 to 3-10
   81-mm mortar, M252 (*see also* 81-mm mortar, M252), 4-12
     example, 4-12
   in the gunner's examination (*see also* gunner's examination)
     ground-mounted mortars, 9-10 to 9-11
     track-mounted mortar, M121, 9-17 to 9-19

S

safety card (*see also* safety procedures), 1-7
   example, 1-9 (illus)

safety checks before firing
   60-mm mortar, M224 (*see also* 60-mm mortar, M224), 3-7 to 3-8
     crewman, 3-8
     gunner, 3-7
     mask and overhead clearance, 3-7 to 3-8
   81-mm mortar, M252 (*see also* 81-mm mortar, M252), 4-10
     ammunition bearer, 4-10
     assistant gunner, 4-10
     gunner, 4-10
     mask and overhead clearance, 4-10
   120-mm mortars, M120 and M121, ground-mounted (*see also* 120-mm mortars, M120 and M121), 5-11 to 5-12
   120-mm mortars, M120 and M121, carrier-mounted (*see also* 120-mm mortars, M120 and M121), 5-28 to 5-29
     mask and overhead clearance, 5-28 to 5-29

safety diagram (*see also* safety "T"), 1-8 to 1-11
   example, 1-9 to 1-11
   for M821 HE, 1-11 (illus)
   for M853A1 ILLUM, 1-11 (illus)

safety procedures, 1-5 to 1-13
   duties of safety officer and supervisory personnel, 1-5 to 1-7
   before departing for range, 1-5
   mortar range safety checklist, 1-5 to 1-7
     items to check
       after firing, 1-7
       before firing, 1-5 to 1-7
       during firing, 1-7
   safety card, 1-7

Index

safety "T" (*see also* safety diagram), 1-8
  to 1-11
  example, 1-9 to 1-11
    for M821 HE, 1-11 (illus)
    for M853A1 ILLUM, 1-11 (illus)

short-range training round, M880, B-2
  to B-14, B-4 (illus)
  advantages, B-2
  characteristics, B-2
  clean and service round, B-6 to B-7
  components, B-3
    fuze M775, B-4
    projectile body, B-4
  defuze a dud M880 round, B-8 to
    B-9
  malfunctions and removal of misfire,
    B-13 to B-14
  parts, B-9, B-9 (table)
  preparation, B-5
  range, B-2 to B-3, B-3 (illus)
  recovery procedures, B-7 to B-8
  refurbish round, B-6
  safety, B-13
  scaled map, construction, B-9 to
    B-13, B-10 (illus), B-11 (illus),
    B-12 (illus), B-13 (illus)
  training, B-2
  training considerations, B-9

sightunit, 2-18 to 2-24
  care and maintenance, 2-23
  M64-series, 2-19 to 2-21, 2-20 (illus)
    illumination, 2-20
    major components, 2-19 to 2-20
      dovetail, 2-19
      locking knobs, 2-19
      micrometer knobs, 2-19
      scales, 2-20
    tabulated data, 2-21, 2-21 (table)
  M67, 2-18 to 2-19, 2-18 (illus)
    equipment data, 2-19, 2-19 (table)
    major components, 2-18 to 2-19
      elbow telescope, 2-19
      telescope mount, 2-19
  operation, 2-21 to 2-22
    attaching the sightunit, 2-21
    correct sight picture, 2-22
    laying for direction, 2-22
    placing the sightunit into
      operation, 2-21
    setting the deflection, 2-21
    setting the elevation, 2-22
    replacing the sightunit in the
      carrying case, 2-22
  radioactive tritium gas, 2-23
  identification, 2-23
  storage and shipping, 2-24
  warning label, 2-23 (illus)

signals, arm-and-hand (*see also* loading
  and firing), 2-45 to 2-46,
  2-46 (illus)
  cease firing, 2-45
  fire, 2-45
  ready, 2-45

subcaliber insert
  M303 (*see also* training devices),
    B-14 to B-16, B-14 (illus)
    characteristics, B-14
    maintenance, B-15 to B-16
    misfire procedures, B-16
  M313 (*see also* training devices),
    B-16 to B-19, B-16 (illus)
    assembly, B-17
    characteristics, B-16 to B-17
    maintenance, B-17 to B-19
    minimum and maximum range,
      B-17
    misfire procedures, B-19

surface danger zones, 1-12 to 1-13
  components, 1-12 to 1-13
    area A, 1-12
    area B, 1-13
    buffer zones, 1-12
    firing position, 1-12
    impact area, 1-12
      dispersion area, 1-12
      target area, 1-12
  construction, 1-13

squad and section duties and
  organization
  60-mm mortar, M224 (*see also*
    60-mm mortar, M224)
    duties, 3-1
      ammunition bearer, 3-1
      gunner, 3-1
      squad leader, 3-1
    organization, 3-1
  81-mm mortar, M252 (*see also*
    81-mm mortar, M252)
    duties, 4-1
      ammunition bearer, 4-1
      assistant gunner, 4-1
      gunner, 4-1
      squad leader, 4-1
    organization, 4-1
    positions, 4-2 (illus)
  120-mm mortars, M120 and M121
    (*see also* 120-mm mortars,
    M120 and M121)
    duties, 5-1 to 5-2
      ammunition bearer, 5-2
      assistant gunner, 5-1
      gunner, 5-1
      squad leader, 5-1
    organization, 5-1
    positions, 5-2 (illus)

T

target engagement, 2-42 to 2-43, 2-43
  (table)

training, A-1 to A-10
  at training base (*see also* training, at
    training base), A-1 to A-2
  example annual mortar training
    programs, A-7 to A-8, A-7 (illus),
    A-8 (illus)
  in units (*see also* training, in units),
    A-3 to A-8
  philosophy, A-1
  training evaluation (*see also* training
    evaluation), A-9 to A-10
  unit mortar training, A-1

training, at training base (*see also*
  training), A-1 to A-2
  Basic Noncommissioned
    Officer Course (11C), A-2
  Infantry Basic Officer Leader's
    Course, A-2
  Infantry Career Captain Course, A-2
  Infantry Mortar Leader Course, A-2
  Maneuver Advanced
    Noncommissioned Officer
    Course, A-2
  One-Station Unit Training (11C), A-2
  Pre-Command Course, A-2

training, in units (*see also* training), A-3
  to A-8
  collective training, A-5 to A-7
    collective training resources, A-6
      to A-7
      battalion-level training model,
        A-6 to A-7
      drill books, A-6
      mission training plan, A-6
      section and platoon training, A-6
    squad training, A-5 to A-6
  cross training, A-5

example annual mortar training programs, A-7 to A-8, A-7 (illus), A-8 (illus)
training plan development, A-3
   integrated training strategy, A-4, A-4 (illus)
   common tasks, A-5
   MOS 11C tasks, A-5
     Army Correspondence Course Program, A-5
     training extension courses, A-5
   mission-essential task list tasks, A-3
   initial training, A-3 to A-4
   sustainment training, A-4

training devices, B-1 to B-19
   full-range training cartridge, M931 (*see also* full-range training cartridge, M931), B-1

short-range training round, M880 (*see also* short-range training round, M880), B-2 to B-14, B-4 (illus)
subcaliber insert, M303 (*see also* subcaliber insert) B-14 to B-15, B-14 (illus)
subcaliber insert, M313 (see also subcaliber insert) B-16 to B-19, B-16 (illus)

training evaluation (*see also* training), A-9 to A-10
   collective training, A-9 to A-10
     Army Training and Evaluation program, A-9
     evaluation of the indirect fire team, A-10
     external evaluation, A-10
   individual training, A-9
     commander's evaluation, A-9
     fire direction center certification, A-9
     gunner's examination (see also gunner's examination), A-9
     Warrior task list, A-9

U

United States mortars, 1-3 to 1-4, 1-4 (table)
   ammunition, 1-3
     cartridges, 1-4
     fuzes, 1-4
     propellant charges, 1-4
   crew, 1-3
   equipment, 1-3

This page intentionally left blank.

GUNNER'S EXAMINATION SCORECARD - MORTARS

For use of this form, see FM 3-22.90, the proponent agency is TRADOC.

1. MORTARS (CHECK ONE)
   - ☐ Ground Mounted
   - ☐ Carrier-Mounted
   - ☐ Support Squad

DATA REQUIRED BY THE PRIVACY ACT OF 1974

Authority   10 USC 3012 (g), Executive Order 9397.

Principal Purpose   Evaluates individual's training on mortars.

Routine Uses   Evaluate individual proficiency; SSN is used for positive identification purposes only.

Disclosure   Disclosure of this information is voluntary. However, individuals not providing information cannot be rated/scored on a mass basis.

2. ORGANIZATION
3. DATE
4. NAME (Last, first, MI)
5. GRADE
6. SSN
7. CHECK ONE
   - ☐ M224   ☐ M29A1   ☐ M252   ☐ M30   ☐ XM120

ITEMS	TRIAL a	TIME b	TRIAL SCORE c	CREDIT SCORE d	INITIALS e
8. Ground-Mounting the mortar carrier and support. Placing the mortar in firing position from traveling position.	1st Trial				
	2d Trial				
9. All - Small deflection change.	1st Trial				
	2d Trial				
10. Ground and Carrier - Referring the sight and realigning aiming posts. Support direct laying.	1st Trial				
	2d Trial				
11. All - Large deflection and elevation change	1st Trial				
	2d Trial				
12. Ground and Carrier - Reciprocal laying. Support - Laying by compass.	1st Trial				
	2d Trial				
13. TOTAL CREDIT SCORE					

EXPERT GUNNER   90 to 100
FIRST-CLASS GUNNER   80 to 89
SECOND-CLASS GUNNER   70 to 79
UNQUALIFIED   69 or LESS

14. VERIFIED BY

a. PRINTED/TYPED NAME

d. UNIT

b. SIGNATURE

c. GRADE

e. DATE

DA FORM 5964-R, SEP 90   ☆ U.S. GOVERNMENT PRINTING OFFICE: 1990 — 527-027 20089

www.ingramcontent.com/pod-product-compliance
Lightning Source LLC
Chambersburg PA
CBHW080741250426
43671CB00038B/2649